Digital Signal Processing

Belle A. Shenoi
Magnitude and Delay Approximation of
1-D and 2-D Digital Filters

Springer

*Berlin
Heidelberg
New York
Barcelona
Hong Kong
London
Milan
Paris
Singapore
Tokyo*

Belle A. Shenoi

Magnitude and Delay Approximation of 1-D and 2-D Digital Filters

With 114 Figures

Springer

Series Editors

Prof. Dr.-Ing. Arild Lacroix
Johann-Wolfgang-Goethe-Universität
Institut für Angewandte Physik
Robert-Mayer-Str. 2-4
D-60325 Frankfurt
Germany

Prof. Dr.-Ing. Anastasios Venetsanopoulos
University of Toronto
Dept. of Electrical and Computer Engineering
10 King's College Road
M5S 3G4 Toronto, Ontario
Canada

Author

Professor Dr. Belle A. Shenoi
Electrical Engineering Department
Wright State University
3640 Colonel Glenn Highway
Dayton, Ohio 45435-0001
USA

Cataloging-in-Publication Data applied for

Die Deutsche Bibliothek - CIP-Einheitsaufnahme
Shenoi, Belle A.: Magnitude and delay approximation of 1-D and 2-D digital filters /
Belle A. Shenoi. - Berlin; Heidelberg; New York; Barcelona; Hong Kong; London;
Milan; Paris; Singapore;Tokyo: Springer, 1999
   (Digital Signal Processing)
   ISBN 3-540-64161-0

ISBN 3-540-64161-0 Springer-Verlag Berlin Heidelberg New York

This work is subject to copyright. All rights are reserved, whether the whole or part of the material is concerned, specifically the rights of translation, reprinting, reuse of illustrations, recitation, broadcasting, reproduction on microfilm or in other ways, and storage in data banks. Duplication of this publication or parts thereof is permitted only under the provisions of the German Copyright Law of September 9, 1965, in its current version, and permission for use must always be obtained from Springer-Verlag. Violations are liable for prosecution act under German Copyright Law.

© Springer-Verlag Berlin Heidelberg 1999
Printed in Germany

The use of general descriptive names, registered names, trademarks, etc. in this publication does not imply, even in the absence of a specific statement, that such names are exempt from the relevant protective laws and regulations and therefore free for general use.

Typesetting: Camera-ready copy from author
Cover-Design: de'blik, Berlin
SPIN 10670637   62/3020 5 4 3 2 1 0   Printed on acid-free paper

This book is dedicated

to my mother Amba Bai Shenai

and

to my teacher Mac E. Van Valkenburg

Preface

There are more than 100 books on Circuit Analysis, Network Synthesis, Analog and Digital Filters and Signal Processing written at the undergraduate and graduate level and a few more written as Reference and Handbooks. When and if they discuss the design of analog and digital filters, they treat mainly the approximation of the magnitude response of the filters and very little of their phase or group delay response. There is hardly any discussion of designing filters that simultaneously approximate the magnitude and group delay response of the filters. Thus most of the books routinely discuss Butterworth, Chebyshev and sometimes the Cauer or elliptic function response of the lowpass prototype filters, followed by the transformations to design highpass, bandpass and bandstop filters-all of them approximating their magnitude response only. Due to the rapid progress from analog to digital communication and data transmission that has taken place in recent years, there is a greater need for designing filters that approximate both the magnitude and group delay requirements. So also is the need to design 2-dimensional digital filters,particularly those used in image processing, that approximate prescribed magnitude as well as constant group delay responses. A lot of research work has been published in professional journals on the design of these filters in the last 10-15 years. But even the books that have been published recently, particularly on digital filters and signal processing, have little information on the approximation of both the magnitude and group delay of 1-D and 2-D digital filters. The purpose of this book is to fill this void.

The outstanding feature of this book, therefore, is that it treats the theory of approximating the magnitude only, group delay only and both the magnitude and group delay of the 1-D as well as the 2-D IIR filters. As a prerequisite background material, the classical methods of approximating the magnitude of analog filters is included at the beginning of the first chapter, followed by the approximation of the magnitude of 1-D IIR filters. The material covered in this chapter may be considered as concise, because most of it is found in many well-known text books-a few of them being listed below.

In Chap.2, we discuss a large number of methods for designing 1-D digital filters that approximate a constant delay and both the magnitude and group delay. The methods included in this chapter, are based on (1) the use of allpass sections connected in cascade (2) the use of mirror image polynomial (3) the use of two allpass sections connected in parallel (4) use of singular value decomposition (5) linear programming (6) theory of commensurate, distributed networks, and (7) theory of eigen filters. These methods, except the first one, are fairly new and very little detail on these methods has been published in any text book so far. For that reason, they are discussed in greater detail than the methods covered in Chap.1.

Chapter 3 is devoted to the design of 2-D IIR filters that approximate the magnitude only. We discuss the filters having a magnitude response in the passband or stopband regions that are classified into three categories: (1) regions bounded by straight lines parallel to the ω_1 and ω_2 axes in the 2-D frequency plane i.e. rectangular regions and union of rectangular regions. (2) regions bounded by straight lines inclined to the ω_1 and ω_2 axes i.e. fan filters, and (3) regions bounded by closed contours i.e. circularly symmetric and elliptically symmetric regions. At least three methods are described under each of the three categories.

In Chap.4, we describe several methods for the design of 2-D IIR filters that approximate both the magnitude and group delay specifications. None of the methods described in Chaps.3 and 4, have been published in any book so far.

It should be obvious, therefore, that the scope of this book is very specific and selective. It can be used as a good reference book by graduate students and researchers who are interested in getting a review of the research on the above topics, which is still found mainly in professional journals and in extending it. It may also be used as a textbook, when supplemented with the book by J.S. Lim cited below, for graduate level courses on two-dimensional signal and image processing.

There are many excellent books that treat a lot of topics in 1-D digital signal processing and only a few books that treat the topics in 2-D digital signal processing. Topics in these two areas such as the analysis of discrete systems, Fourier transform, Discrete Fourier Transform (DFT), Fast Fourier Transform (FFT), realization and synthesis of digital filters by different structures, finite word length effects with respect to the diverse structures and types of quantization, stability analysis and so on are covered in these books. Because of page limitations and the well-defined scope of this book, these topics are not included in it. Another major subject that is contained in these books is the theory and design of 1-D and 2-D FIR filters. They can be easily designed to exhibit a constant delay over the entire frequency range and also to approximate a prescribed magnitude response over a finite bandwidth or a finite passband region. But, regretfully, the extensive and important material on FIR filters had to be deleted from this book. Readers using this book are assumed to be familiar with the topics contained in such excellent books, a few of which are listed below. The author has used the contents of this book as supplementary material to that of the listed text books, while he has been teaching graduate level courses on 1-D and 2-D digital filters and signal processing and continuously improving the class notes during the last five years. The students who took these courses as well as the author himself have worked out many examples to illustrate the large number of design procedures included in the book. A beneficial feature of this book is that it contains lots of design examples and plots of the magnitude and delay response that accompany the discussion of the theoretical material. The design examples and the plots clarify the theoretical material and provide a better insight about the relative merits of the various design procedures. Particularly in Chaps.3 and 4, the merits of the design theory and the design examples published by the different authors have been compared in order to put their research work in proper perspective. But the author of this book admits that this comparison may be found to be partly subjective and partly objective. Indeed he admits that all the design methods have not been treated with equal extent of detail and also that he might have missed the discussion of some very important methods that have been published in professional journals. So he would very much appreciate receiving constructive criticisms and comments on this book from those who have reviewed it or used it diligently-via e-mail to <bshenoi@ieee.org>. He would be very pleased to hear from them about the deficiencies of this edition and suggestions for improving the book in its next edition.

I take great pleasure in thanking a few of my friends and colleagues who have helped me in the writing of this book. I appreciate the timely suggestions made by Dr.M.A.Pai, University of Illinois, Urbana and Dr.Sanjit Mitra, University of California, Santa Barbara about writing and publishing this book. Dr.H.K.Kwan, University of Windsor and Dr.Weiping Zhu, Concordia University reviewed the early versions of the manuscript and provided very good suggestions to improve it. Apart from the students who took my courses and offered their opinions on the draft manuscript, Rajamohana Hegde especially deserves my sincere thanks for his original contributions. His research work on the theory and design procedure for approximating the magnitude and the group delay of 1-D and 2-D filters is very well covered in Chap.2 and 4. I owe many thanks Professor A.N.Venetsanopolous, University of Toronto,Canada and Professor A. Lacroix, J.W.Goethe-Universitat, Germany who decided with fast dispatch and great enthusiasm to include this book in the Springer-Verlag Series on Digital Signal Processing. I am also thankful to Dr Rahul Singh for his valuable help in the last stages of getting all the chapters compiled, previewed and printed together as a single manuscript, under LaTeX. Finally I wish to thank my wife Suman immensely, for her patience, understanding and encouragement during the long hours and years I spent in writing this book-without her love and care I would not have completed this book at all.

B.A.Shenoi
December 23, 1998

A.V.Oppenheim and R.W.Schafer, *Discrete-Time Signal Processing*, Prentice-Hall, 1989
J.G.Proakis and D.G.Manolakis, *Digital Signal Processing: Principles, Algorithms, and Applications*, Prentice-Hall, 1996
S.K.Mitra and J.F.Kaiser,(ed.) *Handbook for Digital Signal Processing*, John Wiley & Sons,1993
D.E.Dudgeon and R.M.Mersereau, *Multidimensional Digital Signal Processing*, Prentice-Hall, 1984
J.S.Lim, *Two-Dimensional Signal and Image Processing*, Prentice-Hall, 1990

Contents

1 **Classical Methods of Approximation**   1
   1.1 Introduction 1
   1.2 Classical Filter Theory 2
   1.3 Types of Approximation 4
   1.4 Butterworth Approximation 9
       1.4.1 Example 1.1 14
       1.4.2 Example 1.2 17
   1.5 Chebyshev Approximation 18
       1.5.1 Example 1.3 22
   1.6 Inverse Chebyshev Approximation 23
       1.6.1 Example 1.4 26
   1.7 Elliptic Approximation 27
   1.8 Least Squares Approximation 31
   1.9 Analog Frequency Transformations 34
   1.10 Highpass Filter 34
       1.10.1 Example 1.5 35
   1.11 Bandpass Filter 36
       1.11.1 Example 1.6 37
   1.12 Bandstop Filter 38
       1.12.1 Example 1.7 40
   1.13 1-D Digital Filters 41
   1.14 Theory of Sampling 41
   1.15 Design Procedures 49
   1.16 Impulse Invariant Transformation 50
   1.17 Bilinear Transformation 54
       1.17.1 Example 1.8 56
       1.17.2 Digital Spectral Transformation 60
       1.17.3 Example 1.9 60
       1.17.4 Example 1.10 64
   1.18 Computer-Aided Optimization 64
   1.19 Conclusion 65

2 Magnitude and Delay of I-D Filters 71
- 2.1 Introduction 71
- 2.2 Properties of an IIR Transfer Function 72
- 2.3 Maximally Flat Group Delay Filters 74
- 2.4 Simultaneous Approximation of Magnitude and Group Delay . . 83
  - 2.4.1 Method 2.1: Nonlinear Optimization 83
  - 2.4.2 Method 2.2 : Use of Mirror Image Polynomial 85
  - 2.4.3 Method 2.3 : Maximally Flat Magnitude and Group Delay in the Passband 87
  - 2.4.4 Design Theory 91
  - 2.4.5 Example 2.1 94
  - 2.4.6 Extension for an Equiripple Stopband Response 95
  - 2.4.7 Remez Algorithm 97
  - 2.4.8 Example 2.2 98
- 2.5 Design of Pulse Shaping Filters 99
  - 2.5.1 Example 2.3 101
  - 2.5.2 Method 2.4: Use of Allpass Sections in Parallel 104
  - 2.5.3 Method 2.5: Use of Singular Value Decomposition . . . 110
  - 2.5.4 Example 2.4 113
  - 2.5.5 Method 2.6: Application of Linear Programming 113
  - 2.5.6 Method 2.7: Theory of Commensurate Distributed Networks 117
  - 2.5.7 Equidistant Linear Phase Approximation 121
  - 2.5.8 Method 2.8: Theory of Eigen Filters 122
  - 2.5.9 Example 2.5 127
- 2.6 Conclusion 129

3 Magnitude of 2-D Filters 137
- 3.1 Introduction 137
- 3.2 Filters with Rectangular Passbands 138
- 3.3 Design of Fan Filters 151
  - 3.3.1 Method 3.1 152
  - 3.3.2 Example 3.1 154
  - 3.3.3 Design of Fan Filters using Method 3.2 165
  - 3.3.4 Design Procedure 166
  - 3.3.5 Example 3.2 167
  - 3.3.6 Method 3.3 168
  - 3.3.7 Example 3.3 170
- 3.4 Design of Circularly Symmetric Filters 173
  - 3.4.1 Survey of the literature 173
  - 3.4.2 Design Procedures 175
- 3.5 Conclusion 182

4 Magnitude and Delay of 2-D Filters   **187**
   4.1 Introduction 187
   4.2 Design using Nonlinear Programming 188
        4.2.1 Statement of the problem 188
        4.2.2 Stability conditions 190
        4.2.3 Design Examples 196
   4.3 Design Using Linear Programming 213
   4.4 Design Using Singular Value Decomposition 216
   4.5 Chebyshev Approximation Theory 217
   4.6 Design Using Digital Spectral Transformation 225
        4.6.1 Example 4.5 233
        4.6.2 Example 4.6 240
   4.7 Conclusion 241

Index   **249**

Chapter 1

Classical Methods of Magnitude Approximation

1.1 Introduction

Electric wave filters, called simply as filters in this book, are circuits that are used to selectively filter out some of the frequency components of an input signal and transmit the remaining components as the output signal in such a way that the quality of information contained in the signal is improved. We consider continuous-time signals, one-dimensional (1-D) and two-dimensional (2-D) discrete-time signals as the input and output of filters and characterize them in the frequency domain by their appropriate Fourier Transform. For example, if $x(t)$ and $y(t)$ are the continuous-time, input and output signals of a linear, time-invariant, continuous-time filter, their Fourier Transforms $X(j\omega)$ and $Y(j\omega)$ are related by the equation $Y(j\omega) = H(j\omega)X(j\omega)$ where $H(j\omega)$ is the Fourier Transform of the unit impulse response $h(t)$ of the continuous-time filter. The relation can also be expressed in terms of these complex valued functions as

$$|Y(j\omega)|\,e^{j\varphi} = \left[|H(j\omega)|\,e^{j\theta}\right]\left[|X(j\omega)|\,e^{j\phi}\right] = |H(j\omega)|\,|X(j\omega)|\,e^{j(\theta+\phi)} \quad (1.1)$$

We see that the magnitude of the output as a function of frequency is $|H(j\omega)|$ times the magnitude $|X(j\omega)|$ of the input signal as a function of the frequency and the phase angle of the output signal is the phase angle of the input increased by that of $H(j\omega)$. Therefore by properly shaping the magnitude of $H(j\omega)$ and the phase angle $\theta(j\omega)$ as a function of frequency, we can design the filter to transmit some frequencies and block out other frequencies as we decide. The decision about the shape of the magnitude and/or the phase response of the filter is based on the frequency response characteristics of $X(j\omega)$ and also the frequency response characteristics of $Y(j\omega)$ that we wish to obtain, which depends on the design application we have under consideration. These decisions are translated as specifications for the design of the filter. The first step in the design of the filter is to find the transfer function $H(s)$ of the filter such that the magnitude and/or phase response of $H(j\omega)$ approximates as best as possible the specified magnitude and/or phase response. Only after the transfer function $H(s)$ is derived, we do the synthesis and hardware implementation of the filter. Though there are additional steps involved in the whole design cycle, the words 'design of the filter' used in this book will actually mean the derivation of the transfer function $H(s)$ that approximates the given specifications in the frequency domain. We will also call the continuous-time circuits as analog filters. Similar meaning will be conveyed when we treat the design of 1-D, and 2-D discrete-time filters also called as 1-D and 2-D digital filters.

The theory of approximating the frequency response specifications of these filters is mainly based on the well-advanced mathematical theory of approximation and it has its origin in the early years of electrical circuit theory. It can be said that the mathematical theory of circuit analysis originated in the *"telegrapher's equation"* which was studied by W.Thomson (Lord Kelvin) [31], as the equation to describe the performance of transmission lines. Perhaps the first example of an 'electric wave filter' that was invented from theory is the loading coil [15] which was later built into cables and long distance transmission lines [23], in order to minimize magnitude distortion of the signal transmitted through them. Thus "Filter theory evolved first from loaded lines" [9]. Ever since the beginning of this century when Heaviside [15] proposed the introduction of both distributed and lumped loading of transmission lines, the theory of filters and the concomitant theory of signal processing has continued to develop as a major part of electrical engineering. The use of filters in electrical/electronics industry is so wide spread that it is difficult to pinpoint any one application where they are not used. Millions and millions of analog and digital filters are used in every branch of electrical technology e.g telecommunication networks, consumer electronics and multimedia, defense and automotive electronics, computers and computer networks.

1.2 Classical Filter Theory

The classical theory of filters has focussed on the magnitude (and group delay) of analog filters that approximates a constant value over a finite frequency interval. The normalized frequency interval of an ideal lowpass filter is shown in Fig. 1.1a, whereas the magnitude response of ideal highpass, bandpass and bandstop filters is shown in Figs. 1.1b-d.

The ideal lowpass filter passes all frequencies of the input signal in the interval $|\omega| \leq \omega_p$ with equal gain and completely filters out all frequencies outside this interval. In practice, the intervals $0 \leq \omega \leq \omega_p$ and $\omega_p < \omega < \infty$ are called respectively as the passband and stopband of the ideal lowpass filter -though the intervals $|\omega| \leq \omega_p$ and $\omega_p \leq |\omega| < \infty$ define the passband and stopband. Similarly the ideal bandpass filter transmits all frequencies in the interval $\omega_1 \leq |\omega| \leq \omega_2$ and completely filters out all frequencies outside this interval. Since such ideal specifications can not be realized in practice, it is common to prescribe tolerances within which these specifications have to be met. For ex-

1.2. CLASSICAL FILTER THEORY

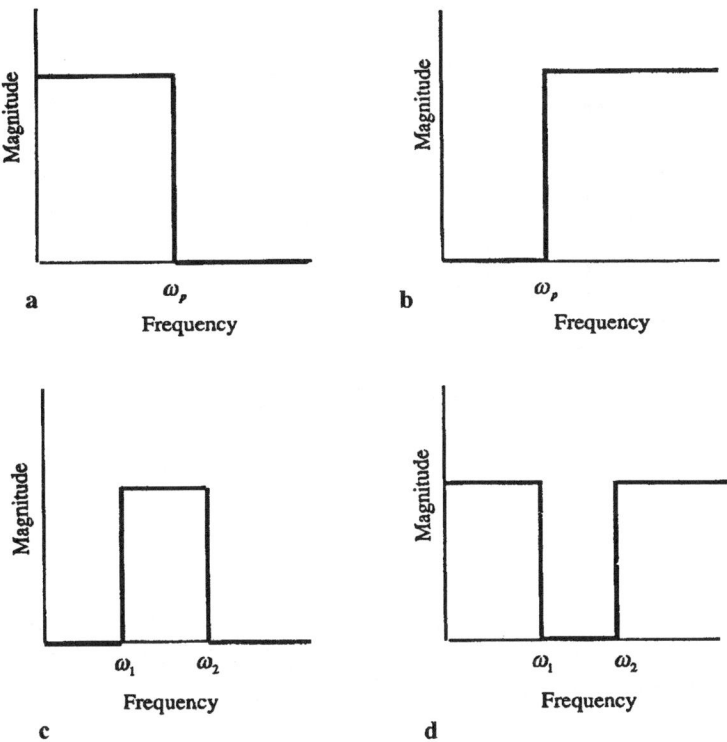

Fig. 1.1. Magnitude response of ideal filters. **a** Lowpass filter **b** Highpass filter **c** Bandpass filter **d** Bandstop filter

ample, the tolerance of δ_p on the ideal value of one for the magnitude in the passband and the tolerance of δ_s for the magnitude of zero in the stopband are shown in Fig. 1.2. A tolerance between the passband and stopband is also provided by a transition band. Similar tolerances are included in the specifications of the other filters.

Besides the class of ideal lowpass, highpass, bandpass and bandstop filters with a piece-wise constant magnitude for the passband and the stopband, there is a wide variety of other filters designed and used for signal processing e.g. comb filters, filters for minimizing intersymbol interference in digital data transmission and so on. There is another equally important class of filter specifications. They are required to compensate for the magnitude distortion and/or to equalize the group delay of a filter or a system, such that the magnitude or group delay of the overall filter (or the system that contains these filters connected in cascade, in parallel or in a feedback loop) approximates a desired response characteristics. Since the magnitude and group delay of the filter or the system may have an arbitrary shape, analytical methods for designing such magnitude compensators

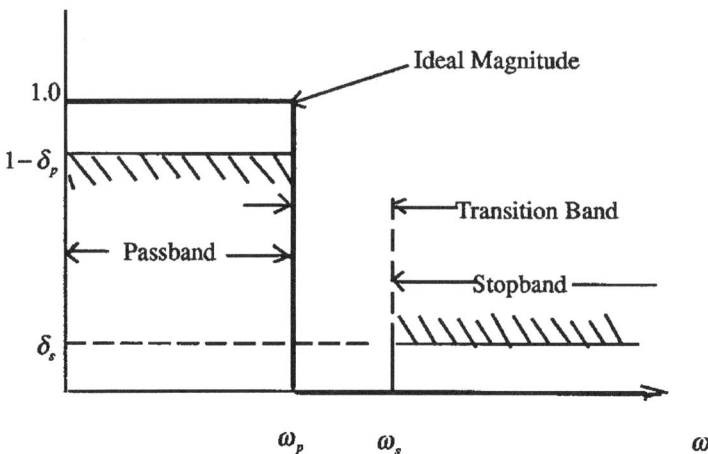

Fig. 1.2. Ideal Lowpass filter with tolerance specifications

and delay equalizers are not well-developed. But for this purpose, efficient numerical methods have been developed which are implemented by use of the very powerful computers currently available.

1.3 Types of Approximation

In this chapter, we will discuss different methods of finding a rational function that approximates a given magnitude response, the type of approximation being defined by the approximation criteria chosen. The most commonly used approximation criteria used for the design of filters in the frequency domain are (1) maximally flat approximation (2) equiripple or min-max approximation and (3) the least p^{th} approximation of which least squares approximation (when $p = 2$) is a special case. These are briefly described below([10, 29]) choosing the frequency response of an analog filter as the example. The design of analog filters satisfying the above three approximation criteria, is treated in the first half of this chapter, because the design of digital filters depends heavily on the design theory of analog filters. The approximation of the magnitude response of the 1-D, digital filters is treated in the second half of this chapter.

So, let us consider the transfer function $H(s)$ of the analog filter denoted by

$$H(s) = K \frac{\sum_{k=0}^{m} b_k s^k}{\sum_{k=0}^{n} a_k s^k} \qquad (1.2)$$

The frequency response of the filter is given by

$$H(s)|_{s=j\omega} = H(j\omega) = |H(j\omega)| e^{j\theta(\omega)}$$

1.3. TYPES OF APPROXIMATION

where $|H(j\omega)|$ is the magnitude response and $\theta(\omega)$ is the phase response. Then we can show that

$$H(-j\omega) = |H(j\omega)| e^{-j\theta(\omega)} \tag{1.3}$$

The square of the magnitude response is

$$|H(j\omega)|^2 = H(j\omega)H(-j\omega) = H(s)H(-s)|_{s=j\omega} \tag{1.4}$$

The phase response of the filter is given by

$$\theta(j\omega) = \frac{1}{2j} \ln \frac{H(s)}{H(-s)}\bigg|_{s=j\omega} \tag{1.5}$$

and the group delay $\tau(j\omega)$ defined as $\tau(j\omega) = -\frac{\partial \theta}{\partial \omega}$ is given by

$$\tau(j\omega) = -\frac{1}{2}\left(\frac{H'(s)}{H(s)} + \frac{H'(-s)}{H(-s)}\right)\bigg|_{s=j\omega} \tag{1.6}$$

Let us take a simple example of the transfer function $H(s)$ to show some other properties of the above functions. Let

$$H(s) = \frac{(s+\gamma)(s^2+\beta_1 s+\beta_o^2)}{(s+\zeta)(s^2+\alpha_1 s+\alpha_o^2)} \tag{1.7}$$

Then directly we can derive

$$|H(j\omega)|^2 = \frac{(\gamma^2+\omega^2)\left\{(\beta_o^2-\omega_o^2)^2+\beta_1^2\omega^2\right\}}{(\zeta^2+\omega^2)\left\{(\alpha_o^2-\omega_o^2)^2+\alpha_1^2\omega^2\right\}} \tag{1.8}$$

$$\theta(j\omega) = \tan^{-1}\left(\frac{\omega}{\gamma}\right) + \tan^{-1}\left(\frac{\beta_1\omega}{\beta_o^2-\omega^2}\right) - \tan^{-1}\left(\frac{\omega}{\zeta}\right) - \tan^{-1}\left(\frac{\alpha_1\omega}{\alpha_o^2-\omega^2}\right)$$

and

$$\tau(j\omega) = \frac{\gamma}{\gamma^2+\omega^2} + \frac{\beta_1(\omega^2+\beta_o^2)}{(\beta_o^2-\omega^2)^2+\beta_1^2\omega^2} - \frac{\zeta}{\zeta^2+\omega^2} - \frac{\alpha_1(\omega^2+\alpha_o^2)}{(\alpha_o^2-\omega^2)^2+\alpha_1^2\omega^2} \tag{1.9}$$

These are functions of the continuous frequency variable ω and we see that $|H(j\omega)|^2$ is an even function of ω whereas the phase response $\theta(j\omega)$ is an odd function of ω. The phase response however involves the sum of arctangent functions. But the group delay $\tau(j\omega)$ is a rational function of ω, is also a real valued function and an even function of ω, just like the magnitude squared response.

Hence we will represent the magnitude squared response by $|H(\omega)|^2$ and the group delay by $\tau(\omega)$, in order to consider the general problem of finding a transfer function $H(s)$ of the analog filter such that its magnitude or group delay approximates the prescribed magnitude or group delay. Let us denote the square of the prescribed magnitude or the prescribed group delay by $g(\omega)$ and the function $|H(\omega)|^2$ or the function $\tau(\omega)$ by $f(\omega)$. The prescribed magnitude (or the group delay) may be given in an analytical form or in the form of a frequency response plot.

For a discussion of the maximally flat approximation, let us first consider the Taylor series expansion of $f(\omega)$ and $g(\omega)$ about some point ω_o, assuming both of them are analytical functions in the frequency range $\omega \in (\omega_1, \omega_2)$. The Taylor series are in the form

$$\begin{aligned} f(\omega) &= f_o + f_1(\omega - \omega_o) + f_2(\omega - \omega_o)^2 + f_3(\omega - \omega_o)^3 + \cdots \\ g(\omega) &= g_o + g_1(\omega - \omega_o) + g_2(\omega - \omega_o)^2 + g_3(\omega - \omega_o)^3 + \cdots \end{aligned} \quad (1.10)$$

The error between the two expansions is given by

$$\mathcal{E} = (f_o - g_o) + (f_1 - g_1)(\omega - \omega_o) + (f_2 - g_2)(\omega - \omega_o)^2 + \cdots \quad (1.11)$$

The function $f(\omega)$ is said to approximate $g(\omega)$ in the Taylor sense at the point $\omega = \omega_o$, if the first k coefficients excluding the constant term $(f_o - g_o)$ in the error function (1.11) are zero so that it reduces to the form

$$\mathcal{E} = (f_{k+1} - g_{k+1})(\omega - \omega_o)^{k+1} + (f_{k+2} - g_{k+2})(\omega - \omega_o)^{k+2} + \cdots \quad (1.12)$$

But for an analytical function, we know that the coefficients of its Taylor series excluding the constant term $(f_o - g_o)$ are its derivatives and if the first k coefficients are zero at $\omega = \omega_o$, it is equivalent to saying that the first k derivatives $f(\omega) - g(\omega)$ are zero at $\omega = \omega_o$. Let us apply this definition to the square of the magnitude function $|H(j\omega)|^2$, expressed in the form of (1.13).

$$|H(j\omega)|^2 = \frac{c_o(1 + c_2\omega^2 + c_4\omega^4 + \cdots + c_{2m}\omega^{2m})}{1 + d_2\omega^2 + d_4\omega^4 + \cdots + d_{2n}\omega^{2n}} \quad (1.13)$$

This is analytic at $\omega = 0$ and when it is expanded in the Maclaurin series by long division, we get

$$|H(j\omega)|^2 = c_0 \left\{ 1 + (c_2 - d_2)\omega^2 + [(c_4 - d_4) - d_2(c_2 - d_2)]\omega^4 + \cdots \right\} \quad (1.14)$$

Note that $|H(j\omega)|^2$ is an even function in ω and in its Maclaurin series expansion (1.14), all odd ordered derivatives are identically zero. If the coefficients $c_{2i} = d_{2i}$, $i = 1, 2, \ldots, (n-1)$, it is clear that all derivatives of order up to $(2n-1)$, of the Maclaurin series are zero at $\omega = 0$. In that case, the lowest non-zero derivative is the coefficient of the term ω^{2n} and therefore the magnitude squared function that is maximally flat is the function that has the highest number of derivatives which are zero at $\omega = 0$. It is therefore given by

$$|H(j\omega)|^2 = \frac{c_o(1 + d_2\omega^2 + d_4\omega^4 + \cdots + d_{2n-2}\omega^{2n-2})}{(1 + d_2\omega^2 + d_4\omega^4 + \cdots + d_{2n-2}\omega^{2n-2}) + d_{2n}\omega^{2n}} \quad (1.15)$$

where we have assumed that $m < n$. Hence it approaches zero as ω approaches ∞. No doubt that the maximally flat approximation, which is of the form (1.15) has the maximum number of derivatives equal to zero at $\omega = 0$ but all

1.3. TYPES OF APPROXIMATION

the degrees of freedom have been used to satisfy this condition and also the condition that the magnitude approaches zero as ω approaches ∞. But as seen from Fig. 1.3, we have no control on the behavior of the magnitude function between these two frequencies.

If however, we satisfy the condition that $d_2 = d_4 = \cdots = d_{2n-2} = 0$, in (1.15), then it corresponds to a transfer function $H(s)$ which has only poles in the finite s-plane and all its zeros are at $s = \infty$. The transfer function (1.2) reduces to the all-pole function in the form (1.16).

$$H(s) = \frac{H_o}{\sum_{k=0}^{n} a_k s^k} \tag{1.16}$$

and the square of its magnitude reduces to the form (1.17).

$$|H(j\omega)|^2 = \frac{H_o^2}{1 + d_{2n}\omega^{2n}} \tag{1.17}$$

The magnitude response not only has the maximum number of its derivatives equal zero at $\omega = 0$ but it is also monotonically decreasing from $\omega = 0$. The special class of all-pole transfer functions which give rise to the magnitude squared function of the form (1.17) are specifically called as Butterworth filter functions. The square of the magnitude of the Butterworth filter function $H(s)$ given in (1.17) now approximates a constant value of H_o^2 in the maximally flat sense at $\omega = 0$ and it is also an all-pole function, whereas (1.15) represents a maximally flat approximation but is not the magnitude squared function of a Butterworth filter. The magnitude response of a Butterworth filter $H_B(s)$ of order 4 and a maximally flat filter $H_M(s)$ of order 4 are plotted in Fig. 1.3. The transfer functions of these two filters and the magnitude squared functions are respectively given by (1.18) and (1.19).

$$H_B(s) = \frac{1}{s^4 + 2.6131s^3 + 3.4142s^2 + 2.6131s + 1}$$
$$|H_B(j\omega)|^2 = \frac{1}{1+\omega^8} \tag{1.18}$$

$$H_M(s) = \frac{(s^2 + .25)}{s^4 + 3.0628s^3 + 4.6904s^2 + 2.2172s + .7906}$$
$$|H_M(j\omega)|^2 = \frac{0.0625 - 0.5\omega^2 + \omega^4}{0.0625 - 0.5\omega^2 + \omega^4 + 0.1\omega^8} \tag{1.19}$$

Both of them are of the form (1.15) and are maximally flat at the origin but only the Butterworth filter has a magnitude which decreases monotonically as ω increases from the origin.

Next we consider the min-max or equiripple approximation. The error function $J_2(\mathbf{a}, \mathbf{b}, \omega)$ to be minimized in order to achieve min-max approximation is the maximum value of the error function $\{W(\omega)|f(\omega) - g(\omega)|\}$ over the frequency range of interest i.e.

$$J_2(\mathbf{a}, \mathbf{b}, \omega) = \max_{\omega \in (\omega_1, \omega_2)} \{W(\omega)|f(\omega) - g(\omega)|\} \tag{1.20}$$

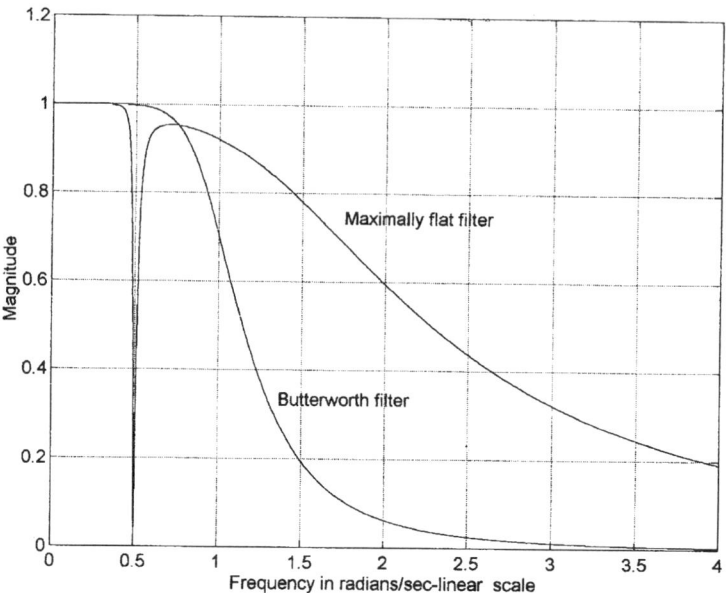

Fig. 1.3. Magnitude of a maximally flat filter and a Butterworth filter compared

When the error function in (1.20) is minimized with respect to the coefficients, it will have the same maxima and the same minima at a number of points in the frequency range $\omega \in (\omega_1, \omega_2)$ and hence the min-max approximation is called the equiripple approximation. It is also called the Chebyshev approximation. It is assumed in the above discussion that $f(\omega)$ and $g(\omega)$ are real valued functions. If we wish to approximate both the magnitude and phase response of the transfer function $H(s)$, then the functions $f(\omega)$ and $g(\omega)$ will be replaced by the corresponding complex valued functions $\overline{f(\omega)}$ and $\overline{g(\omega)}$ and the error function is modified as indicated below:

$$\overline{J_2}(\mathbf{a}, \mathbf{b}, \omega) = \max_{\omega \in (\omega_1, \omega_2)} \left\{ W(\omega) \left| \overline{f(\omega)} - \overline{g(\omega)} \right| \right\} \quad (1.21)$$

The third type of approximation is the least p^{th} approximation. The purpose of finding $f(\omega)$ in the form of a rational function such that it approximates the given function $g(\omega)$ in the least p^{th} sense is achieved by minimizing the error function $J_p(\mathbf{a}, \mathbf{b}, \omega)$ in (1.22), p being an even integer. But p could be chosen as any positive integer when the absolute value $|[f(\omega) - g(\omega)]|$ is chosen in (1.22). When $p = 2$ is chosen, it is called the least squares approximation.

$$J_p(\mathbf{a}, \mathbf{b}, \omega) = \int_{\omega_1}^{\omega_2} W(\omega) \left[f(\omega) - g(\omega) \right]^p d\omega \quad (1.22)$$

In the above two error functions, the unknown coefficients of the transfer function (1.2) are denoted by the vectors $\mathbf{a}$ and $\mathbf{b}$ and $W(\omega)$ is a positive, weighting function that is judiciously chosen over different frequency intervals, if $[f(\omega) - g(\omega)]$ is expected to take different orders of magnitude. Some times the frequency range of interest $[\omega_1, \omega_2]$ is decomposed as the sum of disjoint frequency intervals and the revised form of the error function is expressed in the form (1.23), in which $W_i(\omega)$ is the weighting function chosen for the ith frequency interval $[\omega_{i,1}, \omega_{i,2}]$.

$$J_I(\mathbf{a}, \mathbf{b}, \omega) = \sum_{i=1}^{I} \int_{\omega_{i,1}}^{\omega_{i,2}} W_i(\omega) \left[f(\omega) - g(\omega) \right]^p d\omega \qquad (1.23)$$

So far we have described only three major types of approximation that are used in the design of filters i.e. in finding the coefficients of the transfer function $H(s)$ such that its magnitude squared function approximates a prescribed function in terms of the three approximation criteria. But we have not described the procedure for finding the coefficients when the specifications are prescribed. The procedures are purely analytical in some cases or call for numerical methods of approximation that are implemented by computers in other cases. These procedures will be described briefly in the following sections.

The analytical procedure to be described for designing lowpass filters is based on the following approach. The magnitude squared function $|H(j\Omega)|^2 = H(j\Omega)H(-j\Omega)$ is assumed to be approximated by

$$|H(j\Omega)|^2 = \frac{H_o^2}{1 + \epsilon^2 K^2(j\Omega)} \qquad (1.24)$$

The function $K(s)$ from which $K^2(j\Omega)$ is derived is called the characteristic function. It is either a polynomial or a rational function in its general form. Then $K(j\Omega)K(-j\Omega) = |K(j\Omega)|^2 = K^2(\Omega^2)$ which is an even function in Ω. It approximates a constant value of zero in the frequency range $|\Omega| \leq 1$, in the sense of maximally flat approximation, equiripple approximation or the least squares approximation described above. The function $1 + \epsilon^2 K^2(j\Omega)$ is called the loss function which approximates the value of one over the interval $|\Omega| \leq 1$. It has a value of $(1 + \epsilon^2)$ at $\Omega = 1$, when $K(s)$ is chosen to satisfy the condition that $K^2(\Omega^2)\big|_{\Omega=1} = 1$. The function $10 \, log \left[1 + \epsilon^2 K^2(j\Omega) \right]$ is called the attenuation characteristics of the lowpass filter and it approximates a constant value of 0 dB in $|\Omega| \leq 1$ and at the normalized frequency $\Omega = 1$ it has the maximum attenuation $A_p = 10 \log(1 + \epsilon^2)$ measured in dB. The value of ϵ is chosen to meet the prescribed value for A_p. The design procedure consists of an analytic method for finding $H(s)$ from (1.24), after the appropriate function $K^2(\Omega^2)$ is chosen. This general outline of filter approximation for designing classical, lowpass filters will be elaborated in the next few sections.

1.4 Butterworth Approximation

We have already mentioned that the transfer functions of Butterworth filters constitute one special class of all-pole functions which have a magnitude response that is maximally flat at $\omega = 0$. It is convenient to normalize the gain

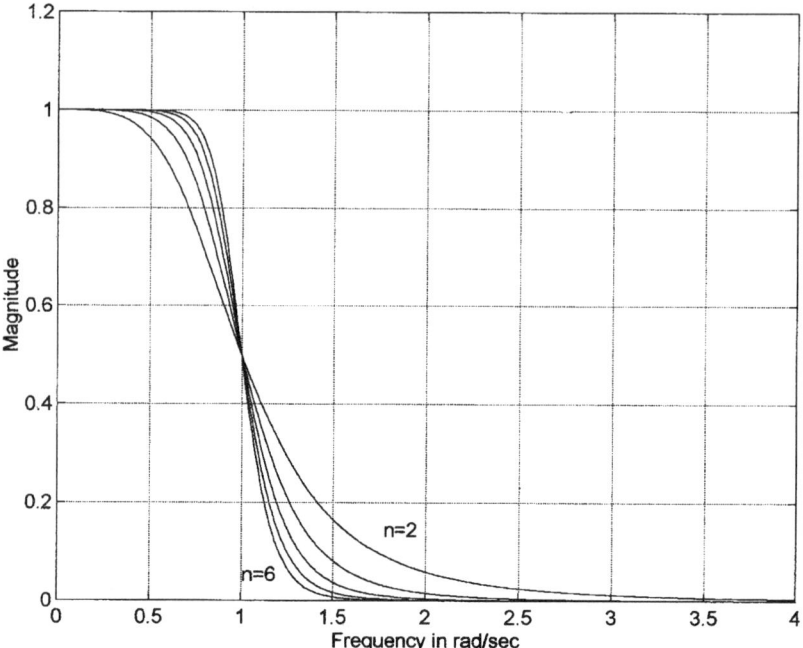

Fig. 1.4. Magnitude response of Butterworth filters

constant of these filters by letting $H_o = 1$ and also to scale the complex frequency $s = \sigma + j\omega$ by a factor ω_p. We will denote the scaled frequency $\frac{s}{\omega_p}$ by $p = \Sigma + j\Omega$. So in (1.17), when we set $H_o = 1$ and $d_{2n} = \left(\frac{1}{\omega_p}\right)^{2n}$, we get the magnitude squared function of the normalized prototype filter, which has the transfer function $H(p)$, in the form given below. (Note that we choose p as the complex frequency variable for the transfer function of the lowpass, *prototype* filter in discussing the design of all classical filters.)

$$|H(j\Omega)|^2 = H(p)H(-p)|_{p=j\Omega} = \frac{1}{1+\Omega^{2n}} \qquad (1.25)$$

The magnitude response of these filters is plotted in Fig. 1.4 for orders of filters $n = 2, 3, 4, 5, 6$.

Comparing (1.25) with (1.17), we see that $K^2(\Omega^2) = \Omega^{2n}$ in the case of Butterworth filters. This function (1.25) is not only a maximally flat approximation to a constant value of one at the origin but also has a magnitude of $\sqrt{\frac{1}{2}}$ at the normalized frequency $\Omega_p = 1$ (or a magnitude of -3.0 dB) for all values of n. The bandwidth of the prototype filter that corresponds to a magnitude of $\sqrt{\frac{1}{2}}$ (or -3.0 dB) is one radian/sec and its magnitude at $\Omega = 0$ is also normalized to one (or 0 dB). It can be easily shown that the magnitude response is maximally

1.4. BUTTERWORTH APPROXIMATION

flat at $\omega = \infty$ also. So we designate the filter with the transfer function $H(p)$ as the normalized prototype, Butterworth, lowpass filter which has a magnitude squared function given by (1.25).

Let us consider the design of a lowpass filter for which the frequency at which the magnitude is 3 dB below the maximum value, and also the magnitude at another frequency in the stopband are specified. These two frequencies are defined as the cutoff frequency ω_c (or the edge of the passband ω_p) and the stopband frequency ω_s respectively. When we normalize the gain constant to unity and the frequency by the scale factor ω_p, we get the cutoff frequency of the normalized prototype filter $\Omega_p = 1$ and the stopband frequency $\Omega_s = \frac{\omega_s}{\omega_p}$, besides having a magnitude of one at the origin. After we have found the transfer function $H(p)$ of this normalized, prototype, lowpass filter, we restore the frequency scale and the magnitude scale to get the transfer function $H(s)$ approximating the prescribed magnitude specification of the lowpass filter.

The analytical procedure to derive $H(p)$ from the magnitude squared function of the prototype lowpass filter is simply carried out by reversing the steps used to derive the magnitude squared function from $H(p)$. First we substitute $\Omega = \frac{p}{j}$ or equivalently $\Omega^2 = -p^2$ in (1.25).

$$\left.\frac{1}{1+\Omega^{2n}}\right|_{\Omega^2=-p^2} = \frac{1}{1+(-1)^n p^{2n}} = H(p)H(-p) \qquad (1.26)$$

The denominator has $2n$ zeros obtained by solving the equation

$$1 + (-)^n p^{2n} = 0 \qquad (1.27)$$

or the equation

$$p^{2n} = \begin{cases} 1 = e^{j2k\pi} & n \text{ odd} \\ -1 = e^{j(2k+1)\pi} & n \text{ even} \end{cases} \qquad (1.28)$$

This gives us the $2n$ poles of $H(p)H(-p)$ which are

$$p_k = e^{j(\frac{2k}{2n})\pi} \qquad k = 1, 2, \ldots, (2n) \qquad \text{when } n \text{ is odd} \qquad (1.29)$$

and

$$p_k = e^{j(\frac{2k-1}{2n})\pi} \qquad k = 1, 2, \ldots, (2n) \qquad \text{when } n \text{ is even} \qquad (1.30)$$

or in general,

$$p_k = e^{j(\frac{2k+n-1}{2n})\pi} \qquad k = 1, 2, \ldots, (2n) \qquad (1.31)$$

It is noticed that in both cases, the poles have a magnitude of one and angle between any two adjacent poles as one goes around the unit circle is equal to $\frac{\pi}{n}$. There are n poles in the left half of the p-plane and n poles in the right half of the p-plane, as illustrated for the case of $n = 2$ and $n = 3$ in Fig. 1.6. For every pole of $H(p)$ at $p = p_a$ which lies in the left half plane, there is a pole of $H(-p)$ at $p = -p_a$ which lies in the right half plane. Because of this property, we identify n poles which are in the left half p-plane as the poles of $H(p)$ so that it is a stable transfer function - the poles which are in the right half plane being assigned as the poles of $H(-p)$. The n poles which are in the left half of the p-plane are given by

$$p_k = \exp\left[j\left(\frac{2k+n-1}{2n}\right)\pi\right] \quad k = 1, 2, \ldots, n \quad (1.32)$$

When we have found these n poles, we construct the denominator polynomial $D(p)$ of the prototype filter $H(p) = \frac{1}{D(p)}$ from (1.33).

$$D(p) = \prod_{k=1}^{n}(p - p_k) \quad (1.33)$$

The only unknown parameter at this stage of design is the order n of the filter function $H(p)$, which is required in (1.32). This is calculated using the specification that at the stopband frequency Ω_s, the magnitude is required to be no more than $-A_s$ dB i.e.

$$10\log|H(j\Omega_s)|^2 = -10\log(1 + \Omega_s^{2n}) \leq -A_s \quad (1.34)$$

from which we derive the formula for calculating n as

$$n \geq \frac{\log(10^{0.1A_s} - 1)}{2\log\Omega_s} \quad (1.35)$$

Since we require that n to be an integer, we choose the actual value of $n = \lceil n \rceil$ which means the next higher integer value or the ceiling of n obtained from the right side of (1.35). We use this integer value for n in (1.32), to calculate the poles and then construct the denominator polynomial $D(p)$ of order n. By multiplying $(p - p_k)$ with $(p - p_k^*)$ where p_k and p_k^* are complex conjugate pairs, the denominator polynomial is reduced to the normal form with real coefficients only. These polynomials are known as Butterworth polynomials and they have many special properties. In the polynomial form, if we represent them as

$$D(p) = 1 + d_1 p + d_2 p^2 + \cdots + d_n p^n \quad (1.36)$$

their coefficients can be computed recursively from ($d_o = 1$)

$$d_k = \frac{\cos\left[(k-1)\frac{\pi}{2}\right]}{\sin\left[\frac{k\pi}{2n}\right]} d_{k-1} \quad k = 1, 2, \ldots, n \quad (1.37)$$

But there is no need to do so, since they can be computed from (1.33). They are also listed in many books for n up to 10 in polynomial form and in some books in a factored form also [10, 24, 29, 33]. We list a few of them below in Table 1.1. The magnitude and delay response of the filters are also given in many books for n up to 10. Fig. 1.5 shows the attenuation characteristics of the Butterworth filters for $n = 2, 3, \ldots 10$ [33].

In the case of lowpass filters, usually the magnitude is specified at $\omega = 0$; hence it is also the magnitude at $\Omega = 0$. Therefore the specified magnitude is

1.4. BUTTERWORTH APPROXIMATION

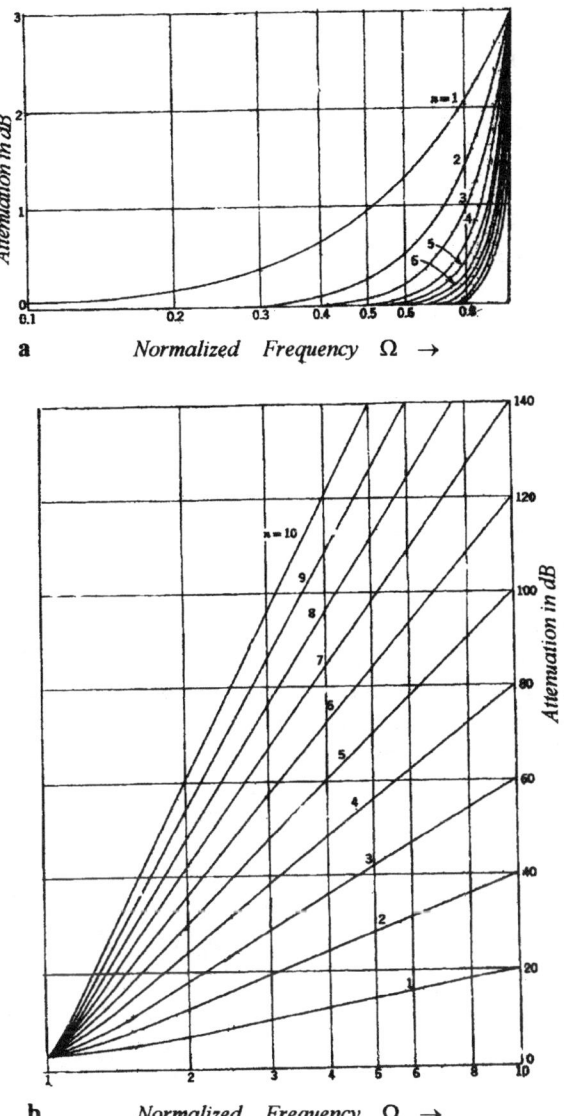

Fig. 1.5. Attenuation characteristics of Butterworth lowpass filters

equated to the value of the transfer function $H(p)$ evaluated at $p = j0$. This is equal to $H(j0) = \frac{H_0}{D(j0)} = H_0$. So we restore the magnitude scale by multiplying the normalized prototype filter function by H_0. To restore the frequency scale by ω_p, we put $p = \frac{s}{\omega_p}$ in $\frac{H_0}{D(p)}$ and simplify the expression to get transfer function $H(s)$ for the specified lowpass filter. This completes the design procedure which

14 CHAPTER 1. CLASSICAL METHODS OF APPROXIMATION

Table 1.1. Butterworth Polynomials.

n	Butterworth Polynomial $D(p)$
1	$(p+1)$
2	$p^2 + \sqrt{2}p + 1$
3	$p^3 + 2p^2 + 2p + 1 = (p+1)(p^2 + p + 1)$
4	$p^4 + 2.613p^3 + 3.414p^2 + 2.613p + 1$
	$= (p^2 + .765p + 1)(p^2 + 1.84p + 1)$
5	$p^5 + 3.236p^4 + 5.236p^3 + 5.236p^2 + 3.236p + 1$
	$= (p+1)(p^2 + .618p + 1)(p^2 + 1.618p + 1)$

will be illustrated below by an example.

1.4.1 Example 1.1

Design a lowpass Butterworth filter with a d.c. gain of 5 dB and a cutoff frequency of 1000 rad/sec at which the gain is 2 dB and a stopband frequency of 5000 rad/sec at which the magnitude is required to be less than -25 dB.

The d.c. gain of 5 dB is the magnitude of the filter function at $\omega = 0$. The passband is the same as the cutoff frequency $\omega_p = 1000$. The magnitude of 2 dB at this frequency means that the filter has a 3 dB bandwidth. The frequency scale factor is chosen as 1000 so that the passband of the prototype filter is $\Omega_p = 1$. The stopband frequency ω_s is specified as 5000 rad/sec and is therefore scaled to $\Omega_s = 5$. The magnitude is normalized so that the normalized prototype lowpass filter function $H(p)$ has a magnitude of one (i.e. 0 dB) at

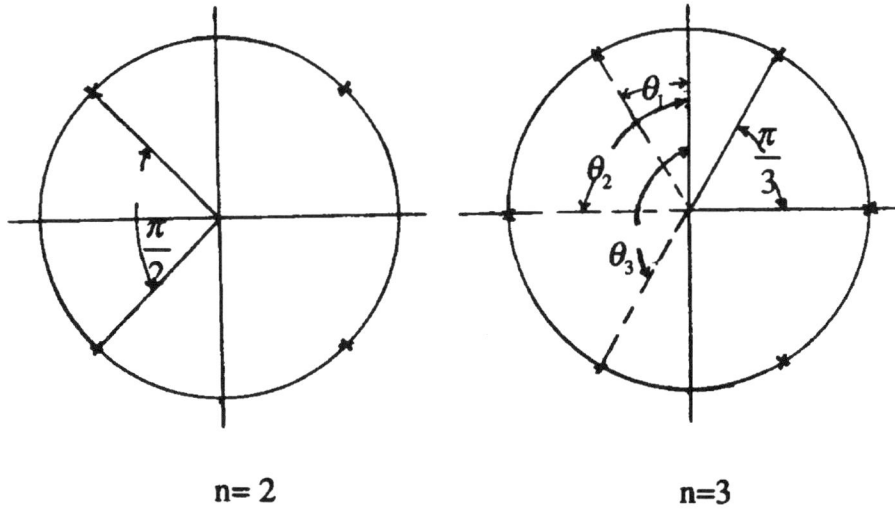

n= 2   n=3

Fig. 1.6. Pole locations of H(p)H(-p)

1.4. BUTTERWORTH APPROXIMATION

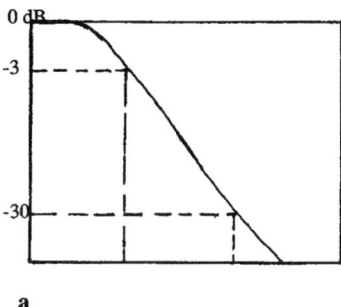
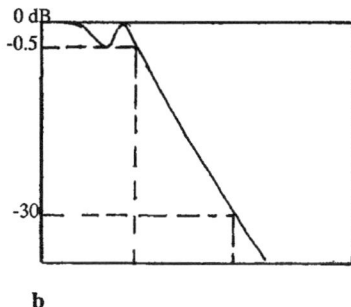

Fig. 1.7. Specifications of normalized prototype filters. **a** Butterworth filter **b** Chebyshev filter

$\Omega = 0$. It is this filter which has a magnitude squared function

$$|H(j\Omega)|^2 = \frac{1}{1+\Omega^{2n}} \qquad (1.38)$$

It is always necessary to reduce the given specifications to the specifications of this normalized prototype filter to which only the the expressions derived above are applicable. The magnitude response of the normalized prototype filter (not to scale) for this example is shown in Fig. 1.7a.

For this example, note that the maximum attenuation in the passband is $A_p = 3$ dB and the minimum attenuation in the stopband is $A_s = 30$ dB. From (1.35) we calculate the value of $n = 2.1457$ and choose $n = \lceil 2.1457 \rceil = 3$. From (1.32), we get the three poles as $p_1 = -0.5 + j\sqrt{.75}$, $p_2 = -1.0$ and $p_3 = -0.5 - j\sqrt{.75}$. Therefore the third order denominator polynomial $D(p)$ is obtained from (1.33) or from Table 1.1.

$$\begin{aligned} D(p) &= (p+0.5-j\sqrt{.75})(p+1)(p+0.5+\sqrt{.75}) \qquad (1.39) \\ &= (p^2+p+1)(p+1) = (p^3+2p^2+2p+1) \end{aligned}$$

Hence the transfer function of the normalized prototype filter of third order is

$$H(p) = \frac{1}{p^3+2p^2+2p+1} \qquad (1.40)$$

To restore the magnitude scale, we multiply the above function by H_o. Now the filter function is

$$H(p) = \frac{H_o}{p^3+2p^2+2p+1} \qquad (1.41)$$

which has a magnitude of H_o at $p = j0$. From the requirement $20\log(H_0) = 5$ dB, we calculate the value of $H_0 = 1.7783$. To restore the frequency scale, we substitute $p = \frac{s}{1000}$ in (1.41) and simplify to get $H(s)$ as shown below:

$$H(p)|_{p=\frac{s}{1000}} = \frac{1.7783}{\left(\frac{s}{1000}\right)^3+2\left(\frac{s}{1000}\right)^2+2\left(\frac{s}{1000}\right)+1} \qquad (1.42)$$

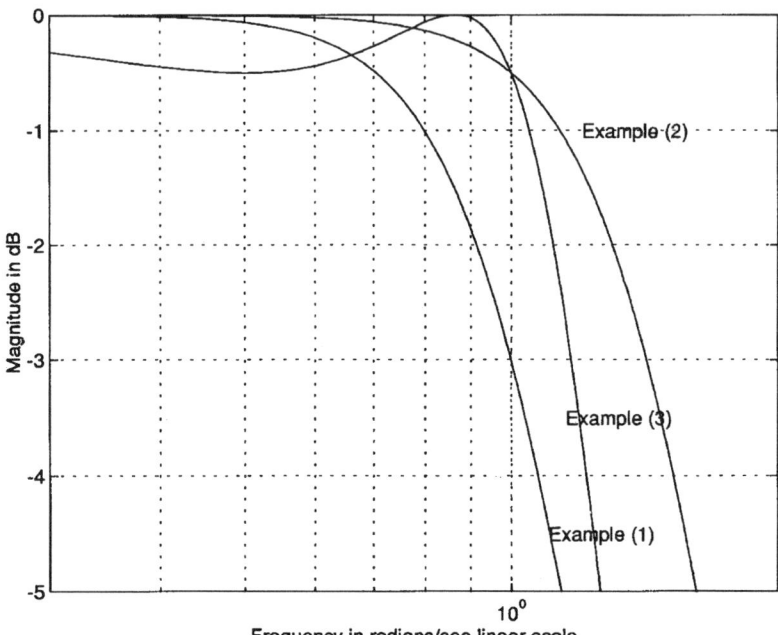

Fig. 1.8. Magnitude response of the prototype filter H(p) in Example 1

$$= \frac{(1.7783)10^9}{s^3 + 2 \times (10^3)s^2 + 2 \times (10^6)s + 10^9}$$
$$= H(s)$$

The magnitude of $H(p)$ is given in Fig. 1.8. It is found that the attenuation at $\Omega = 5$ is about 42 dB which is more than the specified 30 dB.

It must be remembered that in (1.38) $\Omega_p = 1$ is the bandwidth of the filter, and at this frequency, $|H(j\Omega)|^2$ has a value of $\frac{1}{2}$ or a magnitude of -3 dB. But the formulae (1.32) and (1.35) can not be used if the maximum attenuation A_p in the passband is different from 3 dB. In this case, we modify the function to the form (1.43) which is the general case.

$$|H(j\Omega)|^2 = \frac{1}{1 + \epsilon^2 \Omega^{2n}} \qquad (1.43)$$

Now the attenuation at $\Omega = 1$ is given by $10\log(1 + \epsilon^2) = A_p$ from which we get $\epsilon^2 = (10^{0.1A_p} - 1)$. We may also note that $\epsilon^2 = 1$ in the previous case when $A_p = 3$. When A_p is other than 3 dB, the formulae for calculating n and p_k are given below:

$$n \geq \frac{\log\left[(10^{0.1A_s} - 1)/(10^{0.1A_p} - 1)\right]}{2\log \Omega_s} \qquad (1.44)$$

1.4. BUTTERWORTH APPROXIMATION

and

$$p_k = \epsilon^{-\frac{1}{n}} \exp\left[j\left(\frac{2k+n-1}{2n}\right)\pi\right] \qquad k = 1, 2, \ldots, n \qquad (1.45)$$

Comparing (1.45) with (1.32), it is obvious that the poles have been scaled by a factor $\epsilon^{-\frac{1}{n}}$. So the maximum attenuation at $\Omega_p = 1$ is the specified value of A_p; also the frequency at which the attenuation is 3 dB is equal to $\epsilon^{-\frac{1}{n}}$.

1.4.2 Example 1.2

Design a lowpass Butterworth filter with a maximum magnitude of 5 dB, passband of 1000 rad/sec, maximum attenuation in the passband $A_p = 0.5$ dB, minimum attenuation at the stopband frequency of 5000 rad/sec is $A_s = 30$ dB.

First we scale the frequency by $\omega_p = 1000$ so that the frequency ω_s is mapped to $\Omega_s = 5$. Also the magnitude is scaled by 5 dB. The magnitude response for the normalized prototype filter $H(p)$ is similar to that shown in Fig. 1.7a except that now $A_p = 0.5$ dB.

Then we calculate $\epsilon^2 = (10^{0.1A_p} - 1) = 0.1220$ and therefore $\epsilon = 0.3493$. From (1.44), the value of $n = 2.7993$; it is rounded to $n = 3$. Next we compute the three poles from (1.45) as $p_1 = -0.71 + j1.2297$, $p_2 = -1.4199$ and $p_3 = -0.71 - j1.2297$. The transfer function of the filter with these poles is

$$\begin{aligned} H(p) &= \frac{H_0}{(p+1.4199)(p+0.71-j1.2297)(p+0.71+j1.2297)} \\ &= \frac{H_0}{(p+1.4199)(p^2+1.42p+2.0163)} \end{aligned} \qquad (1.46)$$

Since the maximum value is normalized to 0 dB, which occurs at $\Omega = 0$, we equate the magnitude of $H(p)$ evaluated at $p = j0$ to one. Therefore $H_0 = 1.4199 \times 2.0163 = 2.8629$. To raise the magnitude level to 5 dB, we have to multiply this constant by $\sqrt{10^{0.5}} = 1.7783$. The frequency scale is restored by putting $p = \frac{s}{1000}$ in (1.46) to get (1.47) as the transfer function of the filter that meets the given specifications.

$$\begin{aligned} H(s) &= \frac{2.8629 \times 1.7783}{[\frac{s}{1000}+1.4199][(\frac{s}{1000})^2+1.42(\frac{s}{1000})+2.0163]} \\ &= \frac{5.09 \times 10^9}{[s+1419.9][s^2+1420s+2.0163 \times 10^6]} \end{aligned} \qquad (1.47)$$

The magnitude of this filter (1.46) is shown in Fig. 1.8. It has a magnitude of -0.5 dB at $\Omega = 1$ and has a magnitude of about -33 dB at $\Omega = 5$ which exceeds the specified value.

1.5 Chebyshev Approximation

To approximate the ideal magnitude response of the lowpass filter in the equiripple sense, the magnitude squared function of its prototype is chosen to be

$$|H(j\Omega)|^2 = \frac{H_0^2}{1 + \epsilon^2 C_n^2(\Omega)} \qquad (1.48)$$

So, in the case of the Chebyshev filters, we have chosen $K^2(\Omega) = C_n^2(\Omega)$, where $C_n(\Omega)$ is the Chebyshev polynomial of degree n. It is defined by

$$C_n(\Omega) = \cos(n \cos^{-1} \Omega) \quad ; \quad |\Omega| \leq 1 \qquad (1.49)$$

The polynomial $C_n(\Omega)$ approximates a value of zero over the closed interval $\Omega \in [-1, 1]$ in the equiripple sense as shown by examples for $n = 2, 3, 4, 5$ in Fig. 1.9a. These polynomials are

$$\begin{aligned}
C_0(\Omega) &= 1 \\
C_1(\Omega) &= \Omega \\
C_2(\Omega) &= 2\Omega^2 - 1 \\
C_3(\Omega) &= 4\Omega^3 - 3\Omega \\
C_4(\Omega) &= 8\Omega^4 - 8\Omega^2 + 1 \\
C_5(\Omega) &= 16\Omega^5 - 20\Omega^3 + 5\Omega
\end{aligned} \qquad (1.50)$$

Some of the properties of Chebyshev polynomials which are useful for our discussion are described below. Let $\cos(\phi) = \Omega$. Then $C_n(n \cos^{-1} \Omega) = \cos(n\phi)$ and therefore we use the identity

$$\begin{aligned}
\cos(k+1) &= \cos(k\phi)\cos(\phi) - \sin(k\phi)\sin(\phi) \\
&= 2\cos(\phi)\cos(k\phi) - \cos((k-1)\phi)
\end{aligned} \qquad (1.51)$$

from which we obtain a recursive formula to generate Chebyshev polynomials of any order, as

$$\begin{aligned}
C_0(\Omega) &= 1 \\
C_{k+1}(\Omega) &= 2\Omega C_k(\Omega) - C_{k-1}(\Omega)
\end{aligned} \qquad (1.52)$$

To see that $C_n(\Omega) = \cos(n \cos^{-1} \Omega)$ is indeed a polynomial of order n, consider it in the following form:

$$\begin{aligned}
\cos(n\phi) &= Re\left[e^{jn\phi}\right] \\
&= Re\left[\cos(\phi) + j\sin(\phi)\right]^n = Re\left[\phi + j\sqrt{(1-\phi^2)}\right]^n \\
&= Re\left[\phi + \sqrt{\phi^2 - 1}\right]^n
\end{aligned} \qquad (1.53)$$

1.5. CHEBYSHEV APPROXIMATION

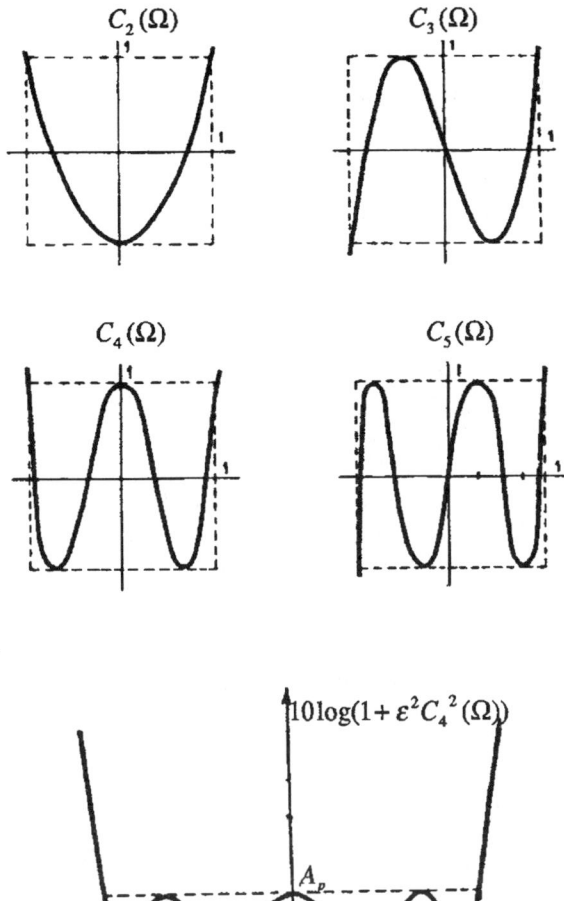

Fig. 1.9. Chebyshev polynomials and Chebyshev filters. **a** Magnitude of Chebyshev polynomials **b** Attenuation of Chebyshev I lowpass filter($n = 4$)

We expand $\left[\phi + \sqrt{\phi^2 - 1}\right]^n$ by the Binomial Theorem and choose the real part, to get the polynomial $C_n(\Omega)$ in the form of

$$\cos(n\phi) = \phi^n + \frac{n(n-1)}{2!}\phi^{n-2}(\phi^2 - 1) + \frac{n(n-1)(n-2)(n-3)}{4!}\phi^{n-4}(\phi^2 - 1)^2 + \cdots$$

Recollect that since n is a positive integer, the above expansion has a finite number of terms and hence we conclude that it is a polynomial (of degree n).

We also notice from (1.50) that

$$C_n^2(0) = \begin{cases} 0 & n \text{ odd} \\ 1 & n \text{ even} \end{cases}$$

But

$$C_n^2(1) = \begin{cases} 1 & n \text{ odd} \\ 1 & n \text{ even} \end{cases}$$

So, we derive the following properties:

$$|H(0)|^2 = \begin{cases} 1 & n \text{ odd} \\ \frac{1}{1+\epsilon^2} & n \text{ even} \end{cases} \quad (1.54)$$

$$|H(1)|^2 = \frac{1}{1+\epsilon^2} \quad n \text{ odd or even} \quad (1.55)$$

The attenuation characteristics of the Chebyshev filter of order $n = 4$ is shown in Fig. 1.9b as an example. The magnitude $|H(j\Omega)|$ shown as Example (3) in Fig. 1.8 has an equiripple response in the passband, with a maximum value of 0 dB and a minimum value of $10 log\left(\frac{1}{1+\epsilon^2}\right)$; it has a magnitude of $10\log\left(\frac{1}{1+\epsilon^2}\right)$ at $\Omega = 1$ for any order n. The magnitude of the ripple can be measured either as $|H(0)| - |H(1)|$ or $|H(0)|^2 - |H(1)|^2 = 1 - \frac{1}{1+\epsilon^2} = \frac{\epsilon^2}{1+\epsilon^2} \approx \epsilon^2$. An unambiguous measure of the ripple in the passband is measured in dB as the value of $A_p = 10\log(1+\epsilon^2)$ and it is shown in Fig. 1.9b. From this we can always calculate $\epsilon^2 = (10^{0.1A_p} - 1)$. It is known that the total number of maxima and minima in the closed interval $[-1,1]$ is $n + 1$.

Typically the specifications for a lowpass Chebyshev filter specify the maximum and minimum value of the magnitude in the passband, the cutoff frequency ω_p which is the highest frequency of the passband, a frequency ω_s in the stopband and the magnitude at the frequency ω_s. As in the case of the Butterworth filter, we normalize the magnitude and the frequency and reduce the given specifications to those of the normalized, prototype, Chebyshev, lowpass filter and follow similar steps to find the poles of $H(p)$.

Since Ω can take real values greater than one in general, let us assume ϕ to be a complex variable $\phi = \varphi_1 + j\varphi_2$. From $1 + \epsilon^2 C_n^2(\Omega) = 0$, we derive

$$\begin{aligned} C_n(\Omega) &= \pm\frac{j}{\epsilon} \quad (1.56) \\ &= \cos(n\phi) = \cos(n(\varphi_1 + j\varphi_2)) \\ &= \cos(n\varphi_1)\cosh(n\varphi_2) - j\sin(n\varphi_1)\sinh(n\varphi_2) \end{aligned}$$

Equating the real and imaginary parts, we get

$$\cos(n\varphi_1)\cosh(n\varphi_2) = 0 \quad (1.57)$$

$$\sin(n\varphi_1)\sinh(n\varphi_2) = \mp\frac{1}{\epsilon} \quad (1.58)$$

1.5. CHEBYSHEV APPROXIMATION

From (1.57) we get $\varphi_1 = \frac{(2k-1)\pi}{2n}$. Substituting this in (1.58), we obtain $\sinh(n\varphi_2) = \pm\frac{1}{\epsilon}$, from which we get

$$\varphi_2 = \frac{1}{n}\sinh^{-1}\left(\frac{1}{\epsilon}\right) \tag{1.59}$$

Now $\Omega = \cos(\phi) = \cos(\varphi_1 + j\varphi_2) = \cos(\varphi_1)\cosh(\varphi_2) - j\sin(\varphi_1)\sinh(\varphi_2)$.
Therefore

$$j\Omega = \sin(\varphi_1)\sinh(\varphi_2) + j\cos(\varphi_1)\cosh(\varphi_2) \tag{1.60}$$

These are the roots in the p-plane which satisfy the condition $1 + \epsilon^2 C_n^2(\Omega) = 0$. Hence the $2n$ poles of $H(p)H(-p)$ are given by

$$p_k = \sinh(\varphi_2)\sin\left[\frac{(2k-1)\pi}{2n}\right] + j\cosh(\varphi_2)\cos\left[\frac{(2k-1)\pi}{2n}\right]$$

$$for \; k = 1, 2, \ldots, (2n) \tag{1.61}$$

The poles in the left half p-plane only, are given by

$$\begin{aligned}p_k &= -\sinh(\varphi_2)\sin\left[\frac{(2k-1)\pi}{2n}\right] + j\cosh(\varphi_2)\cos\left[\frac{(2k-1)\pi}{2n}\right] \\ &= -\sinh(\varphi_2)\sin(\theta_k) + j\cosh(\varphi_2)\cos(\theta_k); \quad k = 1, 2, \cdots, n \end{aligned} \tag{1.62}$$

where φ_2 is obtained from (1.59). Note that θ_k are the angles measured from the imaginary axis of the p-plane and lie in the left half of the p-plane. The $2n$ poles of $H(p)H(-p)$ given by (1.61) are found to lie on an elliptic contour in the p-plane with a major semiaxis equal to $\cosh(\varphi_2)$ along the $j\Omega$ axis and a minor semiaxis equal to $\sinh(\varphi_2)$ along Σ axis. We find that the frequency Ω_3 at which the attenuation of the prototype filter is 3 dB is given by

$$\Omega_3 = \cosh\left[\frac{1}{n}\cosh^{-1}\left(\frac{1}{\epsilon}\right)\right] \tag{1.63}$$

The formula for finding the order n is derived from the requirement that $10\log[1+\epsilon^2 C_n^2(\Omega_s)] \geq A_s$. It is

$$n \geq \frac{\cosh^{-1}\sqrt{[(10^{0.1A_s}-1)/(10^{0.1A_p}-1)]}}{\cosh^{-1}\Omega_s} \tag{1.64}$$

and the value of $\lceil n \rceil$ is chosen for calculating the poles using (1.62). Given ω_p, A_p, ω_s and A_s as the specifications for a Chebyshev lowpass filter $H(s)$, its maximum value in the passband is normalized to one, and its frequencies are scaled by ω_p, to get the values of $\Omega_p = 1$ and $\Omega_s = \frac{\omega_s}{\omega_p}$ for the prototype filter at which the attenuations are A_p and A_s respectively. The design procedure to find $H(s)$ starts with the magnitude squared function (1.48) and proceeds as follows:

1. Calculate $\epsilon = \sqrt{(10^{0.1A_p}-1)}$.

2. Calculate n from (1.64) and choose $n = \lceil n \rceil$.

3. Calculate φ_2 from (1.59).

4. Calculate the poles p_k $(k = 1, 2, \ldots, n)$

5. Compute $H(p) = \frac{H_0}{\prod_{k=1}^{n}(p - p_k)} = \frac{H_0}{\sum_{k=0}^{n} d_k p^k}$.

6. Find the value of H_0 by equating the value of $H(0)$.

$$H(0) = \frac{H_0}{d_0} = \begin{cases} 1 & n \text{ odd} \\ \frac{1}{\sqrt{1+\epsilon^2}} & n \text{ even} \end{cases}$$

7. Restore the magnitude scale.

8. Restore the frequency scale by substituting $p = \frac{s}{\omega_p}$ in $H(p)$ and simplify to get $H(s)$.

A simple example is worked out below to illustrate the above design procedure.

1.5.1 Example 1.3

Let us choose the specifications of a lowpass, Chebyshev filter with a d.c. gain of 5 dB, a bandwidth of 2500 rad/sec, a stopband frequency of 12500 rad/sec, $A_p = 0.5$ dB and $A_s = 30$ dB. For the prototype filter, the maximum value in the magnitude is one (0 dB), and we have $\Omega_p = 1$, $\Omega_s = 5$.

1. $\epsilon = \sqrt{(10^{0.05} - 1)} = .34931$

2. $n \geq \frac{\cosh^{-1}\sqrt{[(10^3-1)/(10^{0.05}-1)]}}{\cosh^{-1}(5)} = 2.2676$; Choose $n = 3$

3. $\varphi_2 = \frac{1}{3}\sinh^{-1}\left(\frac{1}{.34931}\right) = .591378$

4. $p_k = -0.313228 \pm j1.02192$ and -0.626456

5. Obtain

$$H(p) = \frac{H_0}{(p + .31228 - j1.02192)(p + .31228 + j1.02192)(p + 0.626456)}$$
$$= \frac{H_0}{(p^2 + .626456p + 1.142447)(p + 0.626456)}$$

6. $H(0) = \frac{H_0}{(1.142447)(0.626456)} = 1$ (since n=3 is odd). Hence $H_0 = 0.715693$

7. The transfer function $H(p) = \frac{0.715693}{(p^2 + .626456p + 1.142447)(p + 0.626456)}$
The magnitude scale is restored by multiplying $H(p)$ by 1.7783, so that the d.c. gain is raised to 5 dB.

1.6. INVERSE CHEBYSHEV APPROXIMATION

8. The transfer function of the filter is

$$H(p) = \frac{(0.715693)(1.7783)}{(p^2 + .626456p + 1.142447)(p + 0.626456)} \quad (1.65)$$

When we substitute $p = \frac{s}{2500}$ in this $H(p)$ and simplify the expression, we get

$$H(s) = \frac{19.886 \times 10^{12}}{(s^2 + 1566s + 714 \times 10^6)(s + 1566)} \quad (1.66)$$

The magnitude response of the prototype filter (1.66) is shown as Example (3) in Fig. 1.8 in order that the response of the three examples can be compared. It is found that the attenuation of the Chebyshev filter at $\Omega_s = 5$ is found to be 47 dB. The above class of filters with equiripple passband response are sometimes called as Chebyshev I filters to distinguish them from the following class of filters, known as Chebyshev II filters.

1.6 Inverse Chebyshev Approximation

The Chebyshev II filters have a magnitude response that is maximally flat at $\omega = 0$; it decreases monotonically as the frequency increases and has an equiripple response in the stopband. This class of filters are also called the Inverse Chebyshev filters. The first step in deriving them is to apply a frequency transformation $\Omega = \frac{1}{\Omega}$ in $|H(j\Omega)|^2$ of the lowpass normalized prototype filter. This gives the magnitude squared function of the highpass filter $\left|H(\frac{1}{j\Omega})\right|^2$, with an equiripple passband in $|\Omega| > 1$ and a monotonically decreasing response in the stopband $0 < |\Omega| < 1$. As the second step, we subtract $\left|H(\frac{1}{j\Omega})\right|^2$ from one, and get the magnitude squared function (1.68) of the Inverse Chebyshev lowpass filter with a response as described above.

$$\left|H(\frac{1}{j\Omega})\right|^2 = \frac{1}{1 + \epsilon^2 C_n^2(\frac{1}{\Omega})} \quad (1.67)$$

$$1 - \frac{1}{1 + \epsilon^2 C_n^2(\frac{1}{\Omega})} = \frac{\epsilon^2 C_n^2(\frac{1}{\Omega})}{1 + \epsilon^2 C_n^2(\frac{1}{\Omega})} = \frac{1}{\left[1 + \frac{1}{\epsilon^2 C_n^2(\frac{1}{\Omega})}\right]} \quad (1.68)$$

Typical magnitudes of Chebyshev lowpass filters are shown in Figs. 1.10a-b. Next we consider the magnitude squared function $|H(j\Omega)|^2$ of a lowpass Chebyshev I filter and transform its magnitude to that of Chebyshev II lowpass filter, as explained above. The magnitude responses of $|H(j\Omega)|^2$, $\left|H(\frac{1}{j\Omega})\right|^2$ and $1 - \left|H(\frac{1}{j\Omega})\right|^2$ used to arrive at the Chebyshev II lowpass filter from Chebyshev I filter, are shown in Fig. 1.11.

We make two important observations in Fig. 1.11. The normalized cutoff frequency $\Omega = 1$ is the lowest frequency in the stopband of the Inverse Chebyshev filter at which the magnitude is $\frac{\epsilon^2}{1+\epsilon^2}$. Hence the frequencies ω_p and ω_s

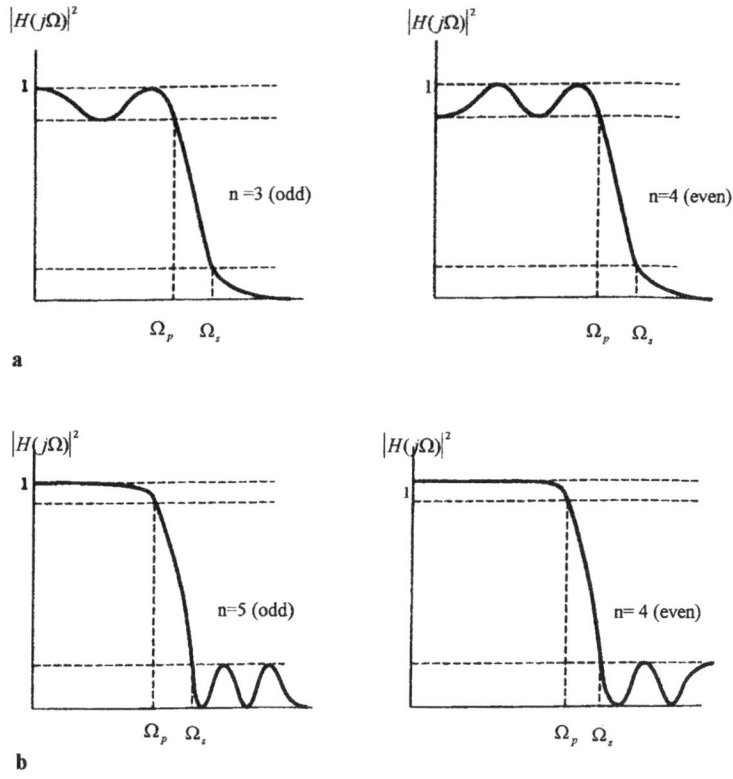

Fig. 1.10. Magnitude response of Chebyshev lowpass filters. **a** Chebyshev I filters **b** Chebyshev II filters

specified for the Inverse Chebyshev filter must be scaled by ω_s and not by ω_p to obtain the prototype of the Inverse Chebyshev filter. We also observe that when n is odd, the number of finite zeros in the stopband is $\frac{n-1}{2} = m$. When n is an odd integer, $\sec(\theta_k)$ attains a value of ∞ when $k = \frac{n+1}{2}$. So one of the zeros is shifted to $j\infty$, the remaining finite zeros appear in conjugate pairs on the imaginary axis and hence the numerator of the Chebyshev II filter is expressed as shown in step 6 below. Note that the value of ϵ_i calculated in step 1 is different from the value calculated in the design of Chebyshev I filters and therefore the value of φ_i used in step 3 and 4 are different from φ_2 used in the design of Chebyshev I filters. Hence it would be misleading to state that the poles of the Chebyshev II filters are obtained as the reciprocals of the poles of the Chebyshev I filters.

Given ω_p, A_p, ω_s, A_s and the maximum value in the passband, we scale the frequencies ω_p and ω_s by ω_s and deduce the specifications for the normalized prototype, lowpass, inverse Chebyshev filter. Equation (1.68) is the magnitude

1.6. INVERSE CHEBYSHEV APPROXIMATION

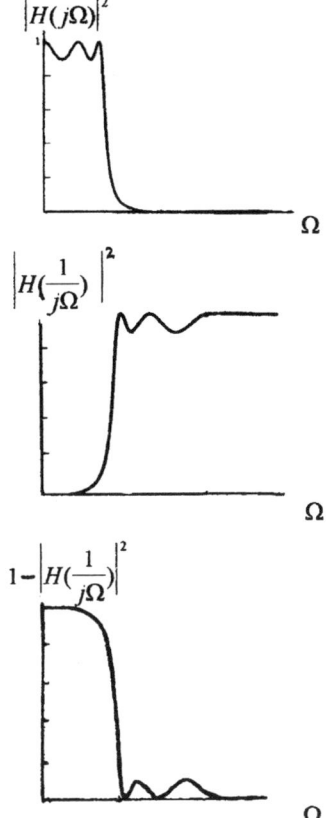

Fig. 1.11. Transformation of Chebyshev I to Chebyshev II filter response

squared function of this Inverse Chebyshev filter and we follow the design procedure as given below:

1. Calculate $\epsilon_i = \dfrac{1}{\sqrt{(10^{0.1A_s}-1)}}$

2. Calculate $n \geq \dfrac{\cosh^{-1}\sqrt{[(10^{0.1A_s}-1)/(10^{0.1A_p}-1)]}}{\cosh^{-1}\Omega_s}$ and choose $n = \lceil n \rceil$

3. Calculate φ_i from $\varphi_i = \dfrac{1}{n}\sinh^{-1}(\dfrac{1}{\epsilon_i})$

4. Compute the poles in the left-half plane p_k

$$p_k = \dfrac{1}{-\sinh(\varphi_i)\sin(\theta_k) + j\cosh(\varphi_i)\cos(\theta_k)}; \quad k = 1, 2, \ldots, n$$

5. The zeros of the transfer function $H(p)$ are calculated.

$$z_k = \pm j\Omega_{ok} = j\sec(\theta_k) \quad k = 1, 2, \ldots m = \lfloor \frac{n}{2} \rfloor$$

and the numerator $N(p)$ of $H(p)$ as $\prod_{k=1}^{m}(p + \Omega_{ok}^2)$. Here $m = \lfloor \frac{n}{2} \rfloor$ is the value obtained by truncating $\frac{n}{2}$ to its integer value

6. Compute $H(p) = \frac{H_0 \prod_{k=1}^{m}(p+\Omega_{ok}^2)}{\prod_{k=1}^{n}(p-p_k)}$ and calculate $H_0 = \frac{\prod_{k=1}^{n}(p_k)}{\prod_{k=1}^{m}(\Omega_{ok})^2}$

7. Restore the magnitude scale

8. Restore the frequency scale by putting $p = \frac{s}{\omega_s}$ in $H(p)$ to get $H(s)$ for the Inverse Chebyshev filter

1.6.1 Example 1.4

Design the lowpass Inverse Chebyshev filter with a maximum gain of 0 dB in the passband, $\omega_p = 1000$, $A_p = 0.5$ dB, $\omega_s = 2000$, and $A_s = 40$ dB. We normalize the frequencies by ω_s and get the lowest frequency of the stopband at $\Omega = 1$, while $\omega_p = 1000$ maps to $\Omega_p = 0.5$. We will have to denormalize the frequency by substituting $p = \frac{s}{2000}$ when the transfer function $H(p)$ of the Inverse Chebyshev filter obtained by the steps given above, is completed. The design procedure gives:

1. $\epsilon_i = \frac{1}{\sqrt{10^4 - 1}} = \frac{1}{99.995}$

2. $n = 5$

3. $\varphi_i = \frac{1}{5}\sinh^{-1}(99.995) = 1.05965847$

4. Poles in the left half plane are $p_k = (-0.155955926 \pm j0.6108703175)$, $(-0.524799485 \pm j0.485389011)$ and (-0.7877702666)

5. Zeros are $z_1 = \pm j1.0515$ and $z_2 = \pm j1.7013$

6. The transfer function of the Inverse Chebyshev filter $H(p)$ is given by

$$\frac{H_0(p^2 + 1.0515^2)(p^2 + 1.7013^2)}{(p^2 + .3118311852p + .3974722176) \times (p^2 + 1.04959897p + .5110169847)(p + .787702666)}$$

7. Calculate $H_0 = 0.049995$

8. Hence we simplify $H(p)$ to the final form

$$\frac{0.049995(p^4 + 4.04p^2 + 3.2002)}{(p^5 + 2.1491328p^4 + 2.30818905p^3 + 1.54997p^2 + 0.65725515p + 0.15999426)}$$

1.7. ELLIPTIC APPROXIMATION

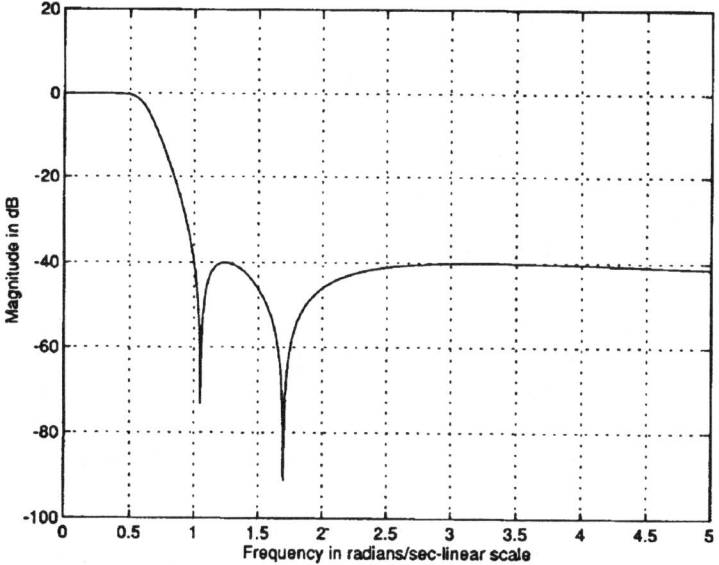

Fig. 1.12. Magnitude response of Chebyshev II lowpass filter

The magnitude response of the prototype Inverse Chebyshev filter is plotted in Fig. 1.12. It is seen that the prototype filter meets the desired specifications. Now we only have to denormalize the frequency by 2000, so that the passband of the specified filter changes from 0.5 to 1000 rad/sec, and it meets the specifications given in Example 1.4.

1.7 Elliptic Approximation

The lowpass elliptic filters exhibit equiripple response in both the passband and the stopband. The magnitude squared function of the prototype filters is

$$|H(j\Omega)|^2 = \frac{1}{1 + \epsilon^2 R_n^2(\Omega, \delta)} \quad (1.69)$$

Hence we have $K^2(\Omega) = R_n^2(\Omega)$ where $R_n(\Omega)$ is called the Chebyshev rational function of order n. It is of the form

$$R_n(\Omega) = C_2 \Omega \prod_{i=1}^{\frac{n-1}{2}} \frac{[\Omega^2 - \Omega_i^2]}{\left[\Omega^2 - \left(\frac{\Omega_s}{\Omega_i}\right)^2\right]}; \quad n \text{ odd} \quad (1.70)$$

and

$$R_n(\Omega) = C_1 \prod_{i=1}^{\frac{n}{2}} \frac{[\Omega^2 - \Omega_i^2]}{\left[\Omega^2 - \left(\frac{\Omega_s}{\Omega_i}\right)^2\right]}; \qquad n \text{ even} \qquad (1.71)$$

It is seen that $R_n(\Omega)$ has poles and zeros which are reciprocal to each other with respect to Ω_s and satisfies the property

$$R_n(\Omega, \delta) = \frac{\delta}{R_n(\frac{\Omega_s}{\Omega}, \delta)} \qquad (1.72)$$

This means that the value of $R_n(\Omega)$ at any frequency Ω_l in the passband $0 \leq \Omega \leq 1$ is the reciprocal of its value at the frequency $\frac{\Omega_s}{\Omega_l}$, multiplied by δ. The frequency $\frac{\Omega_s}{\Omega_l}$ lies in the stopband $\Omega_l \Omega_s \leq \Omega \leq \infty$. As a consequence of this property, we realize that if we find the critical frequencies such that $R_n(\Omega)$ has an equiripple response in the passband, it will have an equiripple response in the stopband, because of (1.72). The problem of finding the Chebyshev rational function that satisfies this property is solved by a transformation $\Omega = sn(\alpha K, m)$ which is analogous to the transformation $\Omega = \cos(n \cos^{-1} \Omega)$ that was used in the theory of Chebyshev polynomials. The function $sn(u, k)$ is called the Jacobian elliptic sine of u, with a modulus k and it is defined by

$$sn(u, k) = \sin(\phi(u, k)) \qquad (1.73)$$

where $u(\phi, k)$, called the elliptic integral of the first kind is the inverse of $\phi(u, k)$. It is defined by the integral

$$u(\phi, k) = \int_0^\phi \frac{dy}{\sqrt{1 - k^2 \sin^2(y)}} \qquad (1.74)$$

When the upper limit in the above integral is $\frac{\pi}{2}$, we get $u(\frac{\pi}{2}, k)$ which is a function of only the modulus k and this function $u(\frac{\pi}{2}, k)$ is called the complete elliptic integral $K(k)$. The properties of these elliptic functions and elliptic integrals and their application for the design of elliptic filters were first investigated by Cauer and hence the elliptic filters are also known as the Cauer filters. We will mention only a few of these properties in the following paragraphs.

Equation (1.73) suggests that $sn(u, k)$ is periodic and is a generalized form of the trigonometric sine function. Indeed its period is equal to $4K(k)$ and hence $K(k)$ is called the quarter period of $sn(u, k)$. A plot of $sn(u, k)$ for some values of the modulus k is plotted in Fig. 1.13.

When $k = 0$, K has a value of $\frac{\pi}{2}$ and hence $sn(u, 0) = \sin(u)$ is a strictly sinusoidal function of u only as shown, with a period of 2π. As k increases from 0, the quarter period increases, approaching the value ∞ as k approaches the value of one. Consequently, the $sn(u, k)$ becomes ' flat and more flat' as shown in Fig. 1.13.

The Chebyshev rational function $R_n(\Omega, \delta)$ has the following properties in addition to (1.72). It is obvious from (1.71) that $R_n(\Omega, \delta)$ is odd when n is

1.7. ELLIPTIC APPROXIMATION

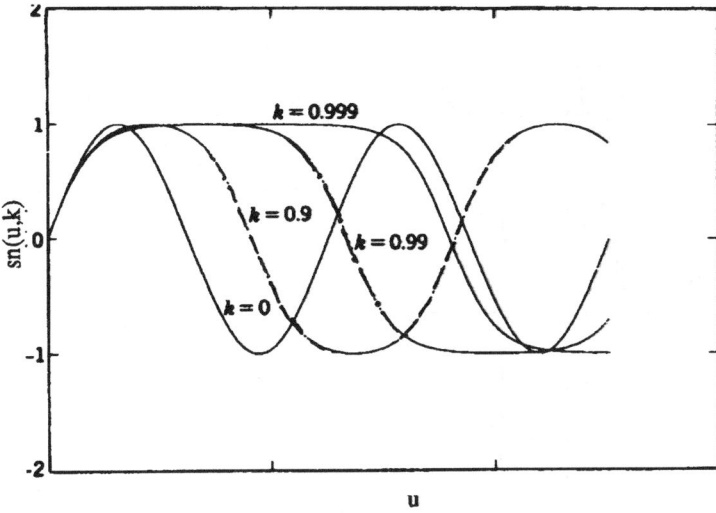

Fig. 1.13. Plot of Jacobian elliptic sine function sn(u,k)

odd and is even when n is even. We have plotted $R_5(\Omega, \delta)$ in Fig 1.14a as an example. We note that $R_n(\Omega, \delta)$ oscillates between +1 and -1 i.e. $|R_n(\Omega, \delta)| \leq 1$ for $|\Omega| \leq 1$. We also notice that $|R_n(\Omega, \delta)| \geq \delta$ for $|\Omega| \geq \Omega_s$, thereby satisfying (1.72). Fig. 1.14b shows the attenuation $10\log(1 + \epsilon^2 R_n^2(\Omega, \delta))$ for $n = 5$. The zeros of $R_n(\Omega, \delta)$ are at Ω_i, $i = 0, 1, 2$ and are called the reflection zeros, denoted as Ω_{ri} in Fig. 1.14b. The poles of $R_n(\Omega, \delta)$ at $\frac{\Omega_s}{\Omega_i}$ are the zeros of $|H(j\Omega)|^2 = \frac{1}{1+\epsilon^2 R_n^2(\Omega,\delta)}$ which lie on the imaginary axis and are called the transmission zeros. They are also called as loss poles and are denoted as $\Omega_{\infty 1}$ and $\Omega_{\infty 2}$. The frequency Ω_s is the product of the reflection zeros and the corresponding transmission zeros. In Fig. 1.14b, the magnitude of $|R_n(\Omega, \delta)|^2 = R_n^2(\Omega, \delta)$ is equal to one at the frequencies Ω_{m1}, Ω_{m2} and $\Omega_p = 1$. Therefore these are the frequencies in the passband at which the attenuation is A_p. According to (1.72), the magnitude of $|R_n(\Omega, \delta)|^2$ at the frequencies $\frac{\Omega_s}{\Omega_{m1}}$, $\frac{\Omega_s}{\Omega_{m2}}$ and Ω_s is δ^2, so we have $10\log \delta^2 = A_s$. Therefore Ω_s is the lowest frequency at which $|R_n(\Omega, \delta)|^2 = \delta^2$. We also have the property of $R_n^2(\Omega, \delta)$ that the sum of the number of its maxima and minima in the closed interval $\Omega \in [0,1]$ is $n+1$ - which is also true of $C_n^2(\Omega)$. The loss function $1 + \epsilon^2 R_n^2(\Omega, \delta)$ also has the same property.

The zeros of the Chebyshev rational function $R_n(\Omega, \delta)$ which has the equiripple behavior in the passband and consequently in the stopband also, are given by

$$\Omega_i = sn(u_i, \delta) \tag{1.75}$$

where

$$u_i = \begin{cases} (2i-1)K(m)/n & n \text{ even} \\ 2iK(m)/n & n \text{ odd} \end{cases} \tag{1.76}$$

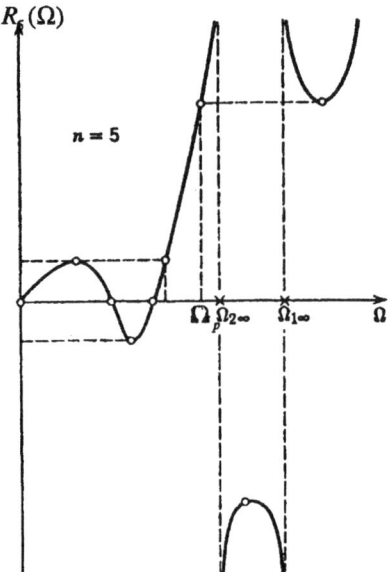

a

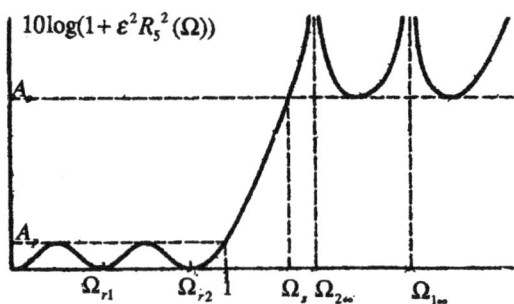

b

Fig. 1.14. a Chebyshev Rational Function b Attenuation of an elliptic lowpass filter (n = 5)

and
$$m = \frac{1}{\Omega_s^2} \tag{1.77}$$

and the poles are given by $\frac{\Omega_s}{\Omega_i}$. Since $R_n(\Omega, \delta) = 1$ when $\Omega = 1 = \Omega_p$, the constants C_1 and C_2 in (1.70) and (1.71) can be easily evaluated as $\frac{1}{R_n(1,\delta)}$.

To find $H(p)$ from the expression for $|H(j\Omega)|^2 = \frac{1}{1+\epsilon^2 R_n^2(\Omega,\delta)}$ in which we have substituted the values for Ω_i from (1.75-1.77), we follow the same procedure as in the case of the Butterworth and Chebyshev filters i.e. substitute $\Omega = \frac{p}{j}$ in $|H(j\Omega)|^2 = \frac{1}{1+\epsilon^2 R_n^2(\Omega,\delta)}$, find the $2n$ poles of $H(p)H(-p)$, choose the n poles in the left half p-plane as those of $H(p)$. In the case of elliptic filters, we also have to find the double zeros of $H(p)H(-p)$ which lie on the imaginary axis and retain only the single poles at these locations as those of $H(p)$. The procedure is analytical but requires a good understanding of the theory of elliptical functions and elliptical integrals. There are only a few text books e.g. [14, 25] in which the procedure is discussed in varying detail. However, [2] is one book that gives a step-by-step design procedure which does not require the calculation of the Jacobian elliptic sine functions. The books [7] and [33] are two valuable sources which list the poles and zeros of elliptic filters of many orders with different values for the parameters corresponding to different values of A_p and A_s. Then there are many softwares available to design these filters[12, 14, 17] The Signal Processing Toolbox running under MATLAB$^{(R)}$ [17] is one of the most popular software available for designing all of these filters as well as the filters that will be discussed in the remaining part of this chapter. Those who might be still interested in understanding the theory and design procedure for elliptic filters should read the references cited above, as the first step.

There are a few more classes of analog filters e.g. Class-L filters [20], published in the literature on classical filter theory but are not used very often. The criteria for comparing and choosing the type of filter can vary but given A_p, A_s and the order n, we find that the elliptic filters have the narrowest transition band. Another way of comparing is the order of the filter, when A_p, A_s and the lowest frequency Ω_s in the stopband are specified; then the order of the elliptic filter is the lowest, next comes the Chebyshev filter and the Butterworth filter needs the highest order to meet the specifications. The Chebyshev I filters and also the Butterworth filters have a poor group delay response near the cutoff frequency, compared to the elliptic and Inverse Chebyshev filters that meet the same specifications. However the order of the Inverse Chebyshev filter is higher than the elliptic filters. The group delay response and the order of the filters that meet the specifications $A_p = 0.4$ dB, $A_s = 25$ dB and $\Omega_s = 1.25$ may be compared by looking at the plots in Fig. 1.15.

But there are cases where filters with ripples in the passband of the magnitude response or in the group delay response may not be preferable over those with a maximally flat response. For example, as pointed in [32], when most of the energy in a signal is concentrated near the origin, maximally flat filters or Inverse Chebyshev filters would be preferred over Chebyshev I or elliptic filters. Since the Inverse Chebyshev filters have a maximally flat passband and also a narrower transition band than the Butterworth filters, they may be the best compromise in such applications.

1.8 Least Squares Approximation

The design of filters that approximate a constant magnitude of the ideal lowpass filter in the least squares sense, is not very common, because the above methods

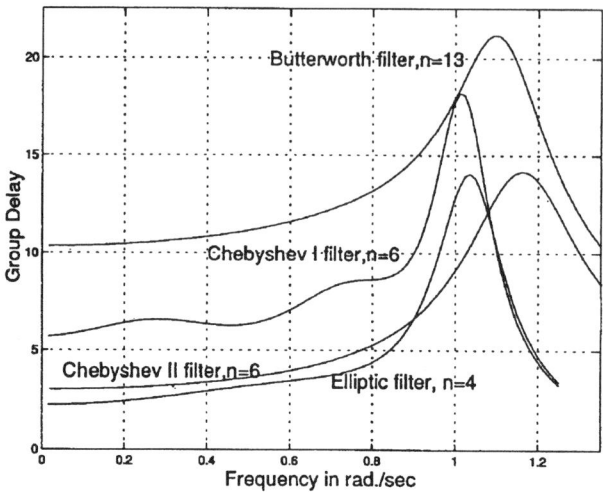

Fig. 1.15. Comparison of the delays of Butterworth, Chebyshev I and II filters, and Elliptic lowpass filters, meeting the same specifications

which are analytical give better results. However, when the magnitude or group delay response is specified arbitrarily over a frequency range, the above methods are not applicable and in such cases, least squares approximation or least p^{th} approximation is a common approach used, not only for designing filters but other systems in general ([3]- [6]). For a brief discussion of this method, let us consider the magnitude squared response of the prototype filter in the form

$$|H(j\Omega)|^2 = \frac{1}{1+\epsilon^2 L^2(\Omega)} \qquad (1.78)$$

where

$$L(\Omega) = \begin{cases} a_o + a_2\Omega^2 + \cdots + a_n\Omega^n & n \text{ even} \\ a_1\Omega + a_3\Omega^3 + \cdots + a_n\Omega^n & n \text{ odd} \end{cases} \qquad (1.79)$$

The author in [16] chooses to minimize the error between $L^2(\Omega)$ and zero over the range $|\Omega| \leq 1$ i.e. to minimize the error

$$J_p(\mathbf{a}, \Omega) = \int_{-1}^{1} W(\Omega) L^2(\Omega) d\Omega$$

subject to the constraint that $L(1) = 1$. This approach is similar to that followed in the above three filter designs. The minimum is found by differentiating the error function with respect to Ω, and equating the derivatives to zero. When we get linear equations, as in this case, it is easy to solve the equations; however, if the derivatives are non-linear functions of the unknown coefficients in $\mathbf{a}$, we have to resort to numerical methods to find the solution for the coefficients. The

1.8. LEAST SQUARES APPROXIMATION

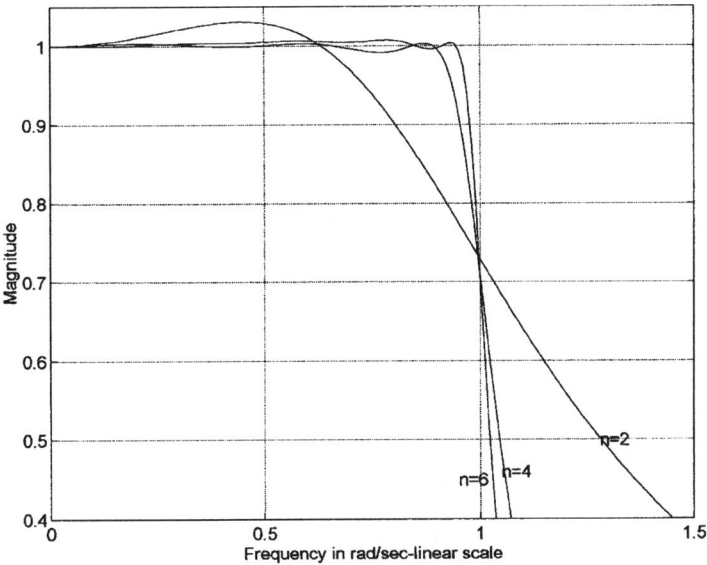

Fig. 1.16. Magnitude of Least Squares filters

author lists the polynomials $L_n(\Omega)$ for $n = 2, 3, \cdots, 9$, which are given below in Table 1.2. The design procedure is completed by substituting $-\Omega^2 = p^2$ in $1 + \epsilon^2 L^2(\Omega)$ and finding the $2n$ zeros, and picking the n poles which lie in the left half p-plane. But compared to the analytical methods of Butterworth, Chebyshev and elliptical filters, the $2n$ poles of (1.80) must be found by numerical methods only.

$$H(p)H(-p) = \frac{1}{1 + \epsilon^2 L^2(\Omega)}\bigg|_{\Omega^2 = -p^2} \quad (1.80)$$

When A_p, A_s and Ω_s are specified, we do not have a formula for finding the order of the filter either; hence we need to plot the magnitude response of the filters for many values of the order n and pick the value of n that meets the specifications of Ω_s and A_s, from the plots. For examples, the magnitude response of the Least Squares Error filters using the polynomials $L_n(\Omega)$ listed in Table 1.2, is shown for $n = 2, 4, 6$ in Fig. 1.16. (It is found, however, that $L_8(\Omega)$ is not equal to one at $\Omega = 1$ and gives inconsistent results.)

It has been shown that as p approaches ∞, the minimization of the error function (1.22) leads to the equiripple approximation [30]. A practical choice of $p = 6, 8$, or 10 is found to give fairly satisfactory equiripple response in filter design [4]. The magnitude response of the Least Squares filters shown in Fig. 1.16 is seen to approach an equiripple response as n increases.

Table 1.2. Least Squares polynomials.

n	$L_n(\Omega)$
2	$\frac{5}{4}\Omega^2 - \frac{1}{4}$
3	$\frac{7}{4}\Omega^3 - \frac{3}{4}\Omega$
4	$\frac{21}{8}\Omega^4 - \frac{7}{4}\Omega^2 + \frac{1}{8}$
5	$\frac{33}{8}\Omega^5 - \frac{15}{4}\Omega^3 + \frac{5}{8}\Omega$
6	$\frac{429}{64}\Omega^6 - \frac{495}{64}\Omega^4 + \frac{135}{64}\Omega^2 - \frac{5}{64}$
7	$\frac{715}{64}\Omega^7 - \frac{1001}{64}\Omega^5 + \frac{385}{64}\Omega^3 - \frac{35}{64}\Omega$
8	$\frac{243}{128}\Omega^8 - \frac{1001}{32}\Omega^6 + \frac{1001}{64}\Omega^4 - \frac{77}{32}\Omega^2 + \frac{7}{128}$
9	$\frac{4199}{128}\Omega^9 - \frac{1989}{32}\Omega^7 + \frac{2457}{64}\Omega^5 - \frac{273}{32}\Omega^3 + \frac{63}{128}\Omega$

1.9 Analog Frequency Transformations

Once we have learned the methods of approximating the magnitude response of the ideal lowpass prototype filter, the design of filters that approximate the ideal magnitude response of highpass, bandpass and bandstop filters is easily carried out. This is done by using well-known analog frequency transformations $p = g(s)$ that map the magnitude response of the lowpass filter $H(j\Omega)$ to that of the specified highpass, bandpass or bandstop filters $H(j\omega)$. The parameters of the transformation are determined by the cutoff frequency (frequencies) and the stopband frequency specified for the highpass, bandpass or bandstop filter so that frequencies in their passband(s) are mapped to the passband of the normalized, prototype filter and the frequencies in the stopband(s) of the highpass, bandpass or bandstop filters are mapped to frequencies in the stopband of the prototype filter. After the normalized prototype filter $H(p)$ is designed according to the methods discussed in earlier sections, the frequency transformation $p = g(s)$ is applied on $H(p)$ to get the transfer function $H(s)$ of the specified filter. With this general outline, let us consider the design of each filter is some more detail.

1.10 Highpass Filter

It is easy to describe the design of a highpass filter, by choosing an an example. Suppose a highpass filter with an equiripple passband $\omega_p \leq |\omega| < \infty$ is specified, along with a stopband frequency ω_s. The magnitude at ω_p which is the cutoff frequency of the passband and also the magnitude at ω_s (or A_p and A_s) are given. The LP to HP frequency transformation to be used in designing the

1.10. HIGHPASS FILTER

highpass (HP) filters is

$$p = \frac{\omega_p}{s} \tag{1.81}$$

It is seen that when $s = j\omega_p$, the value of $p = -j$ and when $s = -j\omega_p$, the value of $p = j$. It can also be shown that under this transformation, all frequencies in the passband of the highpass filter, map into the passband frequencies $-1 \leq \Omega \leq 1$ of the lowpass prototype filter. We calculate the frequency Ω_s to which the specified stopband frequency ω_s maps, by putting $s = j\omega_s$ in (1.81). The stopband frequency is found to be $\Omega_s = \frac{\omega_p}{\omega_s}$. So the specified magnitude response of the highpass filter is transformed to that of the lowpass, prototype, equiripple filter. We design the prototype lowpass filter to meet these specifications and then substitute $p = \frac{\omega_p}{s}$ in $H(p)$ to get the transfer function $H(s)$ of the specified highpass filter.

1.10.1 Example 1.5

The cutoff frequency of a Chebyshev highpass filter is $\omega_p = 2500$ which is the lowest frequency in the passband, and the maximum attenuation in the passband $A_p = 0.5$ dB. The maximum gain in the passband is 5 dB. At the stopband frequency $\omega_s = 500$, the minimum attenuation required is 30 dB. Design the highpass filter $H(s)$.

When we apply the LP to HP transformation $p = \frac{2500}{s}$, the cutoff frequency $\omega_p = 2500$ maps to $\Omega_p = 1$ and the stopband frequency ω_s maps to $\Omega_s = 5$. In the lowpass prototype filter, we have $\Omega_p = 1$, $\Omega_s = 5$, $A_p = 0.5$ dB, $A_s = 30$ dB and the maximum value of 5 dB in the passband. This filter has been already designed under Example 1.3 and has a transfer function given by (1.65) which is repeated below.

$$H(p) = \frac{(0.715693)(1.7783)}{(p^2 + .626456p + 1.142447)(p + 0.626456)}$$

Next we substitute $p = \frac{2500}{s}$ in this transfer function and when simplified, the transfer function of the specified highpass, Chebyshev filter becomes

$$H(s) = \frac{(0.715693)(1.7783)}{(p^2 + .626456p + 1.142447)(p + 0.626456)}\Big|_{p=\frac{2500}{s}}$$

$$= \frac{1.7783s^3}{[s^2 + 1370.9s + 5.4707 \times 10^6][s + 3990]} \tag{1.82}$$

The magnitude response of (1.82) is plotted in Fig. 1.17 and is found to exceed the specifications of the given highpass filter. The design of highpass filter with a maximally flat passband response or with an equiripple response in both the passband and the stopband is carried out in a similar manner.

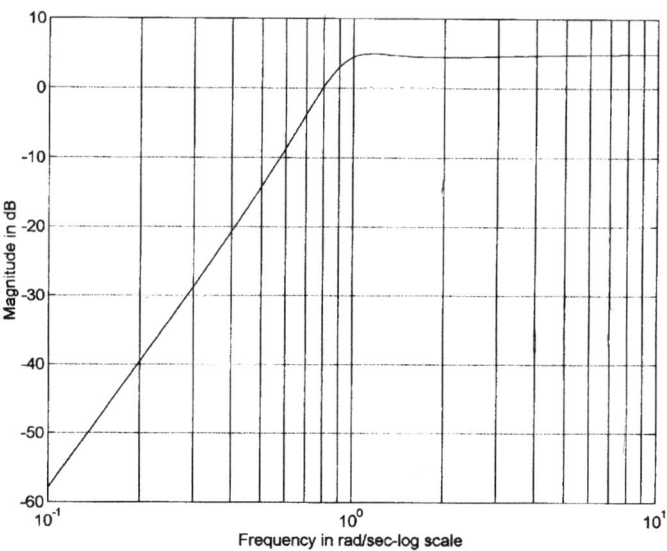

Fig. 1.17. Magnitude response of a highpass filter

1.11 Bandpass Filter

The normal specifications of a bandpass filter $H(s)$ are shown in Fig. 1.18. The cutoff frequencies ω_1 and ω_2, the maximum value of the magnitude in the passband between the cutoff frequencies, the maximum attenuation in this passband or the minimum magnitude at the cutoff frequencies, a frequency ω_s ($= \omega_3$ or ω_4) in the stopband at which the minimum attenuation or the maximum magnitude are specified. The type of passband response may be maximally flat, equiripple or elliptic response.

The LP to BP frequency transformation that is used for the design of a specified bandpass filter is

$$p = \frac{1}{B}\left(\frac{s^2 + \omega_o^2}{s}\right) \tag{1.83}$$

where $B = \omega_2 - \omega_1$ is the bandwidth of the filter and $\omega_o = \sqrt{\omega_1 \omega_2}$ is called the mean frequency of the bandpass filter.

A frequency $s = j\omega_k$ in the bandpass filter, is mapped to a frequency $p = j\Omega_k$ under this transformation which is obtained by

$$j\Omega_k = \frac{j}{B}\left(\frac{\omega_o^2 - \omega_k^2}{\omega_k}\right) \tag{1.84}$$

$$= \frac{j\omega_o}{B}\left(\frac{\omega_k}{\omega_o} - \frac{\omega_o}{\omega_k}\right) \tag{1.85}$$

Therefore the frequencies ω_1 and ω_2 map to $\Omega = \mp 1$ and the frequencies $-\omega_1$ and $-\omega_2$ map to $\Omega = \pm 1$. Similarly the positive value of the stopband frequency

1.11. BANDPASS FILTER

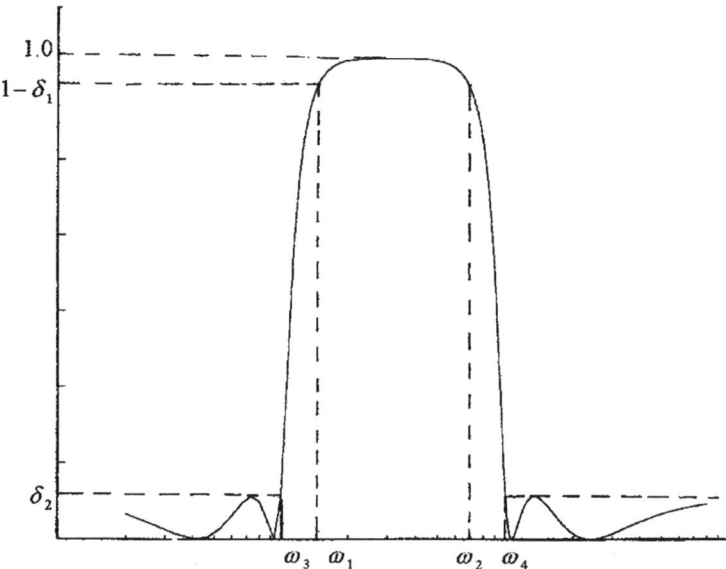

Fig. 1.18. Typical specifications of a bandpass filter

Ω_s to which the frequency ω_s maps, is calculated from

$$\Omega_s = \left| \frac{\omega_o}{B} \left(\frac{\omega_o^2 - \omega_s^2}{\omega_s} \right) \right| \qquad (1.86)$$

The magnitude or the attenuation at the frequencies $\Omega = 1$ and Ω_s for the prototype filter are the same as those at the corresponding frequencies of the bandpass filter. From the specification of the lowpass prototype filter, we obtain its transfer function $H(p)$, following the appropriate design procedure discussed earlier. Then we substitute (1.83) in $H(p)$ to get the transfer function $H(s)$ of the bandpass filter specified.

1.11.1 Example 1.6

The specifications of a Chebyshev bandpass filter are the following: $\omega_1 = 10^4$, $\omega_2 = 10^5$, $\omega_s = 2 \times 10^5$, $A_p = 0.8$ dB and $A_s = 30$ dB and the maximum magnitude in the passband $= 10$ dB. We use the following procedure to design the filter.

1. $B = \omega_2 - \omega_1 = 9 \times 10^4$ and $\omega_o = \sqrt{\omega_2 \omega_1} = \sqrt{10^9} = 31.62 \times 10^3$

2. The LP to BP transformation is $p = \frac{1}{9 \times 10^4} \left(\frac{s^2 + 10^9}{s} \right)$

3. Let $s = j\omega_s = j2 \times 10^5$. From the above transformation, we get $\Omega_s = 2.1667$

4. The lowpass Chebyshev, prototype filter has a magnitude response as shown in Fig. 1.7

5. Calculate $\epsilon = \sqrt{10^{0.1A_p} - 1} = \sqrt{10^{.08} - 1} = 0.4497$

6. Calculate n from (1.64). Choose $n = \lceil 3.5 \rceil = 4$

7. Calculate φ_2 from (1.59). We get $\varphi_2 = 0.3848$

8. Calculate the poles from (1.62). We get $p_k = -.15093 \pm j.9931$ and $-.36438 \pm j.41137$

9. The transfer function of the lowpass, prototype, Chebyshev filter is derived from $H(p) = \frac{H_o}{\prod_{k=1}^{4}(p-p_k)}$ where H_o is fixed to match the gain = 10 dB at $\Omega = 0$

$$H(p) = \frac{.8788}{(p^2 + .3018p + 1.009)(p^2 + .7287p + .302)} \quad (1.87)$$

10. Now substitute $p = \left(\frac{s^2 + 10^9}{9 \times 10^4 s}\right)$ in $H(p)$ and simplify to get $H(s)$.

$$H(s) = \frac{5.7658 \times 10^{19} s^4}{D(s)}$$

where $D(s) = (s^4 + 2.7162 \times 10^4 s^3 + 101.729 \times 10^8 s^2 + 2.7162 \times 10^{13} s + 10^{18}) \times (s^4 + 6.5583 \times 10^4 s^3 + 44.462 \times 10^8 s^2 + 6.5583 \times 10^{13} s + 10^{18})$

To verify the design, we have plotted the magnitude response in Fig. 1.19.

1.12 Bandstop Filter

The normal specification of a bandstop (band reject) filter is shown in Fig. 1.20. The passband of this filter is given by $0 \leq \omega \leq \omega_1$ and $\omega_2 \leq \omega \leq \infty$ whereas the frequencies ω_3 or ω_4 define a frequency in the stopband. These frequencies and the corresponding magnitudes are normally specified. Note that in Fig. 1.18 showing the magnitude response of a bandpass filter, the stopband is equiripple and the passband between ω_1 and ω_2 has a maximally flat response; so it is a Chebyshev II type of filter.

But in the bandstop filter response shown in Fig. 1.20, the passband is equiripple and the stopband is maximally flat so it is a Chebyshev I type of bandstop filter. It is pointed out that the parameter B in the LP to BS transformation is chosen as $\omega_2 - \omega_1$ which is actually the width of the stopband. The mean frequency $\omega_o = \sqrt{\omega_2 \omega_1}$ and the LP to BS frequency transformation is given by

$$p = B\left(\frac{s}{s^2 + \omega_o^2}\right) \quad (1.88)$$

This transformation transforms the entire passband of the bandstop filter to the passband $|\Omega| \leq 1$ of the prototype lowpass filter. So we have to find the

1.12. BANDSTOP FILTER

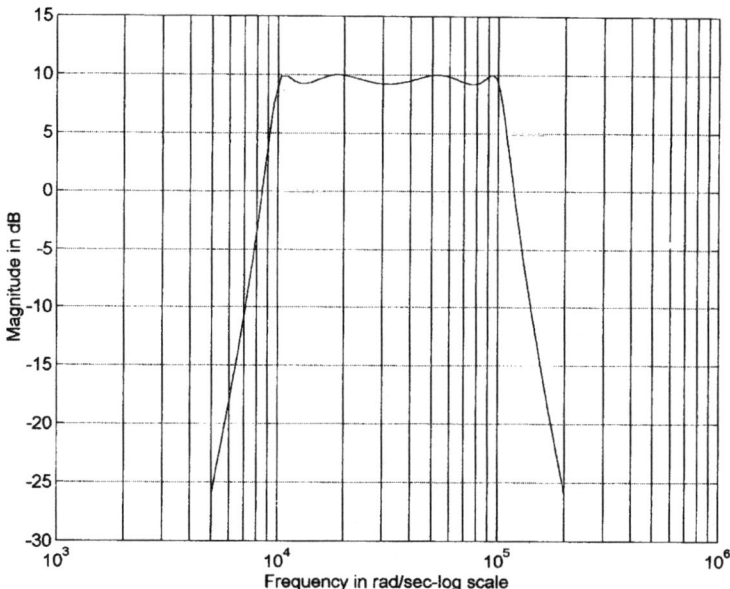

Fig. 1.19. Magnitude response of a bandpass filter

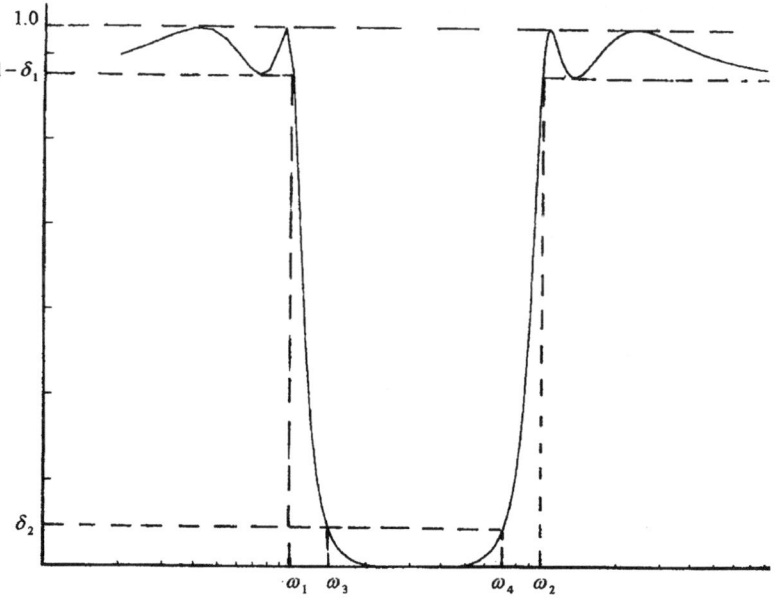

Fig. 1.20. Typical specifications of a bandstop filter

frequency Ω_s to which the stopband frequency ω_s is transformed under the transformation. It is found from

$$\Omega_s = \left| \frac{B\omega_s}{\omega_s^2 - \omega_o^2} \right| \qquad (1.89)$$

Thus we have reduced the specification of a bandstop filter to that of a prototype lowpass filter. It is designed by the procedures discussed earlier. When the transfer function $H(p)$ of the prototype filter is completed, the transformation $p = B\left(\frac{s}{s^2+\omega_o^2}\right)$ is used to transform $H(p)$ to $H(s)$. The design of the bandstop filter is illustrated by the following example.

1.12.1 Example 1.7

Suppose we are given the specification of a bandstop filter as shown in Fig. 1.20. In this example, we are given $\omega_1 = 1500$, $\omega_2 = 2000$, $\omega_s = \omega_4 = 1800$, $A_p = 0.2$ dB and $A_s = 55$ dB. The passband is required to have a maximally flat response. With these specifications, we design the bandstop filter following procedure given below:

1. $B = 2000 - 1500 = 500$ and $\omega_o = \sqrt{(2000)(1500)} = 1732.1$

2. The LP to BS frequency transformation is $p = 500\left(\frac{s}{s^2+3\times 10^6}\right)$

3. Let $s = j\omega_s = j1000$ Then we get $\Omega_s = 3.74$

4. Following the design procedure used in Example 1.2, we get $\epsilon = \sqrt{10^{.02} - 1} = .21709$ and from (1.44), we get $n = 5.946$ and choose $n = 6$

5. The six poles are calculated from (1.45) as $p_k = -0.33385 \pm j1.2459$, $-0.9121 \pm j0.9121$ and $-1.246 \pm j0.3329$

6. The transfer function of the lowpass prototype filter $H(p)$ is constructed from $H(p) = \frac{H_0}{\prod_{k=1}^{4}(p-p_k)}$ as

$$\frac{(1.664)^3}{(p^2+.6677p+1.664)(p^2+1.824p+1.664)(p^2+2.492p+1.664)}$$

7. Next we substitute $p = 500\left(\frac{s}{s^2+3\times 10^6}\right)$ in this $H(p)$ and simplify the expression to get the transfer function $H(s)$ of the specified bandstop filter. This completes the design of the bandstop filter. The magnitude response is found to exceed the given specifications.

The above sections give a summary of the theory of approximating the piecewise constant magnitude of analog filters. It will be required for approximating the magnitude of digital filters which will be treated in the following sections. There is an extensive body of literature on approximating the constant group delay and/or linear phase of analog filters and also on approximating both the magnitude and group group delay of analog filters. But the transformation that will be applied on the classical analog filters for designing digital filters is not useful for approximating the group delay or the magnitude and group delay of digital filters, because of the nonlinear distortion of the group delay under the transformation. If we design an analog filter with the prescribed magnitude and delay response and apply the frequency transformations, the magnitude response matches fairly well but the delay will be severely distorted. Hence the subject of approximating constant group delay and approximating both magnitude and group delay of analog filters is not useful for designing digital filters. Therefore the subject is not discussed in this book.

1.13   1-D Digital Filters

In contrast to the analog filters, the digital filters are described by two types of transfer functions, namely transfer functions of finite impulse response filters and those of infinite impulse response filters. The finite impulse response (FIR) filters can be designed to approximate a piecewise constant magnitude response and they are preferred in applications where a linear phase or a constant group delay is highly desirable. The FIR filters are also used for magnitude compensation and delay equalization. These filters form one very important class of digital filters that are used extensively in many practical applications. However, the scope of this book is limited-due to page limitations - to a discussion of only the infinite impulse response (IIR) digital filters. We refer readers to the many excellent books that treat the theory, design and applications of FIR filters [18].

The transfer functions of the 1-D IIR digital filters are rational functions in the z-transform variable, and they are analogous to the transfer functions of continuous-time (analog) filters. It is necessary to understand not only the theory of analog filter design discussed in sections 1.1 to 1.12, but also the relationship between the frequency domain description of the analog and digital filters, in order to understand the frequency transformation that is used to transform the analog frequency response specifications to that of the digital filters. This relationship is derived from the theory of sampling, which will be treated next.

1.14   Theory of Sampling

Let us first assume a continuous - time (analog) function $x_a(t)$ which satisfies the Dirichlet conditions and hence can be represented by its Fourier Transform $X_a(j\Omega)$

$$X_a(j\Omega) = \int_{-\infty}^{\infty} x_a(t)e^{-j\Omega t}dt \qquad (1.90)$$

whereas the inverse Fourier Transform of $X_a(j\Omega)$ is given by

$$x_a(t) = \frac{1}{2\pi} \int_{-\infty}^{\infty} X_a(j\Omega) e^{j\Omega t} d\Omega \tag{1.91}$$

Now we consider a discrete-time sequence $x(nT)$ obtained as the samples of $x_a(t)$ at $t = nT$.

From (1.91), we can write,

$$x_a(nT) = x(n) = \frac{1}{2\pi} \int_{-\infty}^{\infty} X_a(j\Omega) e^{j\Omega nT} d\Omega \tag{1.92}$$

Next, we define the z-transform of this discrete-time sequence as

$$X(z) = \sum_{n=0}^{\infty} x(n) z^{-n} \tag{1.93}$$

and evaluate it along the unit circle in the z-plane i.e. when $z = e^{j\omega T}$. This gives us the expression

$$X(e^{j\omega T}) = \sum_{n=0}^{\infty} x(n) e^{-j\omega nT} \tag{1.94}$$

If $h(n)$ is the unit pulse response of a linear, time-invariant, discrete-time system, and the input to the system is $e^{j\omega nT}$, then the output $y(n)$, when computed by convolution of this input with $h(n)$, can be derived as $e^{j\omega nT} H(e^{j\omega T})$, where $H(e^{j\omega T})$ is of the same form as (1.94). Hence

$$H(e^{j\omega T}) = \sum_{n=0}^{\infty} h(n) e^{-j\omega nT} \tag{1.95}$$

Because of the above reason, $H(e^{j\omega T})$ is called the frequency response of the discrete-time system or the Discrete-Time Fourier Transform (DTFT) of the sequence $h(n)$. Extending this definition, we will call $X(e^{j\omega T})$ as the DTFT of the discrete-time sequence $x(n)$ and ask this question: What is the relation between the Fourier Transform $X_a(j\Omega)$ of the continuous-time function $x_a(t)$ and the Fourier Transform $X(e^{j\omega T})$ of the discrete-time sequence $x_a(nT)$? To answer this question, we start with the observation that $X(e^{j\omega T})$ is a periodic function of ω with a period $\omega_s = \frac{2\pi}{T}$. Therefore it can be expressed in a Fourier Series which is exactly of the same form as given by (1.94). Therefore, we identify $x(n)$ as the Fourier Series coefficients of the periodic function $X(e^{j\omega T})$. These coefficients are evaluated from (1.96).

$$x(n) = x(nT) = \frac{T}{2\pi} \int_{-\frac{\pi}{T}}^{\frac{\pi}{T}} X(e^{j\omega T}) e^{j\omega nT} d\omega \tag{1.96}$$

1.14. THEORY OF SAMPLING

Let us express (1.92) which involves integration from $\Omega = -\infty$ to $\Omega = \infty$ as the sum of integrals over successive intervals each equal to one period $\frac{2\pi}{T}$ i.e.

$$x(nT) = \frac{1}{2\pi} \sum_{r=-\infty}^{\infty} \int_{\frac{(2r-1)\pi}{T}}^{\frac{(2r+1)\pi}{T}} X_a(j\Omega)e^{j\Omega nT} d\Omega \tag{1.97}$$

But each of the terms in the above summation can be reduced to an integral over the range $-\frac{\pi}{T}$ to $\frac{\pi}{T}$ by a change of variable from Ω to $\Omega + \frac{2\pi r}{T}$, to get

$$x(nT) = \frac{T}{2\pi} \sum_{r=-\infty}^{\infty} \frac{1}{T} \int_{-\frac{\pi}{T}}^{\frac{\pi}{T}} X_a(j\Omega + j\frac{2\pi r}{T})e^{j\Omega nT} e^{j2\pi rn} d\Omega \tag{1.98}$$

Notice that $e^{j2\pi rn} = 1$ for all integer values of r and n. By changing the order of summation and integration, the above equation can be reduced to (1.99).

$$x(nT) = \frac{T}{2\pi} \int_{-\frac{\pi}{T}}^{\frac{\pi}{T}} \left[\frac{1}{T} \sum_{r=-\infty}^{\infty} X_a(j\Omega + j\frac{2\pi r}{T}) \right] e^{j\Omega nT} d\Omega \tag{1.99}$$

For finding a relationship between $X_a(j\Omega)$ and $X(e^{j\omega T})$, now it is appropriate to change the frequency variable Ω to ω, thereby getting (1.100).

$$x(nT) = \frac{T}{2\pi} \int_{-\frac{\pi}{T}}^{\frac{\pi}{T}} \left[\frac{1}{T} \sum_{r=-\infty}^{\infty} X_a(j\omega + j\frac{2\pi r}{T}) \right] e^{j\omega nT} d\omega \tag{1.100}$$

Comparing (1.96) with (1.100), we get the desired relationship, given below

$$X(e^{j\omega T}) = \left[\frac{1}{T} \sum_{r=-\infty}^{\infty} X_a(j\omega + j\frac{2\pi r}{T}) \right] \tag{1.101}$$

In Figs. 1.21a-b, a typical analog signal $x_a(t)$ and the magnitude of its Fourier Transform are sketched and in Fig. 1.22a, the discrete-time sequence generated by sampling $x_a(t)$ is shown. In Fig. 1.22b, the magnitude of a few terms of (1.101) and the magnitude $|X(e^{j\omega T})|$ are shown.

When we use the theory of analog filter approximation for approximating the frequency response specifications of a digital filter, it is important to remember the implications of the above relationship. The two figures show that when a continuous-time function $x_a(t)$ with a Fourier Transform $X_a(j\omega)$ is sampled to get a discrete-time sequence $x(nT)$, the Fourier Transform $X(e^{j\omega T})$ of this sequence is a periodic replication of $\frac{X(j\omega)}{T}$ with a period $\frac{2\pi}{T} = \omega_s$. Ideally the Fourier Transform of $x_a(t)$ does not approach zero for all frequencies above a finite frequency but may do so only as the frequency approaches ∞.

Hence it is seen that, in general, when $\frac{X_a(j\omega)}{T}$ is replicated as shown in Fig. 1.22b, there is an overlap of the frequency responses at all frequencies or the frequency responses of the individual terms in (1.101) add up giving the actual response as shown by the curve for $\|X(e^{j\omega})\|$ in Fig. 1.22b. Because of

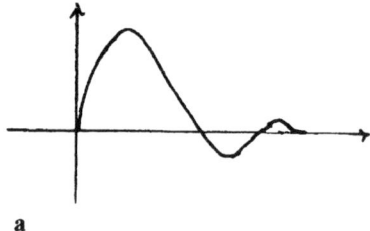

a

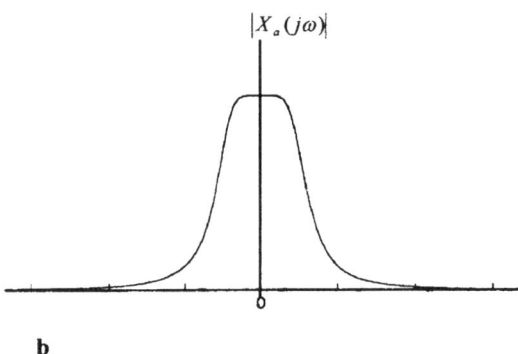

b

Fig. 1.21. a An analog signal b Fourier Transform of an analog filter

this overlapping effect, more famously known as aliasing, there is no way of reconstructing the analog function $x_a(t)$ exactly, from its samples. Aliasing of the Fourier Transform can be avoided if and only if the function $x_a(t)$ is assumed to be bandlimited i.e. it is a function such that its Fourier Transform $X_a(j\omega) \equiv 0$ for $|\omega| > \omega_b$ and the sampling period T is chosen such that $\omega_s = \frac{2\pi}{T} > 2\omega_b$. This is illustrated in Figs. 1.23 and 1.24.

When the analog signal $x_b(t)$ is bandlimited as shown in Fig. 1.23b and is sampled at a frequency $\omega_s \geq 2\omega_b$, the resulting discrete-time signal $x_b(nT)$ and its Fourier Transform $X(e^{j\omega})$ are as shown in Figs. 1.24a and b respectively.

If this signal $x_b(nT)$ is passed through an ideal lowpass filter with a bandwidth of $\frac{\omega_s}{2}$, the output will be a signal with a Fourier Transform equal to $X(e^{j\omega T})H(j\omega) = X_b(j\omega)/T$. This implies that the output signal will be the result of convolving the discrete input sequence $x_b(nT)$ with the unit impulse response $h_{lp}(t)$ of the ideal analog lowpass filter. Using this argument, Shannon[26] derived the formula for reconstructing the continuous-time function $x_a(t)$, from only the samples $x(n) = x_b(nT)$ - under the condition that $x_b(t)$ is bandlimited up to a maximum frequency ω_b and that it is sampled with a period $T < \frac{\pi}{\omega_b}$. This formula (1.102) is commonly called the Reconstruction Formula and the statement that the function $x_a(t)$ can be reconstructed from its samples $x(nT)$

1.14. THEORY OF SAMPLING

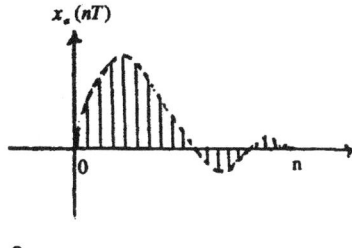

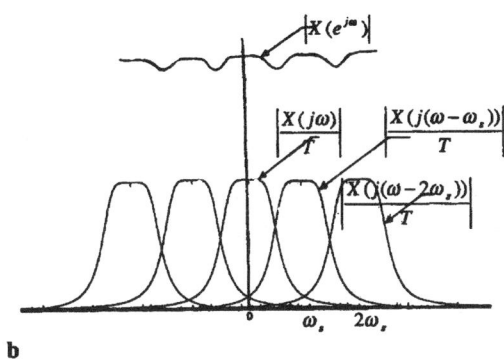

Fig. 1.22. a Discrete-time signal obtained by sampling an analog signal b Fourier Transform of the discrete-time signal and its few components

under the above conditions is known as Shannon's Sampling Theorem.

$$x_b(t) = \sum_{n=-\infty}^{\infty} x_b(nT) \frac{\sin\left[\frac{\pi}{T}(t-nT)\right]}{\left[\frac{\pi}{T}(t-nT)\right]} \qquad (1.102)$$

The reconstruction process is indicated in Fig. 1.25a. An explanation of the reconstruction is also given in Fig. 1.25b, where it is seen that the delayed impulse response $\frac{\sin\left[\frac{\pi}{T}(t-nT)\right]}{\left[\frac{\pi}{T}(t-nT)\right]}$ has a value of $x_b(nT)$ at $t = nT$ and it contributes zero value at all other sampling instants $t \neq nT$ so that the reconstructed, analog signal interpolates exactly between these sample values of the discrete samples.

This revolutionary result implies that the samples $x(nT)$ contain all the information that is contained in the original analog signal $x(t)$, if it is bandlimited and if it has been sampled with a period $T < \frac{\pi}{\omega_b}$. Any given signal can be made almost bandlimited by passing it through an analog lowpass filter of fairly high order. Indeed it is common practice to pass an analog signal through an analog lowpass filter, before it is sampled. Such filters used to precondition the analog signals - for example human speech - are called as the anti-aliasing filters. In the PCM telephone circuits, they are called the channel bank filters and they

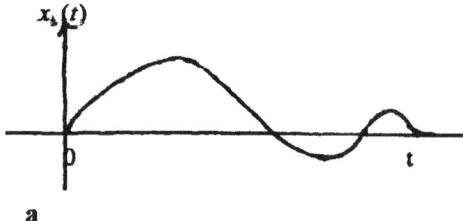

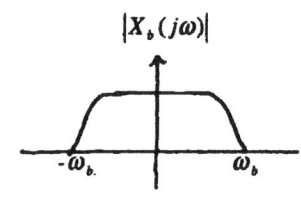

Fig. 1.23. a A bandlimited analog signal b Fourier Transform of the bandlimited analog signal

are usually lowpass filters of order 5 or 6, providing an attenuation of 30 dB at 4 kHz[13]. They are also used to reconstruct the analog signals from the digital signal at the receiving end. It is assumed that the maximum frequency contained in human speech is 3400 Hz and hence the sampling frequency is chosen as 8 kHz. In some other applications the signal may be first sampled at a much higher rate, for example 128 kHz and then this sampling rate reduced to 8 kHz. There is some ambiguity in the published literature, about the definition of what is called the Nyquist frequency. Some books define half the sampling frequency as the Nyquist frequency and a few books define the maximum frequency ω_b as the Nyquist frequency. In our example, when the signal is sampled at 8 kHz, some would consider 4 kHz as the Nyquist frequency and some others would consider 3.4 kHz as the Nyquist frequency. We will choose half the sampling frequency as the Nyquist frequency throughout this book.

Suppose we have an analog signal which is a bandpass signal i.e. it has a Fourier Transform which is zero outside the frequency range $\omega_1 \leq \omega \leq \omega_2$, so the bandwidth of this signal is $B = \omega_2 - \omega_1$ and the maximum frequency of this signal is ω_2. However it is not necessary to choose a sampling frequency $\omega_s \geq 2\omega_2$ in order to assure that we can reconstruct this signal from its sampled values. It has been shown [22] that when ω_2 is a multiple of B, we can recover the analog bandpass signal from its samples obtained with only a sampling frequency $\omega_s \geq 2B$.

We will represent the frequency response of the digital filter either by $H(e^{j\omega T})$ or often by $H(e^{j\omega})$ for convenience. Whenever it is expressed as $H(e^{j\omega})$ - which

1.14. THEORY OF SAMPLING

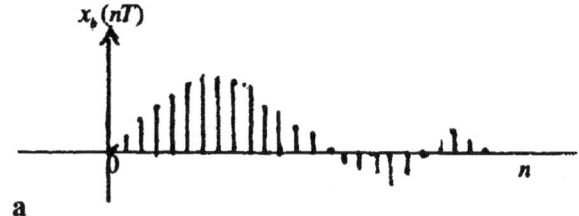

a

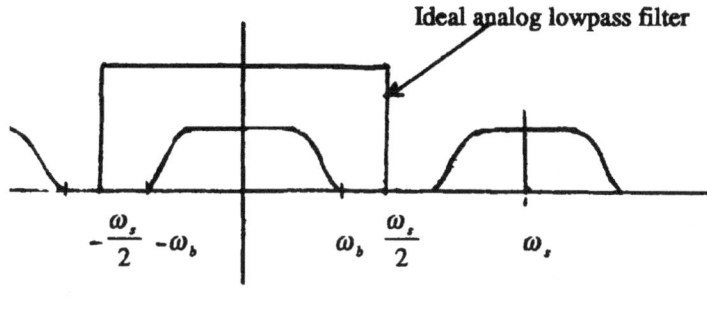

b

Fig. 1.24. a Discrete-time signal obtained by sampling a bandlimited analog signal b Reconstruction of the bandlimited analog signal from its discrete-time samples

is a very common practice in published literature - the frequency variable ω is to be understood as the normalized frequency $\omega T = \omega/f_s$. We may also represent the normalized frequency ωT by θ (radians). Since the frequency response $H(e^{j\omega})$ is periodic and its magnitude is an even function, the magnitude response of a digital filter is completely known for all values of ω over the range $(-\infty, \infty)$, when it is specified over any one half period. Commonly the range $(0, \pi)$ for the the normalized frequency $\omega T = \theta$, is chosen for specifying the magnitude, phase or group delay response of a digital filter, which corresponds to the actual frequency range $(0, \frac{\omega_s}{2})$. So the normalized frequency π corresponds to the Nyquist frequency. Sometimes the frequency ω is normalized by πf_s so that the Nyquist frequency has a value of 1. In Fig. 1.26 we have shown the magnitude response of an ideal lowpass, highpass, bandpass and bandstop filter and also an equiripple approximation of these magnitude responses.

Instead of *simply defining* the z transform $X(z)$ of a sequence $x(n)$ as

$$X(z) = \sum_{n=-\infty}^{\infty} x(n) z^{-n} \qquad (1.103)$$

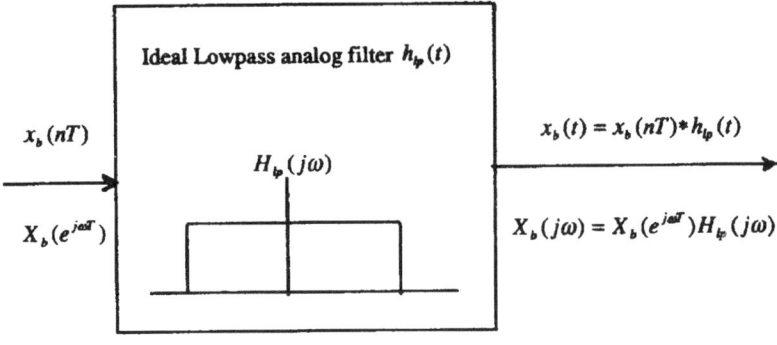

Fig. 1.25. a Reconstruction of the bandlimited analog signal from its samples b Reconstruction of the analog signal signal from its discrete-time samples

it is intuitively meaningful from a signal processing point of view, to consider a continuous-time or analog signal $x_a(t)$ for which a Laplace Transform $X_a(s)$ exists and then generate a discrete-time sequence $x_a(nT)$ which are values of the signal at $t = nT$. These values $x_a(nT)$ are considered as the strengths of impulse functions $\delta(t - nT)$ so that the sequence is represented as a function of the continuous variable t, i.e. $x^*(t) = \sum_{n=0}^{\infty} x_a(nT)\delta(t - nT)$. So we say that $x^*(t)$ is obtained by impulse sampling of $x_a(t)$ and its Laplace Transform $X^*(s)$ is

$$X^*(s) = \sum_{n=0}^{\infty} x_a(nT)e^{-sT} \qquad (1.104)$$

The z transform of $x_a(nT)$ is obtained from $X^*(s)$ when we substitute $z = e^{sT}$ in the above equation, getting $X(z) = \sum_{n=0}^{\infty} x(nT)z^{-n}$. Since T is fixed, $x_a(nT)$ is a function of the integer variable n and $z^{-1} = e^{-sT}$ is interpreted as a pure delay operator with a delay T equal to the sampling period. This leads to the derivation of the same z transform as in (1.103). From this derivation of the z transform, one can see that the inverse z transform of $X(z)$ gives us a sequence

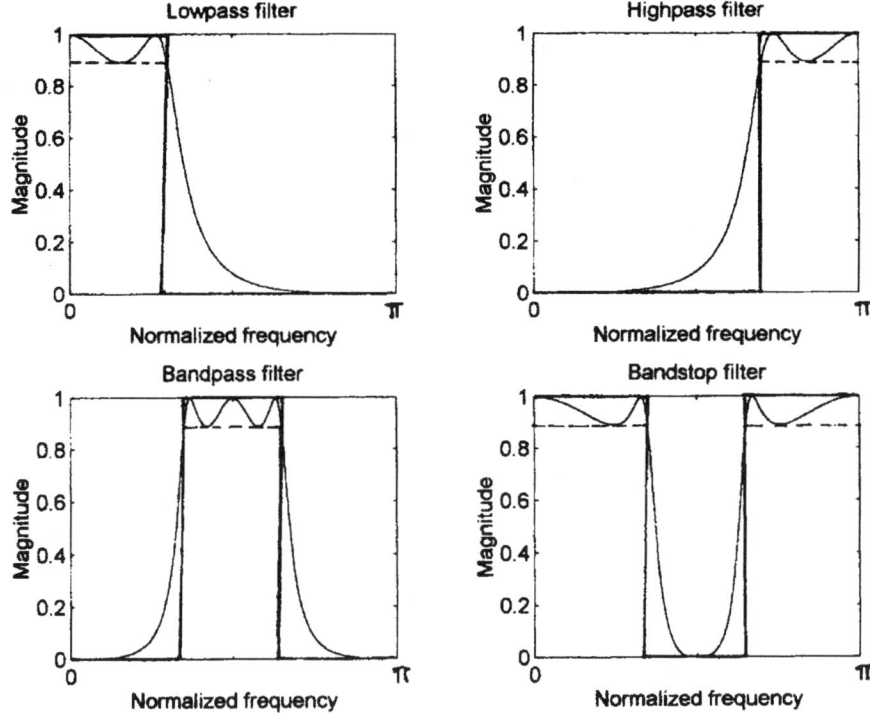

Fig. 1.26. Ideal and approximate magnitude response of digital filters

$x_a(nT)$ which are exactly the same as the values of $x_a(t)$ at $t = nT$, where $x_a(t)$ is the inverse Laplace Transform of $X_a(s)$. For this reason, the z transform derived from the signal processing point of view is called the Impulse Invariant Transform or the standard z transform of $x_a(nT)$. But the impulse invariant transform leads to the same form as (1.103).

1.15 Design Procedures

There are several procedures used for designing IIR filters, some of which make use of the theory of analog filter approximation. These design procedures propose different transformations of the form $s = f(z)$ to transform $H(s)$ to $H(z)$. The transformation $s = f(z)$ must satisfy the requirement that the digital filter transfer function $H(z)$ is stable, when it is obtained from the analog filter transfer functions $H(s)$ - which is easily chosen as a stable transfer function. When $H(s)$ and $f(z)$ are stable in the s and z domain respectively, the poles of $H(s)$ in the left half s-plane map into poles inside the unit circle in the z-plane; therefore $H(z)$ also is a stable transfer function. We also like to have frequencies from $-\infty$ to ∞ on the j ω axis of the s-plane mapped into frequencies on the boundary of the unit circle-without encountering any discontinuities.

One of these transformations is derived by representing the first order derivative of a continuous-time function by the backward difference approximation given below.

$$\frac{dy}{dt} \cong \frac{y(nT) - y(nT - T)}{T} \qquad (1.105)$$

The left side of the above equation is a continuous-time function by assumption and the right side is a difference equation. The transfer function in the Laplace Transform variable for the system that implements the derivative is s whereas the transfer function in the z-transform variable, for the system that implements the backward difference equation shown above is $\frac{1-z^{-1}}{T}$. Substitution of s by $\frac{1-z^{-1}}{T}$ is strictly a mathematical construction that helps in mapping the frequency response of a digital filter to that of an analog filter; but it is different from the transformation $z = e^{sT}$ used in deriving the standard z transform of a discrete-time sequence. The mapping of the s-plane to the z-plane under these two transformations is shown in Figs. 1.27a-b. Though both of them transform a stable transfer function $H(s)$ to a stable $H(z)$, the frequencies on the entire imaginary axis of the s-plane do not map into frequencies on the boundary of the unit circle in the z-plane under the transformation $s = \frac{1-z^{-1}}{T}$. This is one reason why the transformation $s = \frac{1-z^{-1}}{T}$ is not as popular as the standard transformation just mentioned: $z = e^{sT}$ (its inverse function is $s = \frac{1}{T}\ln(z)$).

1.16 Impulse Invariant Transformation

Now let us investigate the mapping of the frequency response $\sum_{n=0}^{\infty} x(nT)e^{-j\omega T}$ obtained under the impulse invariant transformation $z = e^{sT}$. As shown in Fig. 1.27a, strips of the left half plane bounded by $\pm j(\frac{(2r-1)\pi}{T})$ and $\pm j\left(\frac{(2r+1)\pi}{T}\right)$ on the $j\omega$ axis are mapped to the inside and the boundary of the unit circle in the z-plane. Hence $X^*(j\omega) = \sum_{r=-\infty}^{\infty} X(j\omega + jr\frac{2\pi}{T})$ is periodic, and will have no aliasing only if $x_a(t)$ is bandlimited. Approximations to the ideal lowpass and bandpass analog filters map to corresponding approximations for the digital filters without aliasing under this transformation. But the mapping of highpass and bandstop filters will result in aliasing and therefore this transformation is limited to the design of digital, lowpass and bandpass filters only.

Let the magnitude response $H(e^{j\omega T})$ of a digital lowpass filter be specified. Let us choose this as the specification of an analog lowpass filter. It is assumed that for frequencies larger than ω_b, the attenuation of the lowpass filter is so large that any aliasing that would necessarily result when the unit impulse response $h(t)$ of the analog filter is sampled with a sampling rate of $\frac{1}{T}Hz$ is negligible. In order to satisfy this requirement, we may have to increase the attenuation at ω_2 to be more than the specified value. Then we follow the design procedure described earlier in this chapter to obtain the transfer function $H(s)$ of the analog filter which approximates the above magnitude specification. After obtaining the transfer function $H(s)$, it is expressed in its partial fraction form.

1.16. IMPULSE INVARIANT TRANSFORMATION

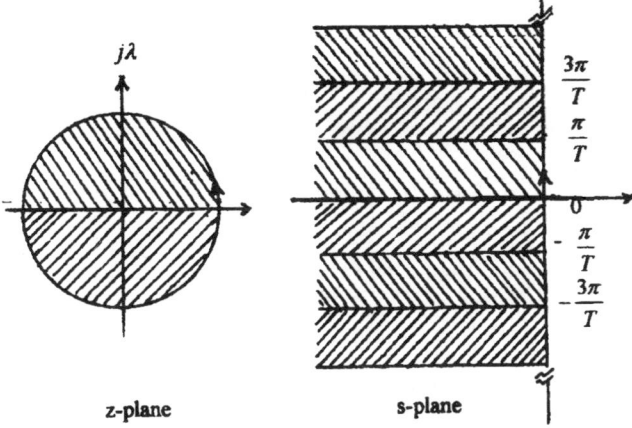

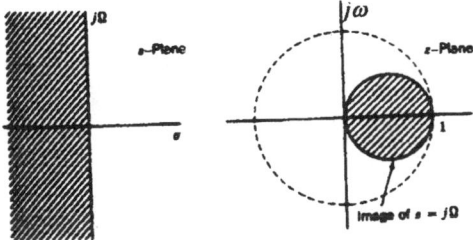

Fig. 1.27. a Mapping of s-plane to z-plane under the transformation $z = e^{sT}$ b Mapping of the s-plane to z-plane under the backward difference approximation

Since the poles of a lowpass analog filters obtained from the classical methods are simple, the partial fraction expansion is of the form

$$H(s) = \sum_{k=1}^{K} \frac{R_k}{s + s_k} \quad (1.106)$$

The unit impulse response $h_k(t)$ of a typical term $\frac{R_k}{s+s_k}$ is $R_k e^{-s_k t}$. When it is sampled with a sampling period T, and the impulse invariant z transform is evaluated, it becomes

$$R_k \sum_{n=0}^{\infty} e^{-s_k nT} z^{-n} = R_k \frac{1}{1 - e^{-s_k T} z^{-1}} = R_k \frac{z}{z - e^{-s_k T}} \qquad (1.107)$$

Hence the impulse invariant z transform $H(z)$ derived from $H(s)$ is given by

$$H(z) = \sum_{k=1}^{K} \frac{R_k z}{z - e^{-s_k T}} \qquad (1.108)$$

Of course, the unit pulse response $h(nT)$ of the digital filter will match the unit impulse response $h(t)$ at the instants of sampling $t = nT$. But the frequency response of $H(z)$ will not match the frequency response of $H(s)$ at all frequencies. If the magnitude response of the analog filter $H(j\omega)$ is very small for frequencies larger than ω_b, the frequency response of the digital filter $H(z)$ obtained from the impulse invariant transformation may give a fairly good approximation to the specified magnitude response.

Let us consider an example. We choose a Chebyshev I lowpass analog filter with a transfer function $H(s)$ given below:

$$H(s) = \frac{1}{s^2 + 1.098s + 1.103} \qquad (1.109)$$

As shown in Fig. 1.28, its magnitude response has a passband ripple of 1 dB and a bandwidth of 1 radian per second. The corresponding impulse invariant transfer function $H(z)$ is constructed by using (1.108).

$$H(z^{-1}) = \frac{1.1178 e^{-.5498T} \sin(.8953T) z^{-1}}{1 - 2e^{-.5498T} \cos(.8953T) z^{-1} + e^{-1.0996T} z^{-2}} \qquad (1.110)$$

The magnitude responses of three filters with the choice of $T = 1.0, 0.8$ and 0.5 are plotted in Fig. 1.29. They are periodic with periods which are respectively their sampling frequencies 2π, 2.5π and 4π. In Fig. 1.29, the magnitude response of the analog filter also is plotted and it decreases monotonically to zero as the frequency ω approaches ∞. But the frequency response of the digital filters is shown for only one period. In this figure, the frequency on the x-axis is the actual frequency for all filters (but the magnitude of all filters at $\omega = 0$ has been normalized to the same value for comparison only). Though the lowpass magnitude response of the analog filter decreases to a small value for $|\omega| \geq 6$, the effect of aliasing as the sampling period T is increased, is very clearly seen in these plots. An analog, lowpass, inverse Chebyshev filter, which has a ripple in the stopband would exhibit more severe aliasing effect even if a low value for T is chosen.

1.16. IMPULSE INVARIANT TRANSFORMATION

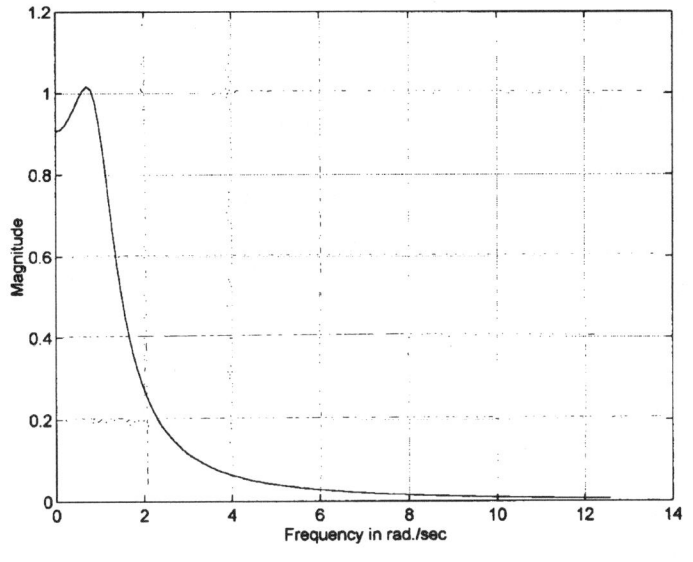

Fig. 1.28. Magnitude response of an analog lowpass filter

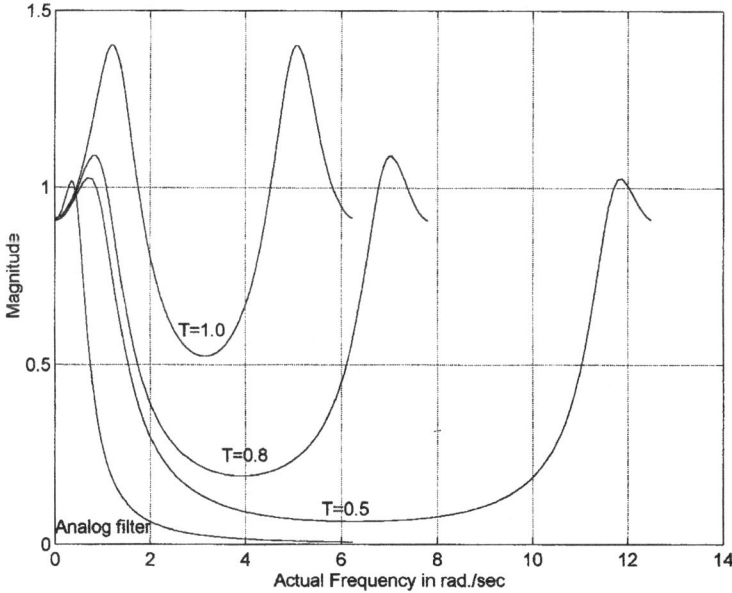

Fig. 1.29. Magnitude response of the analog lowpass filter and the impulse invariant digital filters

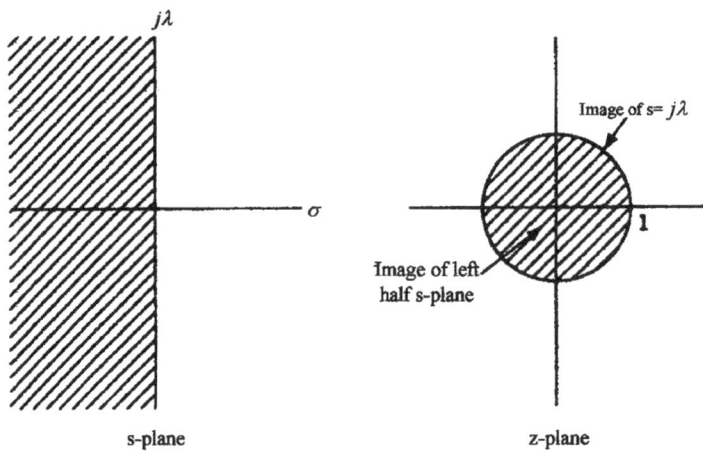

Fig. 1.30. Mapping of the s-plane to z-plane under the bilinear transformation

1.17 Bilinear Transformation

We define the bilinear transformation, which is the most important transformation used in designing IIR filters.

$$s = \frac{2}{T}\left(\frac{z-1}{z+1}\right) \tag{1.111}$$

To find how frequencies on the unit circle in the z-plane map to the s-plane, let us substitute $z = e^{j\omega T}$ in (1.111).

$$\begin{aligned} s &= \frac{2}{T}\left(\frac{e^{j\omega T}-1}{e^{j\omega T}+1}\right) = \frac{2}{T}\left(\frac{e^{j\frac{\omega T}{2}} - e^{-j\frac{\omega T}{2}}}{e^{j\frac{\omega T}{2}} + e^{-j\frac{\omega T}{2}}}\right) \\ &= j\frac{2}{T}\tan = j2f_s \tan\left(\frac{\omega T}{2}\right) = j\lambda \end{aligned}$$

This transformation maps the poles inside the unit circle in the z-plane to the inside of the left half s-plane. It also maps the frequencies on the unit circle in the z-plane to frequencies on the entire imaginary axis of the s-plane, where $s = \sigma + j\lambda$. This mapping is shown in Fig. 1.30.

To understand the mapping in some more detail, let us consider the frequency response of an IIR filter over the interval $(0, \frac{\omega_s}{2})$, where $\frac{\omega_s}{2} = \frac{\pi}{T}$ is the Nyquist frequency. As an example, we choose a frequency response $|H(e^{\omega T})| = |H(e^{j\theta})|$ of a Butterworth, bandpass digital filter as shown in Fig. 1.31a.

In Fig. 1.31, we have also shown the curve showing the relation between ωT and $\lambda = 2f_s \tan\left(\frac{\omega T}{2}\right)$. The value of λ corresponding to any value of $\omega T = \theta$ can be calculated from $\lambda = 2f_s \tan\left(\frac{\theta}{2}\right)$ as illustrated by mapping of a few

1.17. BILINEAR TRANSFORMATION

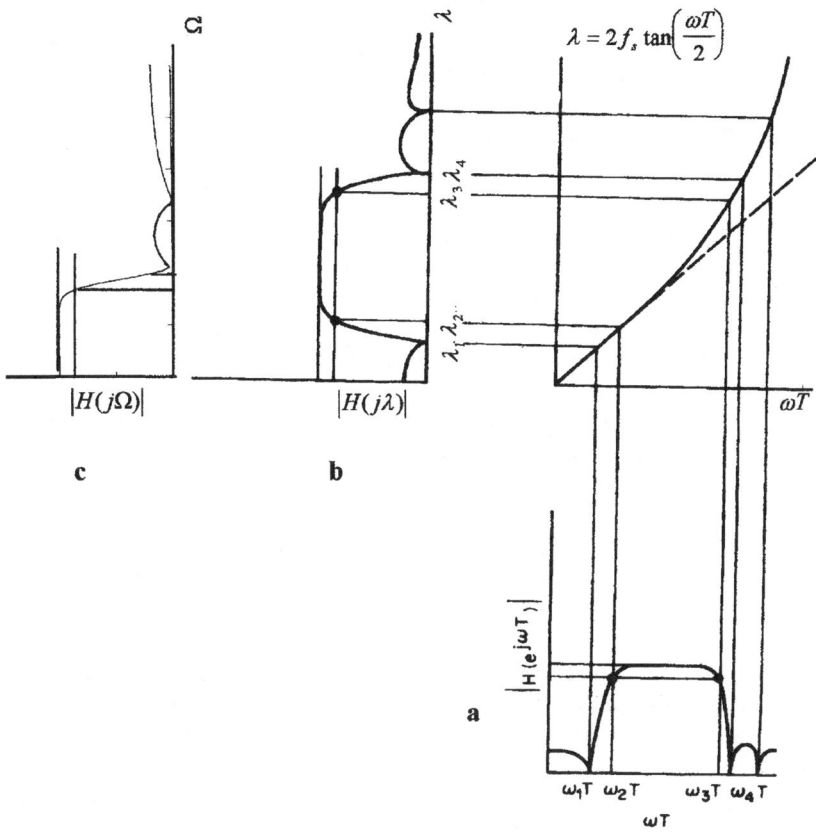

Fig. 1.31. Mapping of the filter response under the bilinear transformation, for IIR filter design

frequencies like $\omega_1 T, \omega_2 T$ in Fig. 1.31. The magnitude of the frequency response of the digital filter at any frequency $j\omega_k T$ is the magnitude of $H(\mathbf{s})$ at the corresponding frequency $\mathbf{s} = j\lambda_k$, where $\lambda_k = 2f_s \tan\left(\frac{\omega_k T}{2}\right)$.

This plot shows that the magnitude response of the digital filter over the Nyquist interval $(0, \pi)$ maps over the entire range $(0, \infty)$ of λ. So there is a nonlinear mapping whereby the frequencies in the ω domain are warped when mapped to the λ domain. Similarly the frequencies in the interval $(0, -\pi)$ are mapped to the entire interval $(0, -\infty)$ of λ. From the periodic nature of the function $tan(.)$, we also see that the periodic replicates of the digital filter frequency response in the ω domain, map to the same frequency response in the λ domain and the transfer function $H(\mathbf{s})$ obtained under the bilinear transform behaves like that of an analog filter. But it is to be pointed out that we use only the mathematical theory of analog filter approximation to solve the problem of finding such a function $H(\mathbf{s})$ and we do not design an analog filter. In other words, the bilinear transformation helps us to reduce the mathematical problem of approximating the frequency response of a digital filter in the variable ω, to the problem of approximating another function (in the variable λ), which is solved by using the mathematical theory of approximation for the frequency response of analog filters. So, if the frequency response $|H(j\lambda|$ is a lowpass frequency response, the frequency s is linearly scaled by λ_c to obtain the frequency response $|H(j\Omega)|$ of a lowpass prototype filter. If it is a highpass, bandpass or bandstop response, then the appropriate analog frequency transformation $\mathbf{p} = g(\mathbf{s})$ is used to convert the specification $|H(j\lambda|$ to that of an analog prototype lowpass filter. We obtain the transfer function $H(\mathbf{p})$ of the prototype filter, in which the complex frequency variable $\mathbf{p}$ is shown in bold in order to differentiate it from $H(p)$ and the magnitude is denoted by $|H(j\Omega)|$. The theory of analog filter approximation is used to find $H(\mathbf{p})$ such that its magnitude $|H(j\Omega)|$ approximates the magnitude response of the lowpass prototype filter. It is important to note that the unit impulse response of the filter $H(\mathbf{p})$ when sampled with a sampling period T does not match the unit impulse response of the digital filter $H(z)$, because the bilinear transformation is not impulse invariant. If on the other hand, we start with an analog filter function $H(s)$, for which the impulse invariant transfer function $H(z)$ is obtained, by using (1.106) and (1.108), the function $H(\mathbf{p})$ obtained by the bilinear transformation has nothing in common with $H(p)$. Hence it is not appropriate to say that we design an analog filter $H(p)$ - as it is likely to mean that we are designing the analog filter described by $H(p)$ from which the impulse invariant transform $H(z)$ was derived. The need to differentiate the frequency variables p and s from $\mathbf{p}$ and $\mathbf{s}$ becomes more evident in the next chapter, when we refer to $H(\mathbf{s})$ as the transfer function of commensurate, distributed networks, which are different from what we understand here as the 'analog filters'.

Once we have designed the lowpass prototype filter function $H(\mathbf{p})$, we apply the appropriate analog frequency transformation $\mathbf{p} = g(\mathbf{s})$ to $H(\mathbf{p})$ to get the function $H(\mathbf{s})$. Then we substitute $\mathbf{s} = 2f_s\left(\frac{z-1}{z+1}\right)$ in $H(\mathbf{s})$ to get $H(z)$ as the transfer function of the digital filter.

1.17.1 Example 1.8

The specified magnitude response of a maximally flat bandpass digital filter has a maximum value of 1.0 in its passband which lies between the cut off frequencies $\omega_1 = 0.4\pi$ and $\omega_2 = 0.5\pi$. The magnitude at these cut off frequencies is specified to be no less than 0.93 and at the frequency $\omega_3 = 0.7\pi$ in the stopband, the magnitude is specified to be no more than 0.004. Design the IIR digital filter that approximates these specifications, using the bilinear transformation.

It is obvious from the above specifications that the frequencies are normalized frequencies. So $\theta_1 = 0.4\pi$ and $\theta_2 = 0.5\pi$ are the normalized cutoff frequencies and $\theta_3 = 0.7\pi$ is the frequency in the stopband. The specified magnitude response is plotted in Fig. 1.32a. The two cut off frequencies ω_1, ω_2 and the

1.17. BILINEAR TRANSFORMATION

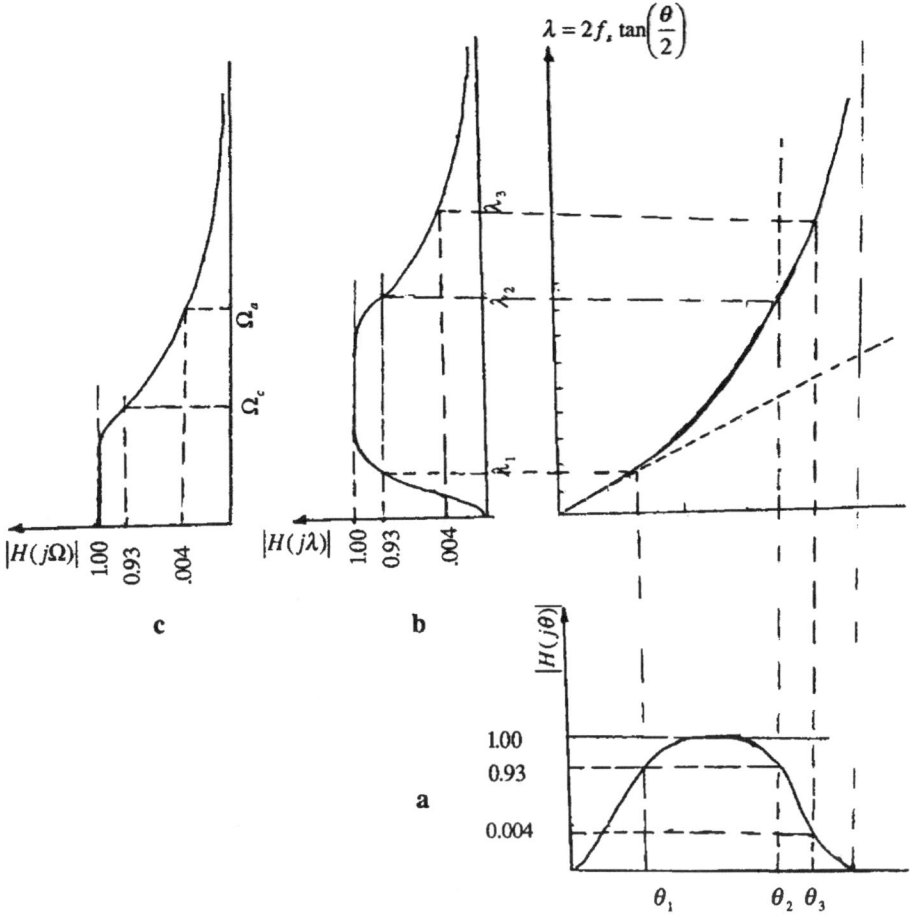

Fig. 1.32. Mapping of the filter under the bilinear transformation, for IIR filter design example

stopband frequency ω_3 map to and λ_1, λ_2 and λ_3 as follows:

$$\begin{aligned}\lambda_1 &= 2\tan(0.2\pi) = 1.453 \text{ radians/sec} \\ \lambda_2 &= 2\tan(0.25\pi) = 2.00 \text{ radians/sec} \\ \lambda_3 &= 2\tan(0.35\pi) = 3.95 \text{ radians/sec}\end{aligned}$$

The frequency response of the 'analog' filter $H(s)$ is plotted in Fig. 1.32b.

Now we find the bandwidth $B = \lambda_2 - \lambda_1 = 0.547$ and the center frequency $\lambda_o = \sqrt{\lambda_2 \lambda_1} = 1.705$ for the pre-warped bandpass filter function $|H(j\lambda)|$. Next

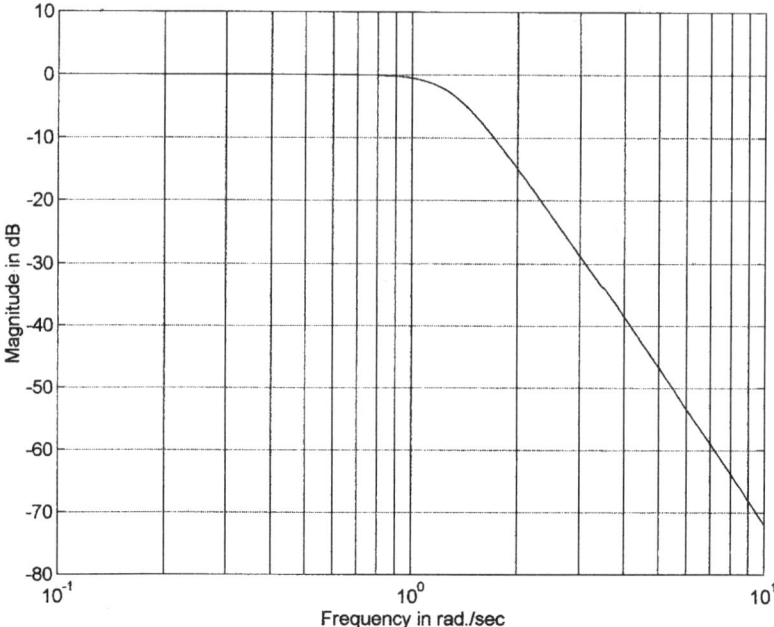

Fig. 1.33. Magnitude response of the analog lowpass filter in example 8

we define the bandpass to lowpass frequency transformation (see Eqn. 1.83).

$$\mathbf{p} = \frac{1}{B}\left(\frac{s^2 + \lambda_o^2}{s}\right) = \frac{1}{0.547}\left(\frac{s^2 + 1.705^2}{s}\right)$$

To find the frequency Ω_3, to which the frequency $\lambda_3 = 3.95$ maps, we substitute $s = j3.95$ in the above transformation and get $\mathbf{p} = j5.876 = j\Omega_3$, whereas the cutoff frequencies map to the normalized frequency $\Omega_c = 1$. Hence the magnitude response of the lowpass, Butterworth, prototype filter function is as shown in Fig. 1.32c. Using the same notations as before, we get $A_p = 0.63$ dB, $A_s = 48$ dB, $\epsilon = 0.395$, and $n = 4$ for this prototype lowpass Butterworth filter.

The four poles in the left half **p**-plane are calculated as

$$\mathbf{p}_1, \mathbf{p}_4 = -0.4827 \pm j1.1654$$
$$\mathbf{p}_2, \mathbf{p}_3 = -1.1654 \pm j0.4827$$

The transfer function of the prototype lowpass filter

$$H(\mathbf{p}) = \frac{2.5317}{\mathbf{p}^4 + 3.296\mathbf{p}^3 + 5.4325\mathbf{p}^2 + 5.24475\mathbf{p} + 2.5317} \quad (1.112)$$

The magnitude response of this lowpass filter is plotted in Fig. 1.33.

1.17. BILINEAR TRANSFORMATION

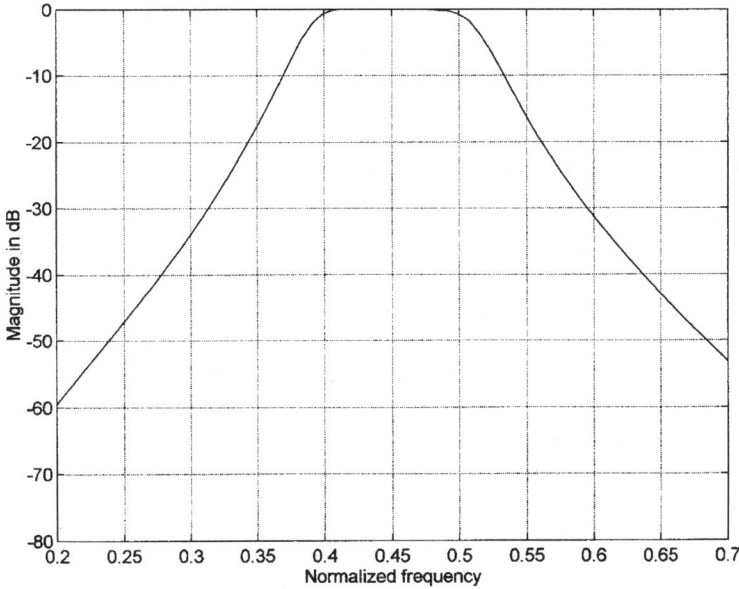

Fig. 1.34. Magnitude responses of the bandpass filters in examples 8 and 9

Next we substitute $\mathbf{p} = \frac{1}{0.547}\left(\frac{s^2+1.705^2}{s}\right)$ in (1.112) and after simplifying, the resulting transfer function is

$$H(s) = \frac{0.2267s^4}{D(s)}$$

where $D(s)$ is given by

$$(s^8 + 1.8030s^7 + 13.2535s^6 + 16.5824s^5 + 60.3813s^4 + \\ 48.205s^3 + 112.0006s^2 + 44.2926s + 71.4135)$$

Now we apply the bilinear transformation $s = 2\left(\frac{z-1}{z+1}\right)$ on this $H(s)$ and simliplify the transfer function $H(z)$ of the digital filter to the form

$$H(z) = \frac{3.6272z^8 - 14.5088z^6 + 21.7632z^4 - 14.5088z^2 + 3.6272}{(3825z^8 - 4221z^7 + 13127z^6 - 9857z^5 + 15753z^4 \\ -7615z^3 + 7849z^2 - 1934z + 1354)} \quad (1.113)$$

A plot of the magnitude response of this function is shown in Fig. 1.34. It is found that the given specifications are met by this transfer function of the digital filter.

The design of lowpass, highpass, and bandstop filters use similar procedures. In contrast with the impulse invariant transformation, we see that the bilinear transformation can be used for designing lowpass, highpass and bandstop filters

60 CHAPTER 1. CLASSICAL METHODS OF APPROXIMATION

as well. Indeed the use of bilinear transformation is the most popular method used for the design of IIR digital filter functions that approximate the magnitude only specifications.

1.17.2 Digital Spectral Transformation

In the design procedure described above, we used the bilinear transformation to convert the magnitude specification of an IIR digital filter to that of $H(j\lambda)$ by prewarping the frequencies on the ω-axis. Then we either scaled the frequencies on the λ-axis or applied the analog frequency transformations $\mathbf{p} = g(\mathbf{s})$ to reduce the frequency response to that of a lowpass, analog prototype filter function. There is an alternative method for designing IIR digital filters which replaces the need for applying the analog frequency transformation by a frequency transformation in the digital domain. Constantinides [8]derived a set of digital spectral transformations (DST) that convert the magnitude of a lowpass, digital filter with an arbitrary bandwidth, say θ_p, to that of highpass, bandpass and bandstop filters or lowpass filters with a different passband. These transformations are similar to the analog frequency transformations and the parameters of the transformation are determined by the cutoff frequencies of these filters just as in the case of the analog frequency transformations. Let us denote the cutoff frequency of the specified lowpass filter and the highpass filter by θ_p'. Let us denote the upper and lower cutoff frequencies of the bandpass and bandstop filters be denoted by θ_u and θ_l respectively - all of them being less than π radians on the normalized frequency basis. Whereas the lowpass, prototype analog filter always has a passband of one radian/sec, the new lowpass digital filter has a passband that is chosen arbitrarily as θ_p; yet, we will call it the lowpass digital, prototype filter, with a transfer function $H(\mathbf{z}^{-1})$. The digital spectral transformations applied on this digital filter are of the form $\mathbf{z}^{-1} = g(z^{-1})$. They map points inside the unit circle in the z-plane to points inside the unit circle in the z-plane, and map the boundary of the unit circle in the z-plane to the boundary of the unit circle in the z-plane. Using these necessary conditions, Constantinides derived the digital spectral transformations (DST) for the $LP \Rightarrow LP, LP \Rightarrow HP, LP \Rightarrow BP$ and $LP \Rightarrow BS$ transformations and they are listed in Table 1.3[1] where the frequencies $\theta = \omega T$ are normalized frequencies in radians.

1.17.3 Example 1.9

We choose the same specifications as in Example 1.8 and illustrate the procedure to design the IIR filter using the digital spectral transformation from the Table 1.3. Let us choose the passband of the lowpass prototype digital filter to be $\theta_p = 0.5\pi$. The values for the cutoff frequencies specified for the bandpass

---

[1]The readers are advised that there are errors in this Table given in several of the well-known text books on digital signal processing-particularly in the expressions for the parameters or the coefficients of the $LP \Leftrightarrow HP$ and the $LP \Leftrightarrow BP$ transformation.

1.17. BILINEAR TRANSFORMATION

filter are $\theta_l = 0.4\pi$, $\theta_u = 0.5\pi$. So we calculate

$$\alpha = \frac{\cos\left(\frac{0.5\pi+0.4\pi}{2}\right)}{\cos\left(\frac{0.5\pi-0.4\pi}{2}\right)} = 0.158$$

$$K = \cot\left(\frac{0.5\pi - 0.4\pi}{2}\right)\tan\left(\frac{0.5\pi}{2}\right) = 6.314$$

$$z^{-1} = -\frac{z^{-2} - 0.273z^{-1} + 0.727}{0.727z^{-2} - 0.273z^{-1} + 1}$$

Now we have to find the frequency θ_s in the lowpass, digital, prototype filter to which the prescribed stopband frequency $\theta'_s = 0.7\pi$ of the bandpass filter maps, by substituting $z = e^{j0.7\pi}$ in the digital spectral transformation given above. The value is found to be $\theta_s = 2.8$ radians $= 0.8913\pi$ radians. Therefore the specification for the lowpass prototype digital filter to be designed is given as Fig. 1.35b

Using the mapping of $\lambda = 2\tan(\frac{\theta}{2})$ vs. θ, we map this lowpass frequency response to the lowpass filter response $|H(j\lambda)|$ as shown in Fig. 1.35c. We calculate $\lambda_1 = 2\tan\left(\frac{\pi}{4}\right) = 1.998$ and $\lambda_2 = 2\tan\left(\frac{2.8}{2}\right) = 11.6$ as the edge of the passband and the edge of the stopband of this filter respectively. So we scale its frequency by 1.998 to get the frequency response of the lowpass, prototype filter $H(j\Omega)$ in order to get a normalized bandwidth $\Omega_p = 1$. The stopband frequency gets scaled down to 5.8, which is slightly different from the value obtained in the Example 1.8 due to numerical inaccuracies. But the order of the lowpass, prototype, analog filter is required to be the same and hence the transfer function is the same as in Example 1.8. The transfer function is repeated below.

$$H(\mathbf{p}) = \frac{2.5317}{\mathbf{p}^4 + 3.2962\mathbf{p}^3 + 5.4325\mathbf{p}^2 + 5.2447\mathbf{p} + 2.5317} \tag{1.114}$$

Next we restore the frequency scale by substituting $\mathbf{p} = \frac{s}{1.998}$ in $H(\mathbf{p})$ to get the transfer function $H(s)$ as

$$H(s) = \frac{40.5072}{s^4 + 6.5924s^3 + 21.73s^2 + 41.9576s + 40.5072} \tag{1.115}$$

and then apply the bilinear transformation $s = 2\left(\frac{z-1}{z+1}\right)$ on this $H(s)$ to get the transfer function of the lowpass prototype digital filter $H(\mathbf{z})$ as

$$\frac{(40.5072\mathbf{z}^4 + 162.0288\mathbf{z}^3 + 243.0432\mathbf{z}^2 + 162.0288\mathbf{z} + 40.5072)}{(280.0816\mathbf{z}^4 + 160.3808\mathbf{z}^3 + 165.2032\mathbf{z}^2 + 35.6768\mathbf{z} + 6.7728)} \tag{1.116}$$

Final step is to apply the digital spectral transformation (1.117) derived earlier, to $H(\mathbf{z})$ in (1.116).

$$\mathbf{z}^{-1} = -\frac{z^{-2} - 0.273z^{-1} + 0.727}{0.727z^{-2} - 0.273z^{-1} + 1} \tag{1.117}$$

Table 1.3. Digital Spectral Transformations.

Transformation	Parameters
$LP \Rightarrow LP$	θ_p =passband of the prototype filter θ'_p =passband of the new LP filter
$z^{-1} \Rightarrow \left(\dfrac{z^{-1}-a}{1-az^{-1}}\right)$	$a = \dfrac{\sin\left[\left(\frac{\theta_p-\theta'_p}{2}\right)\right]}{\sin\left[\left(\frac{\theta_p+\theta'_p}{2}\right)\right]}$
$LP \Rightarrow HP$	θ'_p =cutoff frequency of the HP filter
$z^{-1} \Rightarrow -\left(\dfrac{z^{-1}-a}{1-az^{-1}}\right)$	$a = \dfrac{\cos\left[\left(\frac{\theta_p+\theta'_p}{2}\right)\right]}{\cos\left[\left(\frac{\theta_p-\theta'_p}{2}\right)\right]}$
$LP \Rightarrow BP$ $z^{-1} \Rightarrow$ $-\left(\dfrac{z^{-2}-\frac{2\alpha K}{(K+1)}z^{-1}+\frac{(K-1)}{(K+1)}}{\frac{(K-1)}{(K+1)}z^{-2}-\frac{2\alpha K}{(K+1)}z^{-1}+1}\right)$	θ_l =lower cutoff frequency of the BP filter θ_u =upper cutoff frequency of the BP filter $\alpha = \dfrac{\cos\left[\left(\frac{\theta_u+\theta_l}{2}\right)\right]}{\cos\left[\left(\frac{\theta_u-\theta_l}{2}\right)\right]}$ $K = \cot\left(\frac{\theta_u-\theta_l}{2}\right)\tan\left(\frac{\theta_p}{2}\right)$
$LP \Rightarrow BS$ $z^{-1} \Rightarrow$ $\left(\dfrac{z^{-2}-\frac{2\alpha}{(K+1)}z^{-1}+\frac{1-K}{1+K}}{\frac{1-K}{1+K}z^{-2}-\frac{2\alpha}{(K+1)}z^{-1}+1}\right)$	θ_l =lower cutoff frequency of the BS filter θ_u =upper cutoff frequency of the BS filter $\alpha = \dfrac{\cos\left[\left(\frac{\theta_u+\theta_l}{2}\right)\right]}{\cos\left[\left(\frac{\theta_u-\theta_l}{2}\right)\right]}$ $K = \tan\left(\frac{\theta_u-\theta_l}{2}\right)\tan\left(\frac{\theta_p}{2}\right)$

1.17. BILINEAR TRANSFORMATION

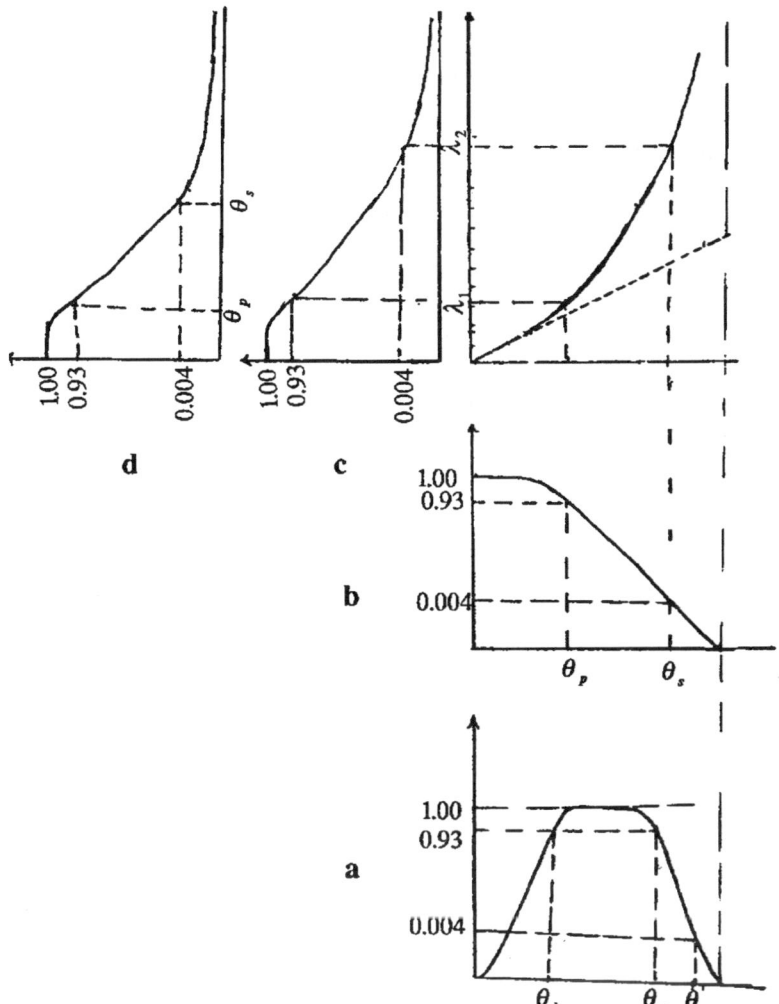

Fig. 1.35. Mapping of the frequency responses under the bilinear transformation, for Example 9

The final result is the transfer function $H(z)$ of the required IIR filter which is found to be the same as (1.113) obtained in Example 1.8. The magnitude is therefore found to be the same as in Fig. 1.34 and is not plotted again. However, it should be obvious that this alternative method does not offer any advantage. Indeed, when compared with method of Example 1.8, the above method requires more computations, particularly in the final step.

We described the procedure for designing an IIR digital filter by two methods in some detail, in order to provide a better understanding of the theory. Now we design the same filter using MATLAB$^{(R)}$.

1.17.4 Example 1.10

MATLAB$^{(R)}$ commands `butter` and `butterap` can design Butterworth filters if the passband edges correspond to an attenuation of 3 dB only. Hence to design the digital filter of Example 1.8, we have to use the command `buttord` and get the value of n, then use this value in the commands `butter` or `buttap`. The MATLAB$^{(R)}$ file used to obtain the transfer function for the bandpass filter given in Example 1.8 is the following.

```
clear
wp=[0.4 0.5];
ws=[0.2 0.7];
Rp=0.63   % Max.attenuation in the passband
          % corresponding to a magnitude = 0.93
Rs= 48    % Min.attenuation in the stopband
          % corresponding to a magnitude = 0.004
[n,wn]=buttord(wp,ws,Rp,Rs);
[b,a]=butter(n,wn);
[h,w]=freqz(b,a,128);
H3=abs(h);
H3db=20*log10(H3);
plot(w/pi,H3db);
axis([0.2 0.7 -80 0]);
grid
title('Magnitude of the BP filter:Example 3');
ylabel('Magnitude in dB');
xlabel('Normalized frequency');
axis('normal')
```

Its magnitude response is plotted along with that of Example 1.8 (and 1.9) in Fig. 1.36. The magnitude response of Example 1.10 in the stopband meets the given specifications but that of Example 1.8 (and 1.9) shows an attenuation in the stopband which is more than the attenuation specified.

1.18 Computer-Aided Optimization

As in every other design effort, the design of IIR digital filters can be carried out using computer-aided methods and if the frequency response is arbitrarily specified, that is the only method available. This method is described in the next chapter, for the design of an IIR filter when both the magnitude and group delay are specified at a number of discrete frequencies θ_i, $i = 1, 2, \ldots, m$.

1.19. CONCLUSION

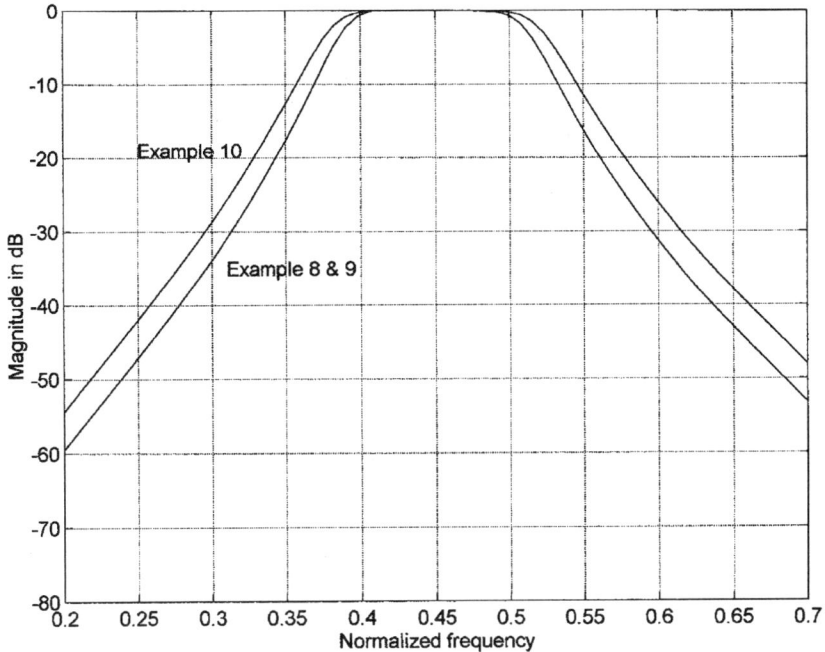

Fig. 1.36. Magnitude of the bandpass filters in Examples 8,9 and 10

If only the magnitude is specified, we prewarp the frequencies by use of the transformation $\lambda_i = 2\tan\left(\frac{\theta_i}{2}\right)$ and the magnitude response in the λ domain is then plotted as a Bode plot. Then using straight line approximation, a rational transfer function $H(s)$ is constructed which has a magnitude closely approximating the prewarped magnitude response. Then we apply the bilinear transformation $s = 2\left(\frac{z-1}{z+1}\right)$ to get $H(z)$. This will have a magnitude response very close to the prescribed magnitude response. If we start the computer-aided optimization procedure with this transfer function as the initial guess, it is easier to reach the global minimum for any error function defined for the purpose. Since this initial guess is a stable transfer function, it is more likely that the final transfer function corresponding to the global minimum value for the error will be a stable transfer function.

1.19 Conclusion

Another approach for the design of IIR filters is aimed at approximating the unit sample response of the digital filter. The length of the unit impulse response is truncated to a long but finite sequence. There are several methods for finding the transfer function $H(z)$ of an infinite impulse response (IIR) filter such that its unit impulse response $h(n)$ approximates the truncated sequence of finite length. The literature on the design of IIR digital filters that approximate prescribed impulse response data is very extensive ([5, 28]) and there are many text books in which the literature is available. Those readers who are interested in this approach are referred to this literature.

In the next chapter, we describe several other methods for designing IIR filters that approximate both prescribed magnitude and group delay response. Some of these methods can therefore be easily adapted for the design of filters which are specified by their magnitude only, for example, by choosing a constant group delay and the magnitude response that is prescribed.

Bibliography

[1] M. Abramowitz and I.A. Stegun, *Handbook of Mathematical Functions*, Dover, NY, 1965.

[2] A. Antoniou, *Digital Filters: Analysis, Design and Applications*, McGraw-Hill, NY, 1993.

[3] J.W. Bandler and B.L. Bardakjian, "Least p^{th} optimization of recursive digital filters," IEEE Trans. on Audio and Electroacoustics, vol. 21, pp. 460-470, October 1973.

[4] J.W. Bandler and R.E. Seviora, "Current trends in network optimization," IEEE Trans. on Microwave Theory and Techniques, MTT-18, pp. 1159-1170, December 1970.

[5] C.S. Burrus and T.W. Parks, "Time domain design of recursive digital filters," IEEE Trans. on Audio and Electroacoustics, AU-18, pp. 137-141, 1970.

[6] C. Charalambous, "Acceleration of least p^{th} algorithm for minimax optimization with engineering applications," Mathematical Programming, vol. 17, pp. 270-297, 1979.

[7] Erich Christian and Egon Eisenmann, *Design Tables and Graphs*, John Wiley, NY, 1966.

[8] A.G. Constantinides, "Spectral transformations for digital filters," Proc. IEE, vol. 117 No. 8, pp. 1585-1590, August 1970.

[9] S. Darlington, "A history of network synthesis and filter theory for circuits composed of resistors, inductors and capacitors," IEEE Trans. on Circuits and Systems, CAS-31, pp. 3-12, January 1984.

[10] R.W. Daniels, *Approximation Methods for Electronic Filter Design*, McGraw-Hill, 1974.

[11] A.G. Evans and R. Fischl, "Optimal least squares time domain synthesis of recursive digital filters," ibid, AU-21, pp. 61-65, 1973.

[12] FILSYN Software for Filter Analysis and Design, DGS Associates, 1353 Sarita Way, Santa Clara, CA 95051.

[13] R.A. Friedenson et al, "RC-Active filters for the D-3 Channel Bank," B.S.T.J. vol. 54, pp. 507-529, 1975.

[14] A.H. Gray, Jr. and J.D. Markel, "A computer program for designing digital elliptic filters," IEEE Trans. on Acoustics, Speech and Signal Processing, ASSP-24, pp. 529-538, June 1976.

[15] O. Heaviside, *Electromagnetic Theory*, 1893 (Reprinted by Dover Publications, New York, 1950).

[16] D.S. Humpherys,The Analysis, Design and Synthesis of Electrical Filters, Prentice-Hall, 1970.

[17] The Signal Processing Toolbox and MATLAB$^{(R)}$ - Software from The Mathworks Inc, Natick, MA.

[18] S.K. Mitra and J.F. Kaiser (Eds.), *Handbook for Digital Signal Processing*, John Wiley & Sons, 1993.

[19] A.V. Oppenheim and R.W. Schafer, *Discrete-Time Signal Processing*, Prentice-Hall, 1989.

[20] A. Papoulis, "A new class of filters," Proc.IRE, vol. 46, pp. 649-653, March 1958.

[21] T.W. Parks and C.S. Burrus,*Digital Filter Design*, John Wiley, NY, 1987.

[22] J.G. Proakis and D.G. Manolakis, *Digital Signal Processing-Principles, Algorithms, and Applications*, Prentice-Hall 1996.

[23] M.I. Pupin, "Wave transmission over non-uniform cables and long distance air lines," Trans. AIEE, vol. 17, pp. 445-507, 1900.

[24] R. Schaumann, M.S. Ghausi and K.R. Laker, *Design of Analog Filters*(Appendix), Prentice-Hall, 1981.

[25] A.S. Sedra and P.O. Bracket, *Filter Theory and Design: Active and Passive*, ISBS Inc, Forest Grove, OR, 1978.

[26] C.E. Shannon, "Communication in the Presence of Noise," Proc. IRE, pp. 10-12, January 1949.

[27] A.K. Shaw, "Optimal identification of discrete-time systems from impulse response," IEEE Trans. on Signal Processing, vol. 42, no. 1, pp. 113-120, January 1994.

[28] A.K. Shaw, "Optimal design of digital IIR filters by model-fitting frequency response data," IEEE Trans. on Circuits and Systems, vol. 42, pp. 702-710, November 1995.

[29] G.C. Temes and J.W. LaPatra, *Introduction to Circuit Synthesis and Design*, McGraw-Hill, 1977.

[30] G.C. Temes and D.Y.F. Zai, "Least p^{th} approximation," IEEE Trans. on Circuit Theory, CT-16, pp. 235-237, May 1969.

[31] W. Thomson, "On the theory of the electric telegraph," Phil. Mag. vol. 11, pp. 146-160, 1856.

[32] P.P. Vaidyanathan, "Optimal design of linear phase FIR digital filters with very flat passband and equiripple stopbands," IEEE Trans. on Circuits and Systems, vol. CAS-32, pp. 904-907, September 1985.

[33] A.I. Zverev, *Handbook of Filter Synthesis*, John Wiley, NY, 1967.

Chapter 2

Magnitude and Delay Approximation of 1-D IIR Filters

2.1 Introduction

It has already been shown in the previous chapter that one of the most well known methods for designing 1-D IIR filters to approximate only a prescribed magnitude response, is based on the application of the bilinear transformation to the corresponding analog filter functions. The classical theory of analog filters that approximates the piece-wise constant magnitude over finite bands of frequency is a highly advanced and mathematically sophisticated part of circuit theory, that has contributed to and benefited from the mathematical theory of approximation. Numerical methods based on both linear and nonlinear optimization techniques have also been developed extensively during the last four decades to design such filters. Many well-known computer software packages are available for the analysis and design of filters using passive elements only or passive elements and active devices like Operational Amplifiers, Transconductance Amplifiers, and Switched Capacitors [14, 45, 46].

In this chapter, we consider the approximation of a constant group delay first and then the simultaneous approximation of both magnitude and group delay of one-dimensional IIR digital filters. Thus we will describe a method of designing an all-pole transfer function $H(z)$ which approximates a constant group delay in the maximally flat sense. Then we will describe the method of designing allpass functions which, when cascaded with a transfer function that approximates the prescribed magnitude, can compensate for the delay distortion of the transfer function without changing its magnitude response. Thus, an approximation for a constant group delay can be achieved. We will then consider the procedures of augmenting an all-pole transfer function that approximates the prescribed group delay, by introducing a numerator polynomial such that it changes the group delay of the all-pole transfer function only by a constant but changes the magnitude such that the overall transfer function approximates the prescribed magnitude. The use of the bilinear transformation to transform the

the problem of designing an IIR filter which approximates both the prescribed magnitude and group delay, to that of designing an analog filter with equivalent magnitude and group delay (or vice-versa) is not successful, because the group delay characteristic is not preserved under such a transformation. That is why the use of the FIR filter became very attractive since such a digital filter can be easily designed to obtain either zero phase or linear phase response, in addition to satisfying the prescribed magnitude specification. But FIR filters designed to approximate a magnitude response that has a narrow transition band between the passband and the stopband usually require a large number of multipliers and they have a large delay also.

Hence in recent years, new methods have been proposed for the approximation of both the magnitude and group delay of IIR digital filters. They have less number of multipliers than their corresponding FIR counterparts and yet provide the required group delay response. Several such methods will be described in detail in the later part of this chapter.

2.2 Properties of an IIR Transfer Function

In its most general form, the transfer function of a one-dimensional IIR filter is of the form

$$H(z) = A \frac{\sum_{i=0}^{M} b_i z^{-i}}{\sum_{i=0}^{N} a_i z^{-i}} = \frac{N(z)}{D(z)} \qquad (2.1)$$

We assume that it has at least one finite pole so that it represents an IIR filter and also without loss of generality we assume $a_0 = 1$, unless stated otherwise.

The magnitude and phase response of the above filter function evaluated on the unit circle $z = e^{j\omega T}$ are given below:

$$\begin{aligned} H(e^{j\omega T}) &= A \frac{\sum_{i=0}^{M} b_i \cos(i\omega T) - j \sum_{i=0}^{M} b_i \sin(i\omega T)}{\sum_{i=0}^{N} a_i \cos(i\omega T) - j \sum_{i=0}^{N} a_i \sin(i\omega T)} \\ &= \left| H(e^{j\omega T}) \right| e^{j\theta(\omega T)} \end{aligned} \qquad (2.2)$$

where the magnitude response is

$$\left| H(e^{j\omega T}) \right| = |A| \left| \frac{\left[\sum_{i=0}^{M} b_i \cos(i\omega T) \right]^2 + \left[\sum_{i=0}^{M} b_i \sin(i\omega T) \right]^2}{\left[\sum_{i=0}^{N} a_i \cos(i\omega T) \right]^2 + \left[\sum_{i=0}^{N} a_i \sin(i\omega T) \right]^2} \right|^{1/2} \qquad (2.3)$$

and the phase response is

$$\theta(\omega T) = -\tan^{-1} \frac{\sum_{i=0}^{M} b_i \sin(i\omega T)}{\sum_{i=0}^{M} b_i \cos(i\omega T)} + \tan^{-1} \frac{\sum_{i=0}^{N} a_i \sin(i\omega T)}{\sum_{i=0}^{N} a_i \cos(i\omega T)} \qquad (2.4)$$

Then the group delay $\tau(\omega T) = -\frac{\partial \theta}{\partial \omega}$ is given by

$$\tau(\omega T) = \frac{1}{1+u^2} \frac{du}{d\omega} - \frac{1}{1+v^2} \frac{dv}{d\omega} \qquad (2.5)$$

2.2. PROPERTIES OF AN IIR TRANSFER FUNCTION

where
$$u = \frac{\sum_{i=0}^{M} b_i \sin(i\omega T)}{\sum_{i=0}^{M} b_i \cos(i\omega T)} \qquad (2.6)$$

and
$$v = \frac{\sum_{i=0}^{N} a_i \sin(i\omega T)}{\sum_{i=0}^{N} a_i \cos(i\omega T)} \qquad (2.7)$$

If the phase response $\theta(\omega T)$ is expressed in the form
$$\theta(\omega T) = Im\left\{\ln H(e^{j\omega T})\right\} \qquad (2.8)$$

then the group delay can be expressed in another form as

$$\begin{aligned} \tau(\omega T) &= -Re\left\{ z\frac{H'(z)}{H(z)}\bigg|_{z=e^{j\omega T}} \right\} \\ &= -\frac{1}{2}\left[\left\{z\frac{H'(z)}{H(z)} + z^{-1}\frac{H'(z^{-1})}{H(z^{-1})}\right\}\bigg|_{e^{j\omega T}}\right] \end{aligned} \qquad (2.9)$$

To derive (2.9), let $H(e^{j\omega T}) = |H(e^{j\omega T})| e^{j\theta(\omega T)}$. Then
$$H(e^{-j\omega T}) = |H(e^{j\omega T})| e^{-j\theta(\omega T)}$$

Therefore
$$\frac{H(e^{j\omega T})}{H(e^{-j\omega T})} = e^{j2\theta(\omega T)} \qquad (2.10)$$

Then $\ln\left\{H(e^{j\omega T})\right\} - \ln\left\{H(e^{-j\omega T})\right\} = j2\theta(e^{j\omega T})$. Therefore, if we let $z = e^{j\omega T}$, we can write $\frac{dz}{d\omega} = jTz$ and $\ln H(z) - \ln H(z^{-1}) = \ln\{j2\theta(z)\}$. Differentiating both sides of this equation with respect to ω and using $\tau(e^{j\omega T}) = -\frac{d\theta}{d\omega} = -\left\{\frac{d\theta}{dz}\frac{dz}{d\omega}\right\}$, we can derive the result (2.9). Similar results were given for analog filters, in section 1.3.

Now consider the special case when $H(z)$ is an allpass function. Then (2.1) reduces to the form
$$H_a(z) = A\frac{\sum_{i=0}^{N} a_{N-i} z^{-i}}{\sum_{i=0}^{N} a_i z^{-i}} = Az^{-N}\frac{\sum_{i=0}^{N} a_i z^{i}}{\sum_{i=0}^{N} a_i z^{-i}} \qquad (2.11)$$

The phase and group delay of this allpass function (which has a constant magnitude equal to A for all frequencies) can be derived as

$$\theta(\omega T) = -N\omega T + 2\tan^{-1}\frac{\sum_{i=0}^{N} a_i \sin(i\omega T)}{\sum_{i=0}^{N} a_i \cos(i\omega T)} \qquad (2.12)$$

and
$$\tau(\omega T) = NT - 2\frac{1}{1+u^2}\frac{du}{d\omega} \qquad (2.13)$$

From (2.6), we get

$$\frac{du}{d\omega} = \frac{\sum_{i=0}^{N} a_i T i \cos(i\omega T)}{\sum_{i=0}^{N} a_i \cos(i\omega T)} + \frac{\sum_{i=0}^{N} a_i \sin(i\omega T)}{\left\{\sum_{i=0}^{N} a_i \cos(i\omega T)\right\}^2} \sum_{i=0}^{N} a_i T i \sin(i\omega T) \quad (2.14)$$

The final expression for $\tau(\omega T)$ can be reduced to the following, by using (2.13).

$$\tau(\omega T) = NT - 2T \frac{\left\{\begin{array}{c}\left(\sum_{i=0}^{N} i a_i \cos(i\omega T)\right)\left(\sum_{i=0}^{N} a_i \cos(i\omega T)\right) + \\ \left(\sum_{i=0}^{N} i a_i \sin(i\omega T)\right)\left(\sum_{i=0}^{N} a_i \sin(i\omega T)\right)\end{array}\right\}}{\left(\sum_{i=0}^{N} a_i \cos(i\omega T)\right)^2 + \left(\sum_{i=0}^{N} a_i \sin(i\omega T)\right)^2} \quad (2.15)$$

It is seen that the allpass transfer function (2.11) has zeros in the z-plane which are reciprocals of the poles and hence the zeros of $H_a(z)$ are outside the unit circle. Let the poles be represented in their polar form $z_i = r_i e^{j\phi_i}$. Then the transfer function can be written in an alternative form as

$$H_a(z) = \prod_{i=1}^{N/2} \frac{(z - r_i^{-1} e^{-j\phi_i})(z - r_i^{-1} e^{j\phi_i})}{(z - r_i e^{j\phi_i})(z - r_i e^{-j\phi_i})} \quad (2.16)$$

Then the group delay $\tau(\omega T)$ can be derived as given below.

$$T \sum_{i=1}^{N/2} \left\{ \frac{1 - r_i^2}{1 + r_i^2 - 2r_i \cos(\omega T - \phi_i)} + \frac{1 - r_i^2}{1 + r_i^2 - 2r_i \cos(\omega T + \phi_i)} \right\} \quad (2.17)$$

It is seen from (2.12) that $\theta(\omega T)$ decreases monotonically from 0 at $\omega = 0$ to $-N\pi$ as ω approaches π. We can also derive

$$\int_0^\pi \tau(\omega T) d(\omega T) = (-\theta(\omega T))|_0^\pi = NT\pi \quad (2.18)$$

Hence an ideal allpass digital filter realizing a constant group delay of NT over the frequency range $0 \le \omega T \le \pi$, can only be of order N. When a constant group delay of NT seconds is specified (which is also measured as N sampling periods) and we plan on approximating it by an optimization technique, a good choice for the initial value for the order of the allpass filter is $(N-1)$; then we decrease it by one in each successive iterative cycle till the order is $N/2$. The best choice is the one at which the error function reaches its minimum value.

2.3 Maximally Flat Group Delay Filters

Thiran [47] developed an analytical method for deriving the all-pole transfer function of the digital filter that approximates a constant group delay in the

2.3. MAXIMALLY FLAT GROUP DELAY FILTERS

maximally flat sense. Let τT be the prescribed group delay, and let the all-pole transfer function be chosen as

$$\mathbf{H}(z^{-1}) = \frac{\sum_{i=0}^{n} a_i}{\sum_{i=0}^{n} a_i z^{-i}} \qquad (2.19)$$

Then the error in the phase is

$$\delta(\omega T) = -\omega\tau - \tan^{-1} \frac{\sum_{i=0}^{n} a_i \sin(i\omega T)}{\sum_{i=0}^{n} a_i \cos(i\omega T)} \qquad (2.20)$$

Another form of the error derived from (2.20) is given as

$$\epsilon(\omega T) = -\tan(\omega\tau) - \frac{\sum_{i=0}^{n} a_i \sin(i\omega T)}{\sum_{i=0}^{n} a_i \cos(i\omega T)} \qquad (2.21)$$

which can be rewritten as

$$\epsilon(\omega T) = \frac{-\sin(\omega\tau) \sum_{i=0}^{n} a_i \cos(i\omega T) - \cos(\omega\tau) \sum_{i=0}^{n} a_i \sin(i\omega T)}{\cos(\omega\tau) \sum_{i=0}^{n} a_i \cos(i\omega T)} \qquad (2.22)$$

Let us assume in the sequel that the sampling period T is normalized to one second so that $\omega_s = 2\pi$ and τ denotes the delay which is the number of sampling periods. Hence ω will be the normalized frequency. Then (2.22) reduces to

$$\epsilon(\omega) = -\frac{\sum_{i=0}^{n} a_i \sin(i+\tau)\omega}{\cos(\omega\tau) \sum_{i=0}^{n} a_i \cos(i\omega)} \qquad (2.23)$$

The numerator is an odd function and the denominator is an even function and therefore their expansion in power series gives

$$\epsilon(\omega) = \frac{\sum_{k=0}^{\infty} p_k \omega^{2k+1}}{\sum_{k=0}^{\infty} q_k \omega^{2k}} \qquad (2.24)$$

Since $\epsilon(\omega)$ is an odd function, its Taylor series contains only odd powers of ω and hence is in the form,

$$\epsilon(\omega) = \sum_{k=0}^{\infty} c_k \omega^{2k+1} \qquad (2.25)$$

where the coefficient c_k is the kth derivative of $\epsilon(\omega)$ evaluated at $\omega = 0$. The coefficients can also be generated from the recursive relation

$$c_k = \frac{1}{q_0}\left[p_k - \sum_{j=1}^{k} c_{k-j} q_j\right] \qquad (2.26)$$

For getting a maximally flat approximation of a constant group delay τ, we need to make the first n derivatives of $\epsilon(\omega)$ at $\omega = 0$ to be zero i.e. $c_k = 0$ for $k = 0, 1, 2, \ldots, (n-1)$. From ((2.24), (2.25) and 2.26), we see that the equivalent condition to be satisfied is $p_k = 0$ for $0 \leq k \leq n-1$. Using the Taylor series expansion

$$\sin x = \sum_{k=0}^{\infty} (-1)^k \frac{x^{2k+1}}{(2k+1)!} \tag{2.27}$$

on (2.23) we get

$$\epsilon(\omega) = \sum_{k=0}^{\infty} (-1)^k \left[\frac{\sum_{i=0}^{n} a_i (i+\tau)^{2k+1}}{(2k+1)!} \right] \omega^{2k+1} \tag{2.28}$$

Hence the coefficient

$$c_k = (-1)^k \left[\frac{\sum_{i=0}^{n} a_i (i+\tau)^{2k+1}}{(2k+1)!} \right] \tag{2.29}$$

is zero when $\sum_{i=0}^{n} a_i(i+\tau)^{2k+1} = 0$. From the condition that the coefficients c_k in (2.26) are zero for $k = 0, 1, 2, \ldots, (n-1)$ (with $a_0 = 1$), the condition for maximally flat approximation of a constant group delay becomes

$$\tau^{2k+1} + \sum_{i=1}^{n} a_i(i+\tau)^{2k+1} = 0 \tag{2.30}$$

So, we get the conditions

$$\sum_{i=1}^{n} a_i(i+\tau)^{2k+1} = -\tau^{2k+1} \quad \text{for } k = 0, 1, 2, \ldots, n-1 \tag{2.31}$$

Solving these linear equations for the coefficients a_i in terms of two Vandermonde determinants, Thiran shows that the coefficients are given by

$$a_k = (-1)^k \binom{n}{k} \prod_{i=0}^{n} \frac{2\tau + i}{2\tau + k + i} \tag{2.32}$$

Using the Gamma functions, he also shows that the polynomial $P_n(z^{-1}, \tau) = \sum_{k=0}^{n} a_k z^{-k}$ in the denominator of the maximally flat delay filter function (2.19) can be expressed as

$$P_n(z^{-1}, \tau) = \sum_{k=0}^{n} \frac{\Gamma(-n+k)}{\Gamma(-n)} \frac{\Gamma(2\tau+n+1)}{\Gamma(2\tau)} \frac{\Gamma(2\tau+k)}{\Gamma(2\tau+k+n+1)} \frac{z^{-k}}{k!} \tag{2.33}$$

The author also derives the numerator of (2.19) from its denominator evaluated on $|z| = 1$ and shows that

$$P_n(1, \tau) = \sum_{i=0}^{n} a_i = \frac{2n!}{n!} \frac{1}{\prod_{i=n+1}^{2n}(2\tau+i)} \tag{2.34}$$

2.3. MAXIMALLY FLAT GROUP DELAY FILTERS

Table 2.1. Coefficients of the Denominator polynomial in Thiran's filter

2	4.6667	-5.333	1.6670		
3	6.0000	-9.000	5.0000	-1.000	
4	7.0714	-12.5714	9.4286	-3.4286	0.5000
5	7:9440	-15.8889	14.444	-7.2222	1.9444
	-0.2222				
6	8.6667	-18.9091	19.697	-12.1212	4.5455
	-0.9697	0.0909			
7	9.2727	-21.6364	24.9650	-17.8322	8.3217
	-2.4965	0.4406	-0.0350		
8	9.7879	-24.0932	30.1166	-24.0932	13.1760
	-4.9604	1.2401	-0.1865	0.0128	
9	10.2308	-26.3077	35.0769	-30.6923	18.9570
	-8.4253	2.6606	-0.5701	0.0747	-0.0045
10	10.6154	-28.3077	39.8077	-37.4661	25.4977
	-12.8831	4.8312	-1.3146	0.2465	-0.0286
	0.0015				

Finally Thiran's expression for the digital transfer function that approximates a maximally flat group delay is given by

$$\mathbf{H}(z^{-1}) = \frac{\left\{\frac{2n!}{n!}\frac{1}{\prod_{i=n+1}^{2n}(2\tau+i)}\right\}}{\sum_{k=0}^{n}\left[(-1)^k \binom{n}{k} \prod_{i=0}^{n} \frac{2\tau+i}{2\tau+k+i}\right] z^{-k}} \qquad (2.35)$$

Thiran has also shown that the above transfer function is stable for all finite positive values of τ. Using the above formula, the coefficients of the denominator polynomials of $\mathbf{H}(z^{-1})$ for $n = 2, 3, \ldots, 10$ (and $\tau = 2$) and their roots are tabulated in Tables 2.1 and 2.2. respectively.

Let us consider the transfer function with the numerator constant normalized to unity i.e. a function $H_0(z^{-1}) = \frac{1}{\sum_{k=0}^{n} a_k z^{-k}}$. Using (2.9), we can derive an expression for the group delay of this transfer function at $\omega = 0$ as

$$\tau(\omega T)|_{\omega=0} = -T \frac{\sum_{k=0}^{n} k a_k}{\sum_{k=0}^{n} a_k} \qquad (2.36)$$

Equation (2.36) can also be expressed as

$$\tau(\omega T)|_{\omega=0} = -T \left\{ \frac{1}{P_n(z^{-1}, \tau)} \frac{dP_n(z^{-1}, \tau)}{dz} \right\}\Bigg|_{z=1} \qquad (2.37)$$

Thiran points out that $P_n(z^{-1}, \tau)$ in (2.33) can be expressed in terms of the Legendre function of the first kind as $P_n(z^{-1}, \tau) = F(-n, 2\tau; 2\tau + n + 1; z^{-1})$.
From equations (15.1.20) and (15.2.1) in [1], we use the following two results:

$$F(a, b, c; 1) = \frac{\Gamma(c)\Gamma(c-a-b)}{\Gamma(c-a)\Gamma(c-b)}$$

Table 2.2. Poles of Thiran's maximally flat group delay filter

n=2	0.5714±j0.1751		
n=3	0.5000±j0.2887	0.5000	
n=4	0.4365± j0.3703	0.4524±j0.1056	
n=5	0.3783±j0.4315	0.4119±j0.1810	0.4197
n=6	0.3247±j0.4784	0.3750±j0.2394	
	0.3911±j0.0741		
n=7	0.2753±j0.5153	0.3407±j0.2845	
	0.3620±j0.1332	0.3773	
n=8	0.2297±j0.5441	0.3060±j0.3214	
	0.3439±j0.1885	0.3512±j0.0219	
n=9	0.1872±j0.5673	0.2831±j0.3509	
	0.2879±j0.2375	0.4039±j0.1134	0.2474
n=10	0.1477±j0.5857	0.2627±j0.3881	
	0.3989±j0.2114	0.2136±j0.2626	0.4551
	0.1660		

$$\text{for } c \neq 0, -1, -2, \ldots \text{ and } Re(c - a - b) > 0 \qquad (2.38)$$

and

$$\frac{dF(a,b,c;1)}{dz} = \frac{ab}{c} \frac{\Gamma(c+1)\Gamma(c-a-b-1)}{\Gamma(c-a)\Gamma(c-b)} \qquad (2.39)$$

When these two equations are applied to (2.37), we get

$$\tau(\omega T)|_{\omega=0} = -T \left[\frac{ab}{c} \frac{\Gamma(c+1)\Gamma(c-a-b-1)}{\Gamma(c)\Gamma(c-a-b)} \right] \qquad (2.40)$$

Substituting $a = -n$, $b = 2\tau$, $c = 2\tau + n + 1$ in the above result, we get the following.

$$\tau(\omega T)|_{\omega=0} = T \left[\frac{2\tau n}{(2\tau + n + 1)} \frac{\Gamma(2\tau + n + 2)\Gamma(2n)}{\Gamma((2\tau + n + 1)\Gamma(2n+1))} \right] \qquad (2.41)$$

But the Gamma function for nonnegative integers is just the factorial i.e. $\Gamma(k) = (k-1)!$. So, (2.41) is reduced to

$$T \left[\frac{2\tau n}{(2\tau + n + 1)} \frac{(2\tau + n + 1)!(2n - 1)!}{(2\tau + n)!(2n)!} \right] = T\tau \qquad (2.42)$$

This result (2.42) proves that the group delay at $\omega = 0$ for the Thiran's filter function is equal to the specified value[1]. The magnitude and group delay responses of Thiran's maximally flat group delay (MFD) filters chosen from Table 2.1 are plotted for n = 2, 4, 6, 8 in Figs. 2.1 and 2.2, respectively.

---

[1] The author is thankful to Dr. George Szentirmai for the help received in deriving this result.

2.3. MAXIMALLY FLAT GROUP DELAY FILTERS

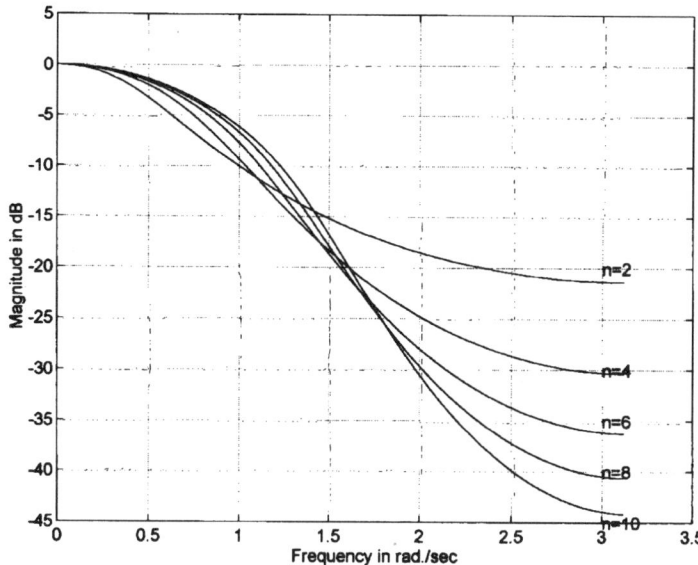

Fig. 2.1. Magnitude of Thiran's Maximally Flat Delay (MFD) filters

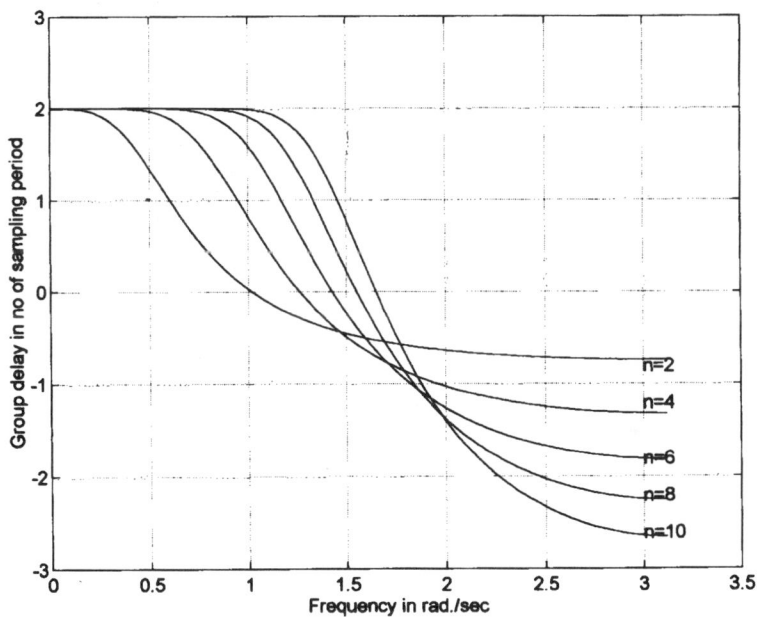

Fig. 2.2. Group Delay of Thiran's Maximally Flat Delay (MFD) filters

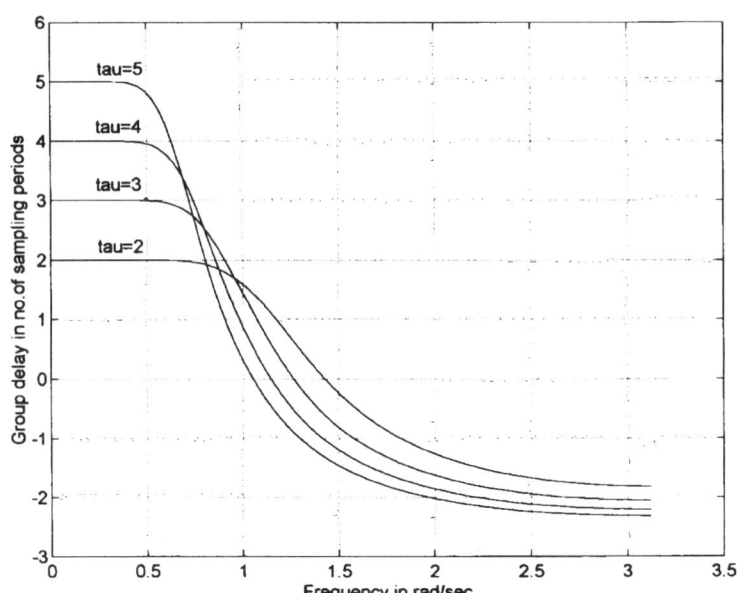

Fig. 2.3. Group Delay of MFD filters for $tau = 2,3,4,5$

In Fig. 2.3, we have plotted the group delay response of Thiran's filters of order 6, with the values of $\tau = 2, 3, 4$ and 5 sampling periods. The two figures show the variation of group delay response as the order n changes, with a fixed value of the group delay and the variation of the response as the group delay changes, with a fixed value for the order of the filter.

The MATLAB$^{(R)}$ [2] program that was used to generate the above data and the plots of magnitude and group delay are given below. It can be used to generate the transfer functions with any other group delay entered as the value of the input variable $\tau = $ tau.

```
clear all
N= input ('Please enter the order of the filter')
tau = input ('Please enter the desired group delay in integers')
k1=fa(N)/fa(2*N);
p1=1;
    for i=N+1:2*N
       temp1=(2*tau +i);
        p1=p1*temp1;
end
    for k=0:N
        p2=1;
         for i=0:N
```

---

[2] MATLAB$^{(R)}$ is a registered trademark of The Mathworks, Inc. Natick, MA.USA

2.3. MAXIMALLY FLAT GROUP DELAY FILTERS

```
            temp2=(2*tau+i)/(2*tau+k+i);
        end
    coeff(k+1)=(-1)^k*(fa(N)/(fa(k))*fa((N-k)))*p2;
end
Q=k1*p1*coeff;
% Q gives the coefficients of the denominator polynomial
disp('Value of the group delay desired');
disp(tau);
disp('Coefficients of the denominator polynomial');
disp(Q);
qr=roots(Q);
disp('Roots of the denominator polynomial');
disp(qr);
% Calculate the Magnitude and Group Delay
N=1;
[h,w]=freqz(N,Q,128);
mag=abs(h)
g=20*log10(mag);
plot(w,g);
grid;
title('Magnitude of Thiran(MFD) filters')
xlabel('Frequency in rad/sec');
ylabel('Magnitude in dB')
[gd,w]=grpdely(N,Q,128);
figure
plot(w,gd);
grid
title('Group Delay of Thiran (MFD)filters')
xlabel('Frequency in rad./sec.')
ylabel('Group delay in no.of sampling period')
%end
```

Thiran [48] has investigated the conditions under which the all-pole transfer functions approximate a constant group delay in the equiripple sense also, following a procedure similar to that used by Ulbrich and Piloty [50] for the design of analog filters. But the conditions give rise to nonlinear equations which can not be solved in general. Deczky [16], however, gave an alternate way of formulating this problem that can be solved by using Remez Exchange Algorithm [5]. In his paper, Deczky assumes the transfer function in the form

$$H(z) = \frac{k_o z^n}{\prod_{i=1}^{n}(z - r_i e^{j\theta_i})} \quad (2.43)$$

in which k_o, r_i and θ_i are the unknown parameters. The corresponding group delay can be derived from (2.9) as

$$\tau(\theta) = \sum_{i=1}^{n} \frac{1 - r_i \cos(\theta - \theta_i)}{1 - 2r_i \cos(\theta - \theta_i) + r_i^2} - n \quad (2.44)$$

Table 2.3. List of poles for the all-pole digital filter with equiripple group delay

$\epsilon = 0.1 \qquad \theta_c = 0.2$

n =2	n =4	n =6	n =8
.4972±j.2655	.7048±j.1517	.7884±j.1055	.8342±j.0807
	.6258±j.4426	.7439±j.3089	.8070±j.2385
		.6799±j.5005	.7590±j.3863
			.7092±j.5273

$\epsilon = 0.1 \qquad \theta_c = 0.4$

n =2	n =4	n =6	n =8
.2467±j.3071	.5290±j.2278	.6527±j.1730	.7231±j.1382
	.2403±j.5834	.4951±j.4771	.6262±j.3952
		.2433±j.6990	.4544±j.6015
			.2490±j.7606

$\epsilon = 0.1 \qquad \theta_c = 0.6$

n=2	n=4	n=6	n=8
.0954±j.2808	.3983±j.2661	.5460±j.2181	.6332±j.1810
	-.0785±j.5238	.2575±j.5415	.4469±j.4895
		-.1632±j.6349	.1374±j.6625
			-.2081±j.6996

$\epsilon = 0.1 \qquad \theta_c = 0.8$

n=2	n=4	n=6	n=8
.0206±j.2409	.3102±j.2797	.4662±j.2450	.5620±j.2109
	.2529±j.3641	.0677±j.5289	.2874±j.5300
		.4103±j.4000	-.1247±j.5997
			-.5004±j.4233

In order to approximate the constant delay τ_o in an equiripple sense, we have to add three more unknown parameters, i.e. θ_c representing the frequency range $[0, \theta_c]$ in the lowpass interval $[0, \pi]$, over which the approximation is achieved, ϵ is the maximum ripple in the range $[0, \theta_c]$ and the total group delay τ_o. These unknown parameters are obtained such that the following conditions are satisfied at $n+1$ points θ_k, $k = 1, 2 \ldots, (n+1)$ in the closed interval $[0, \theta_c]$.

$$E(\theta_k) = \tau(\theta_k) - \tau_o - (-1)^k \epsilon \qquad (2.45)$$

This set of equations and also the method for solving them are similar to those given by Humpherys [25]. Deczky has listed the coefficients of the second order transfer functions connected in cascade to get the all-pole transfer functions of order $n = 2, 4, 6$ and 8, such that they have a maximum ripple $\epsilon = 0.1$ in the group delay. In Table 2.3, we list the poles of the corresponding transfer functions for $\theta_c = 0.2, 0.4, 0.6$ and 0.8. A plot of the equiripple group delay for $n = 6$, and $\epsilon = 0.1$, $\theta_c = 0.6$ is shown in Fig. 2.4. But if values for ϵ or θ_c

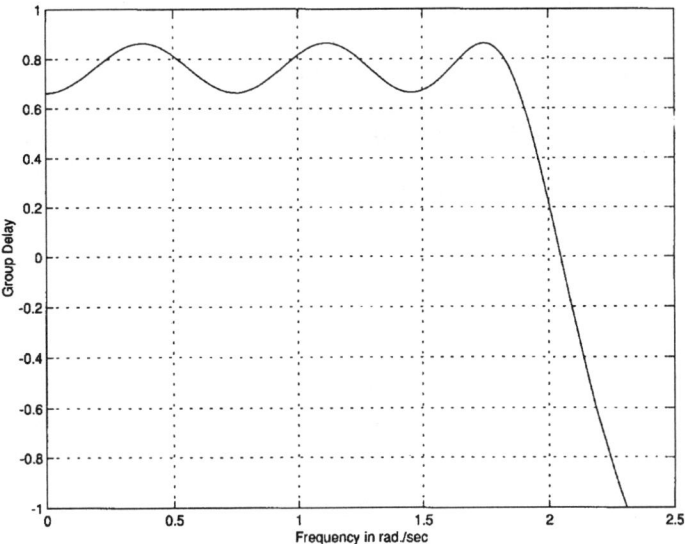

Fig. 2.4. Equiripple group delay of Deczky's filter

are different from what are given in the above table, one will have to solve the approximation problem for that case, since neither a closed form formula like that of Thiran nor a computer program is available to solve the problem with such values.

2.4 Simultaneous Approximation of Magnitude and Group Delay

So far in this chapter we have described how to obtain the all-pole transfer function of an IIR digital filter which approximates only a constant group delay or a linear phase characteristic. Next we consider several methods that have been proposed for simultaneously approximating both the magnitude and the constant group delay response specified for a digital filter. First two of them are listed below.

- Method 2.1: Non-Linear Optimization (Allpass filters in cascade)
- Method 2.2: Use of Mirror Image Polynomial

2.4.1 Method 2.1: Nonlinear Optimization

The first method is the classical method of finding a transfer function that approximates the specified magnitude response and then finding an allpass transfer function which when cascaded with that transfer function gives rise to an overall group delay characteristic that approximates the specified constant group delay. This method is very much similar to the corresponding approach of designing analog filters that approximate both magnitude and group delay and it involves the minimization of an objective function. If $H(z)$ is the transfer function that approximates the specified magnitude response, it can be cascaded with an allpass filter function $H_a(z)$ such that $H(z)H_a(z)$ does not change the magnitude but only the phase of $H(z)$. The allpass filter function is of the form

$$H_a(z) = \prod_{k=1}^{K} \frac{a_k + b_k z + z^2}{1 + b_k z + a_k z^2} \qquad (2.46)$$

The error (objective function) between the specified group delay and the group delay of $H(z)H_a(z)$ is minimized in the least squares or equiripple sense, by choosing the coefficients a_k, and b_k as the unknown parameters of the error function. The computer-aided minimization of the error function is very general because the specified group delay may be a constant or may have any arbitrary characteristics but the minimization involves a highly nonlinear function of the unknown parameters.

But Deczky [15] considered a very general transfer function in the form

$$H(z) = A \prod_{k=1}^{K} \frac{1 + a_{k1} z^{-1} + a_{k2} z^{-2}}{1 + b_{k1} z^{-1} + b_{k2} z^{-2}} \qquad (2.47)$$

and developed an algorithm for minimizing an error function which contains a weighted sum of the error in the magnitude as well as in the group delay. The error function is given by

$$\begin{aligned} J(\mathbf{b}) &= (1-\lambda) \sum_{i=1}^{Q} W_i \left[\left| H(e^{j\omega_i}) \right| - \left| G(e^{j\omega_i}) \right| \right]^p \\ &+ \lambda \sum_{i=1}^{Q} V_i \left[\tau_H(\omega_i) - \tau(\omega_o) - \tau_G(\omega_i) \right]^p \end{aligned} \qquad (2.48)$$

where ω_i represent a set of Q discrete frequencies, $\tau_H(\omega_i)$ is the delay due to the above $H(z)$, $\tau_G(\omega_i)$ is the desired group delay and $\tau(\omega_o)$ is called "the nominal delay" which is a constant value duly chosen to reach a minimum value for the error function. And the vector $\mathbf{b} = [((a_k, b_k), k = 1, 2, \ldots, K), A]^T$ contains all the unknown coefficients. The algorithm uses the least squares approximation ($p = 2$) and the Fletcher-Powell optimization technique, to minimize $J(\mathbf{b})$. Notice that the minimum value for the error depends on the choice of λ and the weighting functions W_i and V_i. Hence the resulting transfer functions are suboptimal. When $\lambda = 1$ is chosen, the problem reduces to that of approximating the group delay of $H(z)$ only, without considering its magnitude distortion. In this case, the author has shown by means of a few examples, how the delay of a digital filter can be equalized in the least squares sense. In doing so, the delay equalizers are assumed to be allpass filters. In this case, we may comment that Deczky's method becomes analogous to the classical method of cascading the transfer function of an analog filter with an allpass function to approximate the prescribed group delay. The authors of [44] have used a new nonlinear programming algorithm called the recursive quadratic programming to design IIR filters that approximate prescribed magnitude and group delay. However they do not assume a cascade connection of allpass filters with an IIR filter; the coefficients of the numerator and the denominator of the IIR filter are considered as the unknown parameters and they are optimized by this algorithm such that the maximum value of the group delay response, (which normally occurs at the passband cutoff frequency when least squares approximation is used) is minimized. For this reason, the algorithm is called peak-constrained least squares (PCLS) optimization and has been shown to yield better delay response than the response for the examples given in [15].

We would also like to digress here to point out that Maria and Fahmy [30] have extended Deczky's work for the design of 2-D IIR digital filters that approximate the constant group delay only, in the least p^{th} sense. However, they do not address the issue of stability of 2-D IIR filters in the design of such filters. On the other hand, Hinamoto and Maekawa [24] approximate both the magnitude and group delay of 2-D IIR filters but in such a way that stability of these filters is a' priori assured. The method of Hinamoto and Maekawa [24] does not involve cascading an IIR filter with an allpass filter in order to equalize the delay of the IIR filter. The method augments an all-pole filter that approximates a constant group delay, by using a mirror image (or anti- mirror image polynomial) for its numerator. This way, the augmentation changes the overall delay response only by a constant but changes the magnitude of the all-pole filter to approximate the prescribed magnitude response. We will defer a more detailed discussion of this method to Chap. 4.

2.4.2 Method 2.2 : Use of Mirror Image Polynomial

The mirror image polynomial of even order, used as the numerator, is a real function of ω. Such a polynomial has zeros inside the unit circle $|z| = 1$ and also the reciprocal of these zeros which are outside the unit circle. It may also have zeros on the unit circle. So the overall transfer function is not a minimum phase transfer function. Consequently, there are less restrictions between the magnitude and the phase response. In other words, there is more flexibility for simultaneously shaping the magnitude and phase responses. Therefore there exist many different choices that have been proposed for solving this general problem, none of which can claim optimality.

These methods which have been summarized below, have something in common but differ in the optimization criteria chosen. All of them decouple the problem of simultaneously approximating the magnitude and group delay by first generating an all-pole transfer function that approximates the group delay - in the maximally flat or equiripple sense or least p^{th} sense. Then they add a numerator which is chosen to approximate the magnitude either in the same sense or in a different sense. For example, the authors of [31] and [24] first optimize the coefficients of the all-pole transfer function $H_1(z^{-1}) = \frac{1}{D(z^{-1})}$ such that it approximates the constant group delay in the least p^{th} sense. Then they augment it by adding a mirror image polynomial $N(z^{-1}) = \sum_{i=0}^{m/2} a_i(z^{-i} + z^{-(m-i)})$ as the numerator and optimize its coefficients such that $H(z^{-1}) = \frac{N(z^{-1})}{D(z^{-1})}$ approximates the prescribed magnitude also in the least p^{th} sense.

Next, we refer to some other methods proposed in [52] and [34]. Briefly, the author of [52], presents an analytical procedure to obtain an explicit form for the transfer function of an allpass filter that approximates a maximally flat group delay. Assuming that it is multiplied by a numerator polynomial, another procedure is developed to obtain the coefficients of this polynomial, such that the magnitude response is equiripple in the stopband. In this two-step procedure, the author introduces the frequency transformations $\zeta = \sqrt{z}$ and $W = \frac{1}{2}(\zeta + \frac{1}{\zeta})$ to derive a function in W. It is then approximated to another function in W and the specified phase response is also expressed as a function in W. The authors of [34] start with Thiran's all-pole transfer function that has a constant group delay response in the maximally flat sense. They also use a couple of frequency transformations like those used in [52]. They propose new methods for choosing the zeros of the numerator polynomial such that the augmented transfer function exhibits a maximally flat magnitude response in the passband and an equiripple response in the stopband. In particular, the method involves a suboptimal choice for the zeros inside the unit circle and their reciprocals (which would be outside the unit circle) such that the magnitude response in the passband is approximated in the maximally flat sense. In the second step, a suboptimal choice of zeros on the unit circle beyond the passband is made, such that the magnitude response in the stopband is approximated in the Chebyshev or equiripple sense. In this second step, the magnitude response in the passband would be altered, so again new locations for the zeros inside the unit circle and their reciprocals are determined. After a few iterative steps, the transfer function is found to meet the specified magnitude response in the maximally flat sense over the passband and an almost equiripple response in the stopband. But this method aims at expanding the magnitude squared function in a Taylor series and requires that the first M terms, excluding the constant term, be zero. It is not clear how optimization technique is used to satisfy this requirement for achieving maximally flat magnitude response. To obtain equiripple response in the stopband, the authors propose a transformation of the zeros on the unit circle in the z-plane to another plane such that the stopband region in the z-plane is mapped to the complete boundary of the unit circle in the new plane. Yet the response in the stopband is equiripple but is not optimal in the Chebyshev sense; so trial and error is involved in their design procedure which adds up to computational load. The same design procedure has been explained and used for the design of digital filters which minimize the intersymbol interference in digital data transmission systems [10].

Another method [27], which does not use a mirror image polynomial, may yet be mentioned which expands the digital filter transfer function in a Laurent

2.4. SIMULTANEOUS APPROXIMATION

series and obtains a simultaneous approximation of the prescribed magnitude and phase response. The transfer function does not have any zeros on the unit circle in the z-plane. But this method requires that the log-magnitude and phase response be given in the form of a finite Laurent series expansion or the trigonometric form of its Fourier series i.e. in the form of (2.49) and (2.50).

$$\log |H(e^{j\omega})| = c_0 + \sum_{r=1}^{P} c_r \cos(r\omega) \qquad (2.49)$$

and

$$\theta(e^{j\omega}) = \sum_{r=1}^{Q} d_r \sin(r\omega) \qquad (2.50)$$

Hence if the magnitude and the phase response are specified in the conventional form as in other methods, their Fourier series coefficients have to be calculated first by other means, before this method can be utilized.

2.4.3 Method 2.3 : Maximally Flat Magnitude and Group Delay in the Passband

Recently a new method has been proposed by the authors of [21, 22] for the design of IIR filters approximating both the magnitude and delay characteristics in the passband, in the maximally flat sense. The method can also achieve an equiripple approximation to a constant delay in the passband and an equiripple magnitude response in the stopband. The method is more general as it can achieve a magnitude response in the passband and stopband with different degrees of flatness at $\omega = 0$ and $\omega = \pi$. In this design theory, we choose Thiran's all-pole transfer function that approximates a constant group delay in the maximally flat sense or equiripple sense. This all-pole filter is augmented by a mirror image polynomial for the numerator. It is shown that the coefficients of this numerator can be obtained by an *analytical method*, when a maximally flat magnitude approximation at $\omega = 0$ or a magnitude response with different degrees of flatness at $\omega = 0$ and $\omega = \pi$ is required. It is a much simpler method than those already mentioned above and provides an analytical method. In this context, it may be pointed out that defining a suitable scalar objective function that satisfies the conditions for *maximally flat magnitude* response is not desirable; even if an objective function is defined, minimization of the objective function by least p^{th} approximation or by min-max approximation methods defeats the purpose of the design objective. The conditions for the maximally flat response of a lowpass filter require that the first n derivatives of the magnitude squared function in the frequency variable ω^2, evaluated at $\omega = 0$ (or at any other frequency within the passband) to be all zero. The new method derives a set of linear equations that satisfy the conditions for the magnitude response to be flat with certain degrees of flatness at $\omega = 0$ and $\omega = \pi$. This is a more general method because the degrees of flatness at $\omega = 0$ and $\omega = \pi$ can be traded off arbitrarily, instead of making it maximally flat at $\omega = 0$. This method has been extended to achieve a passband magnitude response with some desired degree of flatness at $\omega = 0$ and an equiripple response in the stopband. For this purpose, we add additional zeros on the unit circle in the stopband and hence the order of the numerator polynomial is increased. The locations of the zeros are optimally chosen by means of Remez Exchange Algorithm - without affecting the maximally flat magnitude response or the delay response in the passband. Before we develop the design theory, we derive the following two Theorems.

Theorem 1 Let

$$H(z) = \frac{N(z)}{D(z)} \tag{2.51}$$

The first L derivatives of $H(z)$ at $z = z_p$ are zero, if $H(z)|_{z=z_p} = 1$ and the first L derivatives of $N(z)$ at $z = z_p$, are equal to the first L derivatives of $D(z)$ at $z = z_p$, i.e.

$$\left.\frac{d^k H(z)}{dz^k}\right|_{z=z_p} = 0 \qquad \text{for } k = 1, 2, \ldots, L \tag{2.52}$$

if

$$H(z)|_{z=z_p} = 1 \tag{2.53}$$

and

$$\left.\frac{d^k N(z)}{dz^k}\right|_{z=z_p} = \left.\frac{d^k D(z)}{dz^k}\right|_{z=z_p} \qquad \text{for } k = 1, 2, \ldots, L \tag{2.54}$$

Theorem 2 The first K derivatives of $H(z)$ at $z = z_q$ are zero, if $H(z)|_{z=z_q} = 0$ and the first K derivatives of $N(z)$ at $z = z_q$ are equal to zero i.e.

$$\left.\frac{d^k H(z)}{dz^k}\right|_{z=z_q} = 0 \qquad \text{for } k = 1, 2, \ldots, K \tag{2.55}$$

if

$$H(z)|_{z=z_q} = 0 \tag{2.56}$$

and

$$\left.\frac{d^k N(z)}{dz^k}\right|_{z=z_q} = 0 \qquad \text{for } k = 1, 2, \ldots, K \tag{2.57}$$

The values of L and K are called the degrees of flatness of the function $H(z)$ at $z = z_p$ and $z = z_q$ respectively.

Proof. From (2.51), we have $H(z)D(z) = N(z)$. Differentiating both sides, we get

$$\frac{d^k[H(z)D(z)]}{dz^k} = \frac{d^k N(z)}{dz^k} \tag{2.58}$$

2.4. SIMULTANEOUS APPROXIMATION

Using Leibnitz's rule for the higher derivatives of products [37], the left side of (2.58) can be expressed as (2.59).

$$\begin{aligned}\mathbf{D}^k(H_z D_z) &= H_z \mathbf{D}^k(D_z) + \binom{k}{1}\mathbf{D}^1(H_z)\mathbf{D}^{k-1}(D_z) + \cdots \\ &+ \binom{k}{k-1}\mathbf{D}^{k-1}(H_z)\mathbf{D}^1(D_z) + D_z \mathbf{D}^k(H_z) \\ &= \mathbf{D}^k(N_z)\end{aligned} \qquad (2.59)$$

where $\mathbf{D}^p(.) = \frac{d^p(.)}{dz^p}$, $H_z = H(z)$, $N_z = N(z)$ and $D_z = D(z)$. Rearranging the terms in (2.59), we get $\mathbf{D}^k(H_z)$ as

$$\frac{1}{D_z}\left[\mathbf{D}^k(N_z) - \left\{\begin{array}{c} H_z \mathbf{D}^k(D_z) + \binom{k}{1}\mathbf{D}^1(H_z)\mathbf{D}^{k-1}(D_z) \\ +\binom{k}{2}\mathbf{D}^2(H_z)\mathbf{D}^{k-2}(D_z)\cdots \\ +\binom{k}{k-1}\mathbf{D}^{k-1}(H_z)\mathbf{D}^1(D_z) \end{array}\right\}\right] \qquad (2.60)$$

Note that in (2.60) the kth derivative of $H(z)$ is expressed in terms of kth derivative of $N(z)$, the first k derivatives of $D(z)$ and the first $k-1$ derivatives of $H(z)$.

We use the method of induction to prove the theorems. First we show that the theorems are true for $k = 1$. Substituting $k = 1$ in (2.59), we get

$$\left.\frac{dH(z)}{dz}\right|_{z=z_p} = \frac{1}{D(z_p)}\left[\left.\frac{dN(z)}{dz}\right|_{z=z_p} - H(z_p)\left.\frac{dD(z)}{dz}\right|_{z=z_p}\right] \qquad (2.61)$$

If $H(z_p) = 1$, we have

$$\left.\frac{dH(z)}{dz}\right|_{z=z_p} = 0 \qquad (2.62)$$

when

$$\left.\frac{dN(z)}{dz}\right|_{z=z_p} = \left.\frac{dD(z)}{dz}\right|_{z=z_p} \qquad (2.63)$$

If $H(z_p) = 0$, in (2.61), we get

$$\left.\frac{dH(z)}{dz}\right|_{z=z_p} = 0 \qquad (2.64)$$

when

$$\left.\frac{dN(z)}{dz}\right|_{z=z_p} \qquad (2.65)$$

Next we assume the Theorem 1 is true for $k = m$, i.e.

$$\left.\frac{d^k H(z)}{dz^k}\right|_{z=z_p} = 0 \quad \text{for } k = 1, 2\ldots, m \qquad (2.66)$$

when $H(z_p) = 1$ and

$$\left.\frac{d^k N(z)}{dz^k}\right|_{z=z_p} = \left.\frac{d^k D(z)}{dz^k}\right|_{z=z_p} \quad ; \quad k = 1, 2, \ldots, m \tag{2.67}$$

Then substituting $k = m + 1$ and using the above results for $k = m$, we get

$$\left.\frac{d^{m+1} H(z)}{dz^{m+1}}\right|_{z=z_p} = \frac{1}{D(z_p)} \left[\left.\frac{d^{m+1} N(z)}{dz^{m+1}}\right|_{z=z_p} - H(z_p) \left.\frac{d^{m+1} D(z)}{dz^{m+1}}\right|_{z=z_p} \right] \tag{2.68}$$

Note that again when $H(z_p) = 1$, and

$$\left.\frac{d^{m+1} N(z)}{dz^{m+1}}\right|_{z=z_p} = \left.\frac{d^{m+1} D(z)}{dz^{m+1}}\right|_{z=z_p} \tag{2.69}$$

we get the result

$$\left.\frac{d^{m+1} H(z)}{dz^{m+1}}\right|_{z=z_p} = 0 \tag{2.70}$$

Therefore the Theorem 1 is true for $k = 1, 2, \ldots, L$

Now we assume that Theorem 2 is true for $k = 1$, i.e., in (2.61), when

$$H(z_q) = 0 \quad \text{and} \quad \left.\frac{dN(z)}{dz}\right|_{z=z_q} = 0$$

we have

$$\left.\frac{dH(z)}{dz}\right|_{z=z_q} = 0$$

Next we assume that the Theorem 2 is true for $k = m$, i.e.

$$\left.\frac{d^m H(z)}{dz^m}\right|_{z=z_q} = 0 \quad \text{for } k = 1, 2 \ldots, m \tag{2.71}$$

when

$$H(z_q) = 0 \quad \text{and} \quad \left.\frac{d^m N(z)}{dz^m}\right|_{z=z_q} = 0 \tag{2.72}$$

When these conditions are satisfied, we get from (2.68) the result

$$\left.\frac{d^{m+1} H(z)}{dz^{m+1}}\right|_{z=z_q} = 0 \tag{2.73}$$

when $H(z_q) = 0$ and

$$\left.\frac{d^{m+1} N(z)}{dz^{m+1}}\right|_{z=z_q} = 0 \tag{2.74}$$

Thus Theorem 2 is true for $k = 1, 2, \ldots, K$.

2.4.4 Design Theory

In (2.51), we choose $D(z)$ as the Thiran's polynomial to approximate a constant group delay in the maximally flat sense (or equiripple sense). This polynomial derived in section 2.3 is given by (2.35). Then we choose $N(z) = z^{-p}N_a(z)$ where $N_a(z)$ is a mirror image polynomial given in the form

$$N_a(z) = b_0 + b_1\left(\frac{z+z^{-1}}{2}\right) + b_2\left(\frac{z^2+z^{-2}}{2}\right) + \cdots + b_p\left(\frac{z^p+z^{-p}}{2}\right) \quad (2.75)$$

Substituting $z = e^{j\omega}$ in (2.75) we get $N(e^{j\omega}) = e^{-j\omega p}N_a(e^{j\omega})$ where

$$N_a(e^{j\omega}) = b_0 + b_1\cos\omega + b_2\cos(2\omega) + \cdots + b_p\cos(p\omega) \quad (2.76)$$

The numerator $N(e^{j\omega})$ adds a pure delay of p samples to that of the Thiran's all-pole filter $\frac{1}{D(z)}$ because $N_a(e^{j\omega})$ is a real valued function. The coefficients $b_0, b_1, b_2 \cdots, b_p$ have to be found such that $H(e^{j\omega})$ has a magnitude response with the desired degrees of flatness at $\omega = 0$ and $\omega = \pi$, besides having a maximally flat (or equiripple) group delay characteristics. Differentiating (2.76) k times, w.r.t. ω, we get

$$\left.\frac{d^k N_a(e^{j\omega})}{d\omega^k}\right|_{\omega=0} = \begin{cases} (-1)^{\frac{k}{2}}[b_1 + b_2(2)^k + \cdots + b_p(p)^k] \;;\; k \text{ even} \\ 0 \quad\quad\quad\quad\quad\quad\quad\quad\quad\quad\quad\quad\quad\quad ;\; k \text{ odd} \end{cases} \quad (2.77)$$

and $\left.\dfrac{d^k N_a(e^{j\omega})}{d\omega^k}\right|_{\omega=\pi} =$

$$\begin{cases} (-1)^{\frac{k}{2}}[(-1)^1 b_1 + (-1)^2 b_2(2)^k + \cdots + (-1)^p b_p(p)^k]; \; k \text{ even} \\ 0 \quad\quad\quad\quad\quad\quad\quad\quad\quad\quad\quad\quad\quad\quad\quad\quad\quad\quad ;\; k \text{ odd} \end{cases} \quad (2.78)$$

It is also noted that since $|D(e^{j\omega})|$ is an even function of ω, its Fourier Series expansion consists of cosine terms only. Therefore all odd order derivatives contain sine terms only and hence are zero at both $\omega = 0$ and $\omega = \pi$, i.e.

$$\left.\frac{d^k |D(e^{j\omega})|}{d\omega^k}\right|_{\omega=0,\pi} = 0; \qquad k \text{ odd} \quad (2.79)$$

We note that if the derivatives of $N_a(e^{j\omega})$ at $\omega = 0$ and/or $\omega = \pi$ are zero, the derivatives of $|N_a(e^{j\omega})|$ are also zero at these frequencies. We have the following results from Theorems 1 and 2.

(Theorem 1): If

$$\left.|H(e^{j\omega})|\right|_{\omega=0} = \left.\left|\frac{N_a(e^{j\omega})}{D(e^{j\omega})}\right|\right|_{\omega=0} = 1 \quad (2.80)$$

and

$$\left.\frac{d^k |D(e^{j\omega})|}{d\omega^k}\right|_{\omega=0} = \left.\frac{d^k N_a(e^{j\omega})}{d\omega^k}\right|_{\omega=0} \quad (2.81)$$

then
$$\left.\frac{d^k|H(e^{j\omega})|}{d\omega^k}\right|_{\omega=0} = 0 \tag{2.82}$$

(Theorem 2): If
$$|H(e^{j\omega})|\big|_{\omega=\pi} = \left|\frac{N_a(e^{j\omega})}{D(e^{j\omega})}\right|_{\omega=\pi} = 0 \tag{2.83}$$

and
$$\left.\frac{d^k N_a(e^{j\omega})}{d\omega^k}\right|_{\omega=\pi} = 0 \tag{2.84}$$

then
$$\left.\frac{d^k|H(e^{j\omega})|}{d\omega^k}\right|_{\omega=\pi} = 0 \tag{2.85}$$

Now since $H(e^{j\omega}) = \frac{e^{-j\omega p}N_a(e^{j\omega})}{D(e^{j\omega})}$ the equations (2.80) and (2.81) become (2.86) and (2.87) respectively.

$$\begin{aligned}|D(e^{j\omega})|\big|_{\omega=0} &= N_a(e^{j\omega})\big|_{\omega=0} \\ &= b_0 + b_1 + b_2 + \cdots + b_p\end{aligned} \tag{2.86}$$

$$\begin{aligned}\left.\frac{d^k|D(e^{j\omega})|}{d\omega^k}\right|_{\omega=0} &= \left.\frac{d^k N_a(e^{j\omega})}{d\omega^k}\right|_{\omega=0} \\ &= \begin{cases} (-1)^{\frac{k}{2}}[b_1 + b_2(2)^k + \cdots + b_p(p)^k]; & k \text{ even} \\ 0 & ; k \text{ odd} \end{cases}\end{aligned} \tag{2.87}$$

Similarly equations (2.83) and (2.84) now become (2.88) and (2.89) respectively.

$$N(e^{j\omega})\big|_{\omega=\pi} = b_0 - b_1 + b_2 + \cdots + (-1)^p b_p = 0 \tag{2.88}$$

$$\left.\frac{d^k N_a(e^{j\omega})}{d\omega^k}\right|_{\omega=\pi} =$$

$$\begin{cases} (-1)^{\frac{k}{2}}[(-1)^1 b_1 + (-1)^2 b_2(2)^k + \cdots + (-1)^p b_p(p)^k]; & k \text{ even} \\ 0 & ; k \text{ odd} \end{cases} \tag{2.89}$$

It is observed that (2.87) and (2.89) are zero for all odd values of k; the relationships for even values of k constitute the remaining conditions for ensuring that the derivatives of $H(e^{j\omega})$ at $\omega = 0$ and $\omega = \pi$ are zero. Let $k = 1, 2, \ldots, L$ in (2.81) and (2.82) where L can be even or odd. If L is even, we see from (2.87) that

$$(-1)^{\frac{k}{2}}[b_1 + b_2(2)^k + \cdots + b_p(p)^k] = \left.\frac{d^k|D(e^{j\omega})|}{d\omega^k}\right|_{\omega=0} ;$$
$$k = 2, 4, 6, \ldots, L$$

2.4. SIMULTANEOUS APPROXIMATION

whereas when L is odd, we have the condition

$$(-1)^{\frac{k}{2}}[b_1 + b_2(2)^k + \cdots + b_p(p)^k] = \left.\frac{d^k |D(e^{j\omega})|}{d\omega^k}\right|_{\omega=0} ;$$
$$k = 2, 4, 6, \ldots, L-1 \quad (2.90)$$

Similar conclusions can be reached about K with respect to (2.89). Therefore, without loss of generality, let us choose L and K to be odd integers. So we have the above condition (2.90) and also (2.91) to be satisfied by the coefficients of the mirror image polynomial.

$$(-1)^{\frac{k}{2}}[(-1)^1 b_1 + (-1)^2 b_2(2)^k + \cdots + (-1)^p b_p(p)^k] = 0 ;$$
$$k = 2, 4, \ldots, K-1 \quad (2.91)$$

Thus we have to satisfy (2.86) and the $\left(\frac{L-1}{2}\right)$ equations given by (2.90) in order that the magnitude response of $H(e^{j\omega})$ has L degrees of flatness at $\omega = 0$ and we have to satisfy (2.88) and the $\left(\frac{K-1}{2}\right)$ equations given by (2.91) in order that the magnitude response has K degrees of flatness at $\omega = \pi$. Note that all of these equations are linear in terms of the coefficients $b_0, b_1, \ldots, b_p$. We will require that these coefficients satisfy another specification i.e. the magnitude of $H(e^{j\omega})$ at a specified bandwidth ω_b is 3 dB below the 0 dB magnitude at $\omega = 0$. This leads to the condition (2.92) which also is a linear constraint in the coefficients.

$$b_0 + b_1 \cos \omega_b + b_2 \cos 2\omega_b + \cdots + b_p \cos p\omega_b = 0.7071 |D(e^{j\omega_b})| \quad (2.92)$$

Hence the equations (2.86), (2.88), (2.92), (2.90) and (2.91) can be expressed in a matrix form (2.93).

$$\mathbf{A}_1 \mathbf{b}_1 = \mathbf{d}_1 \quad (2.93)$$

The matrix $\mathbf{A}_1$ and the vectors $\mathbf{b}_1$ and $\mathbf{d}_1$ are as shown below:

$$\mathbf{A}_1 = \begin{bmatrix} 1 & 1 & 1 & \cdots & 1 \\ 1 & (-1)^1 & (-1)^2 & \cdots & (-1)^p \\ 1 & \cos \omega_b & \cos 2\omega_b & \cdots & \cos p\omega_b \\ 0 & -(1^2) & -(2^2) & \cdots & -(p^2) \\ 0 & (1^4) & (2^4) & \cdots & (p^4) \\ \vdots & \vdots & \vdots & \ddots & \vdots \\ 0 & (-1)^{\frac{M}{2}} & (-1)^{\frac{M}{2}}(2^M) & \cdots & (-1)^{\frac{M}{2}} p^M \\ 0 & 1 & -(2^2) & \cdots & -(-1)^p p^2 \\ 0 & -1 & (2^4) & \cdots & (-1)^p p^4 \\ \vdots & \vdots & \vdots & \ddots & \vdots \\ 0 & -(-1)^{\frac{n}{2}} & -(-1)^{\frac{n}{2}}(2^N) & \cdots & (-1)^p(-1)^{\frac{N}{2}} p^N \end{bmatrix} \quad (2.94)$$

$$\mathbf{b}_1 = \begin{bmatrix} b_0 & b_1 & b_2 & \cdots & \cdots & \ldots b_p \end{bmatrix}^T \quad (2.95)$$

$$\mathbf{d}_1 = \begin{bmatrix} D^{(0)} & 0 & D_{\omega_b} & D^{(2)} & D^{(4)} & \cdots & D^{(M)} & 0 & \cdots & \cdots & 0 \end{bmatrix}^T \quad (2.96)$$

in which
$$D^{(0)} = D(e^{j0}), \qquad D_{\omega_b} = 0.7071 \left|D(e^{j\omega_b})\right|$$

$$D^{(k)} = \left.\frac{d^k \left|D(e^{j\omega})\right|}{d\omega^k}\right|_{\omega=0}$$

and $M = L - 1$ and $N = K - 1$. Note that there are $\frac{L+K+4}{2}$ equations in (2.93). If we choose $p = \frac{L+K+2}{2}$, we can obtain a unique solution for the $(p+1)$ coefficients $b_0, b_1, \ldots, b_p$ from the system of linear equations (2.93). Therefore we have an *analytical solution* for designing the IIR filters which have (1) a maximally flat (or equiripple) group delay characteristic (2) a magnitude response which has the desired degrees of flatness at $\omega = 0$ and $\omega = \pi$ and (3) any arbitrary attenuation at an arbitrary frequency ω_b. It is obvious that we could choose multiple frequencies in the transition band at which the magnitude is equal to the prespecified values. Of course if these values do not lie on a monotonically decreasing, smooth response, the filter response may exhibit large ripples between these frequencies, even though it will pass through these points. If the more common maximally flat lowpass magnitude response is desired, we choose $K = 0$ in the above design procedure - as illustrated by the examples given below. It is found in some cases, that when $K = 0$, for a fixed value of p, as the 3 dB bandwidth is reduced, some of the zeros of $N_a(z)$ move from locations inside the unit circle $|z| = 1$ to its boundary in order to meet the required steep rolloff in the transition band, thereby giving rise to ripples in the stopband. Hence it is necessary in these cases to choose a non-zero value for K.

2.4.5 Example 2.1

We show some examples to illustrate the design theory and its flexibility in meeting a wide range of specifications [20]. In the first example, the 3 dB bandwidth of the IIR filter is fixed at $\omega_b = 0.35\pi$, the order of its denominator (Thiran's polynomial) is n = 6, the maximally flat constant delay to be realized by the filter is $\tau = 0.7$. The values for the degrees of flatness at $\omega = 0$ and $\omega = \pi$ are varied as (L=1, K=1), (L=3, K=3), (L=5, K=7), (L=7, K=9) and (L=9, K=13). The magnitude response of these filters obtained from the design algorithm are plotted in Fig. 2.5, along with the nominal delay of 0.7. In the second example, we choose n = 6, and $\tau = 0.7$, as before. The 3 dB bandwidth ω_b is varied as $0.25\pi, 0.3\pi, 0.35\pi, 0.4\pi$ and 0.45π. The values for the parameters L and K are set to (L=5, K=17), (L=5, K=11), (L=5, K=11), (L=7, K=11) and (L=7, K=9) respectively. In Fig. 2.6, the magnitude of the filters are shown along with the nominal value $\tau = 0.7$ for the delay. It must be pointed out that the design theory does not give any guidelines for choosing the values for the pair of L and K; we simply try their values arbitrarily and change them if necessary.

2.4. SIMULTANEOUS APPROXIMATION

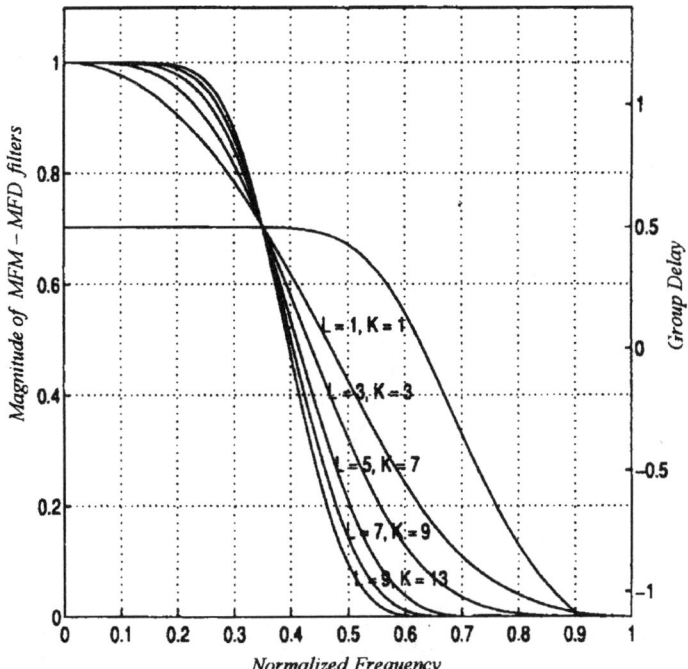

Fig. 2.5. Magnitude and Group Delay of Maximally Flat Magnitude (MFM)-Maximally Flat Delay (MFD) filters

2.4.6 Extension for an Equiripple Stopband Response

For a given value of n and ω_b, the above method can be extended to decrease the transition band and obtain an equiripple magnitude response in the stopband of the filter, while the flat magnitude and group delay response in the passband are maintained. This is done by increasing the order of the mirror image polynomial by purposely adding some zeros in the stopband region on the unit circle.

Let the stopband region of the modified filter be denoted by $\omega_s < \omega < \pi$, where $\omega_s > \omega_b$. The desired magnitude response to be approximated by the modified filter, over this stopband region, is given by

$$\left|H_d(e^{j\omega})\right| = 0 \qquad \omega_s < \omega < \pi \qquad (2.97)$$

In order to obtain an equiripple magnitude response over this stopband region, we require that the magnitude response satisfy the following set of equations:

$$\left[\left|H_d(e^{j\omega_i})\right| - \left|H(e^{j\omega_i})\right|\right] = (-1)^{i+1}\delta \qquad i = 1, 2, \ldots, m \qquad (2.98)$$

where

$$\omega_s \leq \omega_1 < \omega_2 < \cdots < \omega_m \leq \pi \qquad (2.99)$$

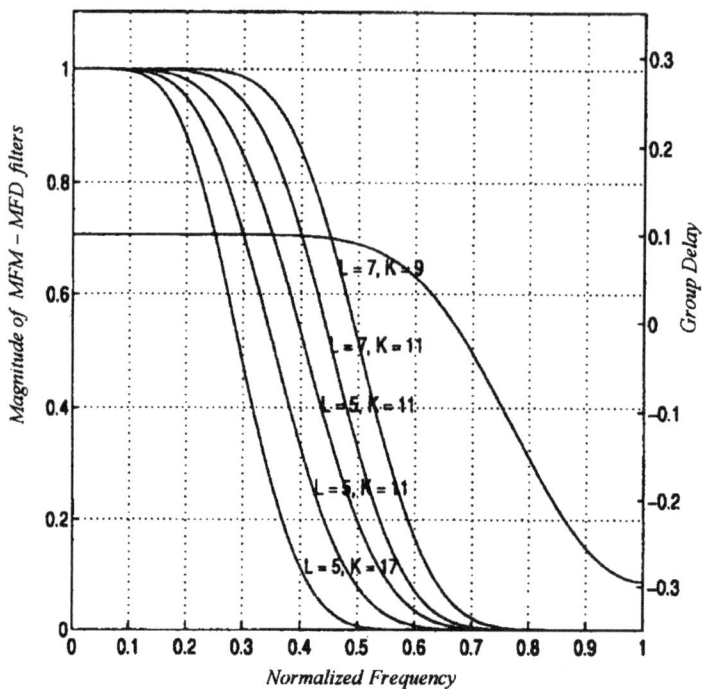

Fig. 2.6. Magnitude and Group Delay of MFM-MFD filters as ω_b is varied

and m is the number of desired extrema in the stopband region. From $H(z) = \frac{z^{-p} N_a(z)}{D(z)}$, and the equations (2.97) and (2.98), we get

$$[|N_a(e^{j\omega_i})| - (-1)^i \delta |D(e^{j\omega_i})|] = 0 \qquad i = 1, 2, \ldots, m \qquad (2.100)$$

Substituting $z = e^{j\omega}$ in (2.100) and in $N_a(z)$, we get

$$b_0 + b_1 \cos \omega_i + b_2 \cos 2\omega_i + \cdots + b_p \cos p\omega_i - (-1)^i \delta |D(e^{j\omega_i})| = 0$$

$$\text{for } i = 1, 2, \ldots, m \qquad (2.101)$$

These equations along with the previously derived equations to obtain a magnitude response with L degrees of flatness at $\omega = 0$ and an attenuation of 3 dB at ω_b can be expressed in a matrix form as shown below:

$$\mathbf{A}_2 \mathbf{b}_2 = \mathbf{d}_2 \qquad (2.102)$$

2.4. SIMULTANEOUS APPROXIMATION

$$\mathbf{A}_2 = \begin{bmatrix} 1 & \cos\omega_b & \cos\omega_b & \cdots & \cos p\omega_b & 0 \\ 1 & 1 & 1 & \cdots & 1 & 0 \\ 0 & -(1^2) & -(2^2) & \cdots & -(p^2) & 0 \\ 0 & (1^4) & (2^4) & \cdots & \vdots & \vdots \\ \vdots & \vdots & \vdots & \ddots & \vdots & \vdots \\ 0 & (-1)^{\frac{M}{2}} & (-1)^{\frac{M}{2}}(2^M) & \cdots & (-1)^{\frac{M}{2}}p^M & 0 \\ 1 & x_{1,1} & x_{1,2} & \cdots & x_{1,p} & D_1 \\ 1 & x_{2,1} & x_{2,2} & \cdots & x_{2,p} & D_2 \\ \vdots & \vdots & \vdots & \ddots & \vdots & \vdots \\ 1 & x_{m,1} & x_{m,2} & \cdots & x_{m,p} & D_m \end{bmatrix} \quad (2.103)$$

$$\mathbf{b}_2 = [b_0 \ b_1 \ b_2 \ \cdots \ b_p \ \delta]^T \quad (2.104)$$

$$\mathbf{d}_2 = [D_{\omega_b} \ D^{(0)} \ D^{(2)} \ D^{(4)} \ \cdots \ D^{(M)} \ 0 \ 0 \ \cdots \ 0]^T \quad (2.105)$$

where $x_{i,n} = \cos n\omega_i$, and $D_i = (-1)^i |D(e^{j\omega_i})|$; $i = 1, 2, \ldots, m$, the other entities being the same as defined in (2.93). There are $\left(\frac{L+2m+3}{2}\right)$ equations in (2.102) which are linear equations in the $(p+2)$ variables $b_0, b_1, \ldots, b_p, \delta$. By choosing $p = \left(\frac{L+2m-1}{2}\right)$ we can obtain a unique solution to (2.102), i.e. the values of the unknown variables $b_0, b_1, \ldots b_p$. But the values of ω at which the maximum deviations or extrema occur are not known á priori, although the interval over which they occur is known as $\omega_s \leq \omega \leq \pi$. Hence the set of equations (2.102) have to be solved recursively, using the Remez Exchange Algorithm [5] by starting with an initial guess for the extremal points: $\omega_1, \omega_2, \ldots, \omega_m$.

2.4.7 Remez Algorithm

The purpose of using the Remez Algorithm is to find the best locations for the zeros on the unit circle $|z| = 1$. This is done by finding the values of ω_i in the stopband at which the equation (2.101) is satisfied. When they are obtained iteratively by Remez Exchange Algorithm, as proposed in [5, 35], we get an equiripple response in the stopband and also the coefficients of the numerator polynomial. The algorithm is briefly outlined below. Given ω_s, we choose initial values for n, p and m. Then,

1. Choose a set of m points, ω_i, $i = 1, 2, \ldots, m$ over the interval $\omega_s \leq \omega \leq \pi$ such that $\omega_s \leq \omega_1 < \omega_2 < \cdots < \omega_m \leq \pi$

2. Solve the set of equations given by (2.102) to find the numerator coefficients $b_0, b_1, b_2, \ldots, b_p$ and δ

3. Compute the magnitude response of the function $H(e^{j\omega})$ at a large number of closely located points over the entire stopband and find the points at which the magnitude attains local maxima

4. If the number of points at which the magnitude of the function exceeds δ is larger than m, then retain m points at which the extrema are the

largest. Repeat step 2 with the choice of these values for ω_i. Check if the number of extrema have changed. If they have not changed, stop the iterative procedure, otherwise, retain the m largest extrema and go back to step 2

2.4.8 Example 2.2

We choose the same specifications as given in [34] i.e. a lowpass filter with flat passband and a bandwidth $\omega_b = 0.2\pi$ at which the maximum attenuation is required to be 3 dB. The stopband cutoff frequency $\omega_s = 0.32\pi$ and an equiripple response with a minimum attenuation of 30 dB in the stopband ($\omega \geq \omega_s$) is also specified. The denominator is chosen to be a 7th order polynomial, whereas the example in [34] has a 12th order polynomial. The all-pole (MFD) transfer function providing a group delay $\tau = 5$ was augmented by using a mirror image polynomial having 10 zeros in the stopband. The magnitude has ten degrees of flatness at $\omega = 0$. The resulting magnitude response and group delay are shown in Fig. 2.7 and Fig. 2.8, respectively. The attenuation in the stopband is found to exceed 30 dB. We have also designed a lowpass filter with a maximally flat magnitude (MFM) in the passband and in the stopband and a maximally flat delay (MFD) in the passband. Its magnitude and delay responses are also shown in these figures, for comparison. In Fig. 2.10 is shown the magnitude and group delay of the filter specified in this example, when the degree of flatness at $\omega = 0$ is changed from ten to nine and the number of zeros in the stopband is changed from ten to eleven-all the other parameters remaining the same. Note that the minimum attenuation in the stopband has increased to about 44 dB. The coefficients of the numerator and denominator polynomial of the MFM-MFD filter and those of the MFD filter with an equiripple stopband are given below.

Numerator Coefficients of the MFM-MFD filter:
-0.0006   0.0026   -0.0021   -0.0042   0.0039   0.0057   0.0039
-0.0042   -0.0021   -0.0026   -0.0006

Denominator Coefficients of the MFM-MFD filter:
0.1250   -0.4000   0.6125   -0.5765   0.3603   -0.1517   0.0417
-0.0068   0.0005

Numerator Coefficients of the MFM-MFD Filter with equiripple stopband:
0.0491   -0.3384   0.9544   -1.4160   1.2095   -0.6473   0.2145   -0.0006
0.0141   -0.0126   -0.0384   -0.0435   -0.0186   0.0278   0.0722   0.0905
0.0722   0.0278   -0.0186   -0.0435   -0.0384   -0.0126   0.0141   -0.0006
0.2145   -0.6473   1.2095   -1.4160   0.9544   -0.3384   0.0491

Denominator Coefficients of the MFM-MFD filter with equiripple stopband:
14.4066   -56.0256   97.3077   -97.3077   60.2381   -23.0000   5.0000
-0.4762

When the variable z^{-1} is replaced by $-z^{-1}$, the magnitude of the lowpass filter with a passband cutoff frequency ω_b changes to that of a highpass filter with a cut off frequency $\omega_B = \pi - \omega_b$. To show that this transformation also preserves

2.5. DESIGN OF PULSE SHAPING FILTERS

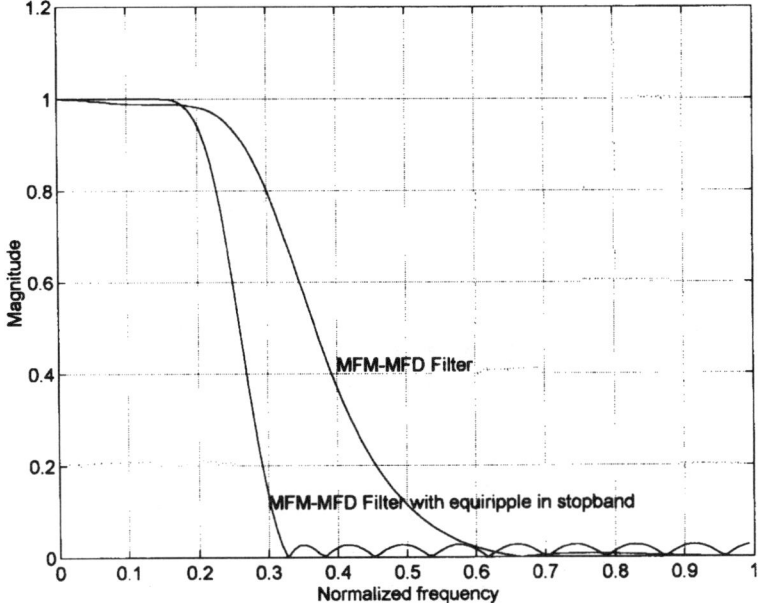

Fig. 2.7. Magnitude response of filters compared

the group delay response characteristics in the passband of the highpass filter, the magnitude and group delay response obtained by use of the above frequency transformation on the MFM-MFD lowpass filter in Example 2.2, are plotted in Fig. 2.9. The transfer function of this 1-D IIR highpass filter, with a maximally flat magnitude and a maximally flat group delay response will also be chosen in Chap. 4, to design a 2-D IIR highpass filter which has both a maximally flat magnitude and maximally flat delay response in a circularly symmetric passband region.

Extensive simulation of the design theory proposed by [22] has shown that the design procedure is mostly analytical. Its implementation is very simple and gives very good results, but we have to try different pair of values for L and K in some cases. The relationship between L, K and the bandwidth ω_b that assures a monotonically decreasing magnitude response has not been obtained and it needs further investigation.

2.5 Design of Pulse Shaping Filters

In the above design theory we chose only one frequency ω_b at which the filter was required to have a specified magnitude (3 dB) and this gave rise to one equation (2.92). But actually we can choose multiple frequencies at which the magnitude is specified and design a filter which will interpolate through all these

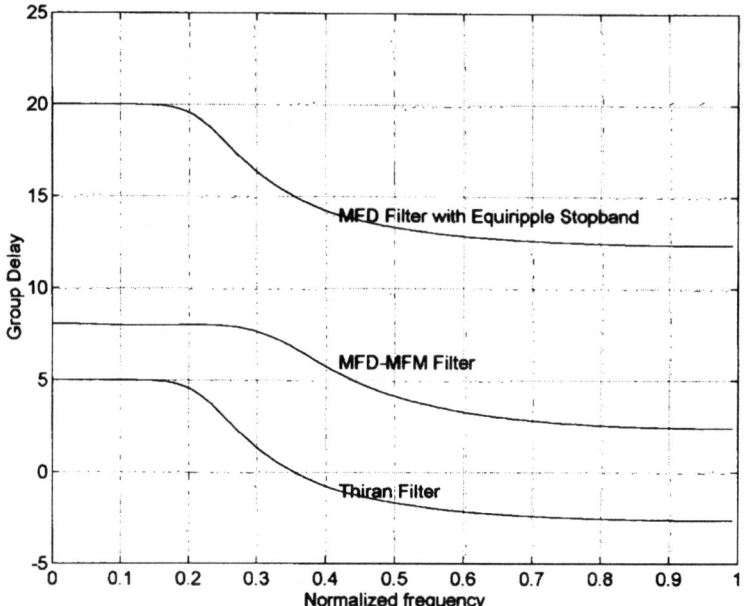

Fig. 2.8. Group Delay of filters compared

points. This indicates that we have more freedom in the choice of the frequency response that can be obtained. In order to illustrate this advantage of the above design theory, we choose the design of pulse shaping filters that are used to minimize intersymbol interference (ISI) in the baseband transmission of binary pulses [23]. Ideally the transfer function of the communication channel cascaded by the transmitting and receiving filter should have a frequency response that is a rectangular function with an ideal cutoff frequency equal to the Nyquist frequency. A more practical alternative is the frequency response which has a raised cosine form as given below:

$$P(f) = \begin{cases} \frac{1}{2f_b} & 0 \leq |f| < f_1 \\ \frac{1}{4f_b}\left\{1 - \sin\left[\frac{\pi(|f|-f_b)}{2f_b - 2f_1}\right]\right\} & f_1 \leq |f| < 2f_b - f_1 \\ 0 & |f| \geq 2f_b - f_1 \end{cases} \quad (2.106)$$

The frequency f_1 and the bandwidth f_b are related by a parameter α called the rolloff factor, according to

$$\alpha = 1 - \frac{f_1}{f_b} \quad (2.107)$$

and the 'transmission bandwidth B_T' becomes $f_b(1 + \alpha)$ i.e. $P(f) = 0$ for $f > B_T$. It is obvious that the magnitude of $P(f)$ is equal to $\frac{1}{2f_b}$ at $f = 0$ and it is equal to half this value at $f = f_b$. Note that the above specification is not an approximation to the ideal lowpass filter in the classical sense of the

2.5. DESIGN OF PULSE SHAPING FILTERS

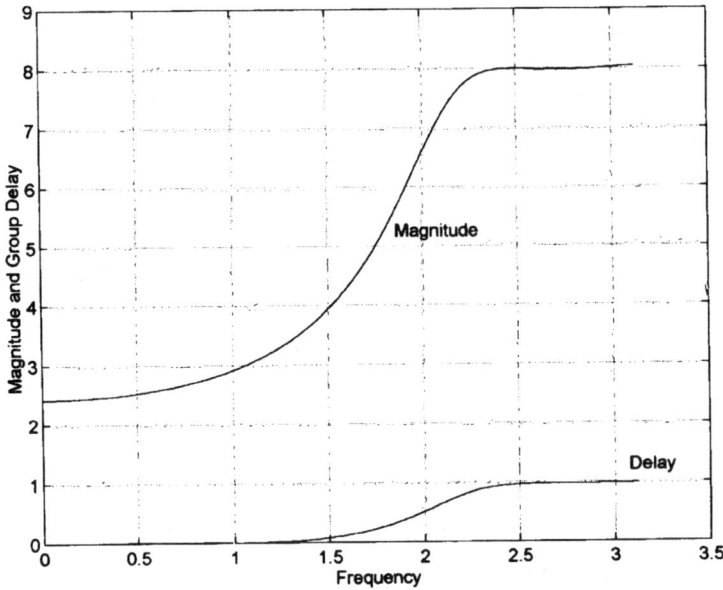

Fig. 2.9. Magnitude and Group Delay of MFM-MFD Highpass filter

maximally flat, equiripple or least squares sense approximation and hence can not be designed by any of the methods that are described in this or the previous chapter.

2.5.1 Example 2.3

We select $\alpha = 0.6$ and $f_b = 0.125$ (i.e. 0.25π radians) and select 20 discrete frequencies $f_k = 0.025k$, $k = 0, 1, 2, \ldots, 19$ (i.e. 0, 0.05π, 0.1π, $0.15\pi, \ldots, 0.95\pi$ radians) at which the magnitude of the desired filter function is required to match the value $P(f_k)$. We generate 20 linear equations to satisfy these constraints and also the additional equations to satisfy the requirement that the magnitude have $L = 7$ degrees of flatness at $f = 0$, and $K = 7$ degrees of flatness at $\omega = \pi$. The order of Thiran's polynomial is chosen as 9 and the group delay τ is chosen as one. The resulting filter has a mirror image polynomial of order 16 that adds 8 sample periods of pure delay, to give a causal filter. The magnitude response matches the value prescribed by (2.106) extremely well. The magnitude and group delay of this filter are plotted in Figs. 2.11 and 2.12 respectively. In these figures, we have also plotted the magnitude and delay responses of the pulse shaping filter designed by the authors of [8]. They have chosen the same value for f_b, α and the order of the filter as in our example. But their filter which will be called the Baher filter has a 10th order numerator polynomial. The overall delay of our filter is 9 sample periods. The authors [8] achieve an approximation to a constant delay of 3 sample periods by an all-pole

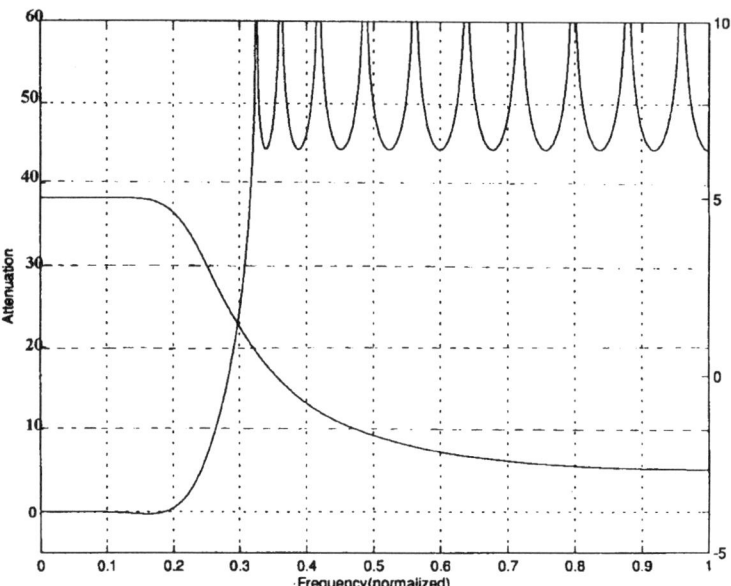

Fig. 2.10. Attenuation and Group Delay of MFM- MFD filter with equiripple stopband

filter and the 10th order mirror image polynomial in the numerator enhances it by 5 sample periods, so that the overall delay of the filter is 8 sample periods. The new type of approximation known as the equidistant approximation used by the authors [8] will be discussed later in this chapter. The delay response of the Baher filter exhibits a small amount of ripple over its passband and a peak at the edge of the passband as shown in Fig. 2.12. The delay of our example filter is maximally flat. Its magnitude shown in Fig. 2.11 matches the magnitude of our example over the passband and shows a larger attenuation over the stopband in comparison. The unit impulse response of our filter is plotted in Fig. 2.13.[3] The response of this filter in the time domain is the same as that given by [8]. But the design procedure we have used is considered to be simpler than the method given by [8].

The coefficients of the filter designed by our method are given below:

The Coefficients of the Numerator are
0.0030 -0.0068   -0.0248 -0.0348   -0.0131   0.0529   0.1489   0.2357
0.2707 0.2357   0.1489   0.0529   -0.0131   -0.0348 -0.0248 -0.0068
0.0030

The Coefficients of the Denominator are

---

[3] In MATLAB, the index n for the plot runs as $1, 2, \ldots$ while the unit impulse response is defined for $n = 0, 1, 2, \ldots$ Hence the plot is centered at n =19 and therefore the unit impulse response has the maximum value at $n = 18$.

2.5. DESIGN OF PULSE SHAPING FILTERS

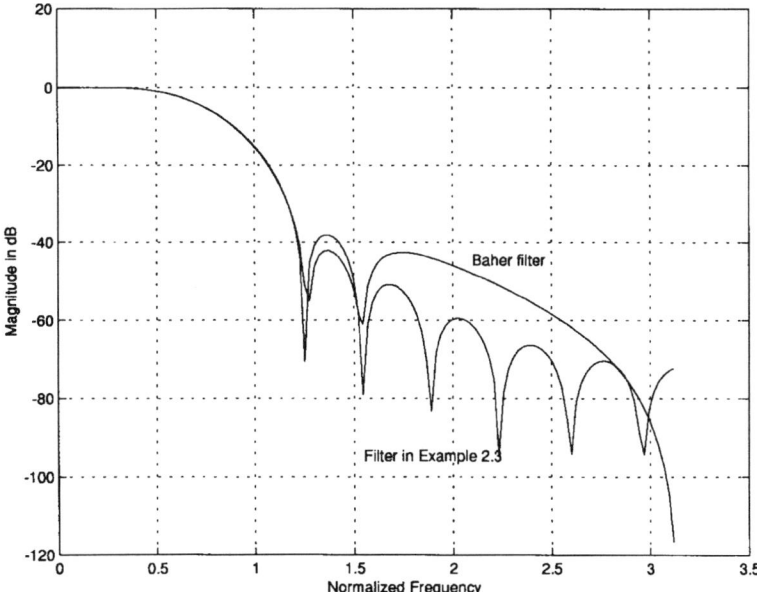

Fig. 2.11. Magnitude response of pulse shaping filters for zero ISI

3.4545   -5.1818   4.7832   -3.1888   1.5944   -0.5979   0.1641
-0.0313   0.0037   -0.0002

The coefficients of the filter in [8] are given below.

Numerator Coefficients
4.6196   281.0891   296.0831   113.5679   -44.7027   847.0423
-44.7027   113.5679   296.0831   -281.0891   4.6196

Denominator Coefficients
38663.07   -162687.2   357297.9   -530424.3   582466.0
-490121.3   318069.5   -156511.2   55727.74   -12945.68
1489.422

Another reason for choosing this filter specification is to point out the importance of designing filters simultaneously approximating both the prescribed magnitude and group delay. The filter designed to minimize intersymbol interference has a frequency response given by (2.106) which has a zero phase response. If it has a constant group delay over the entire frequency range, it will still have the same time domain behavior as shown in Fig. 2.13 but delayed by the same amount of delay. But if the filter has a phase distortion in its passband, the unit impulse response will not be able to minimize the intersymbol interference. This points out the need for developing efficient methods to design filters that approximate not only the prescribed magnitude response but also the prescribed group delay, when the filters are required to meet time domain

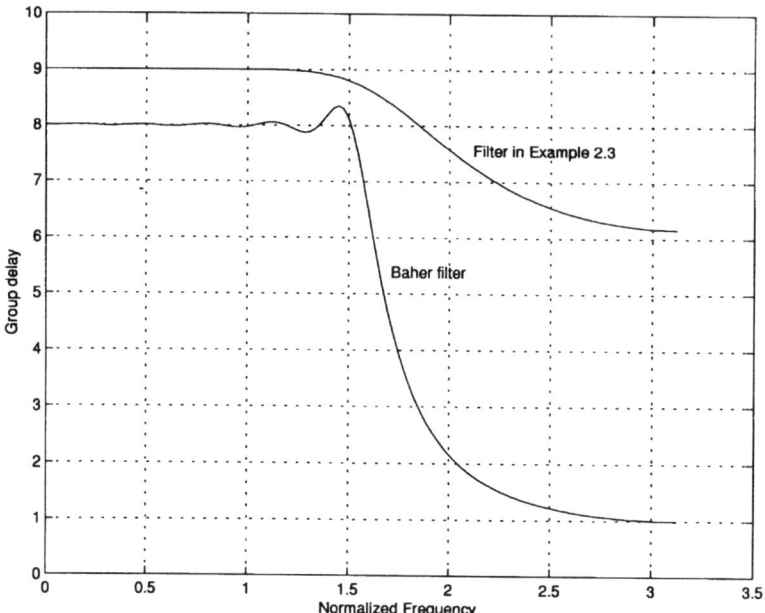

Fig. 2.12. Group Delay of pulse shaping filters for zero ISI

specifications as in high speed digital data transmission.

Several new methods for designing IIR filters that approximate both the prescribed magnitude and the group delay have been reported recently in professional journals and they will be described in the following pages. They are

1. Method 2.4, based on the use of two allpass sections in parallel

2. Method 2.5, based on singular value decomposition

3. Method 2.6, based on linear programming

4. Method 2.7, based on commensurate, distributed networks

5. Method 2.8, based on the theory of eigen filters

2.5.2 Method 2.4: Use of Allpass Sections in Parallel

Let us consider a digital filter which consists of two allpass filters in parallel [53]. Specifically Fig. 2.14 shows the structure which has two outputs $y_1(n)$ and $y_2(n)$ and one input $x(n)$. The transfer functions $A_1(z)$ and $A_2(z)$ are allpass functions and from them, two transfer functions for the structure can be obtained as

$$G(z) = \frac{Y_1(z)}{X(z)} = \frac{1}{2}[A_1(z) + A_2(z)]$$

2.5. DESIGN OF PULSE SHAPING FILTERS

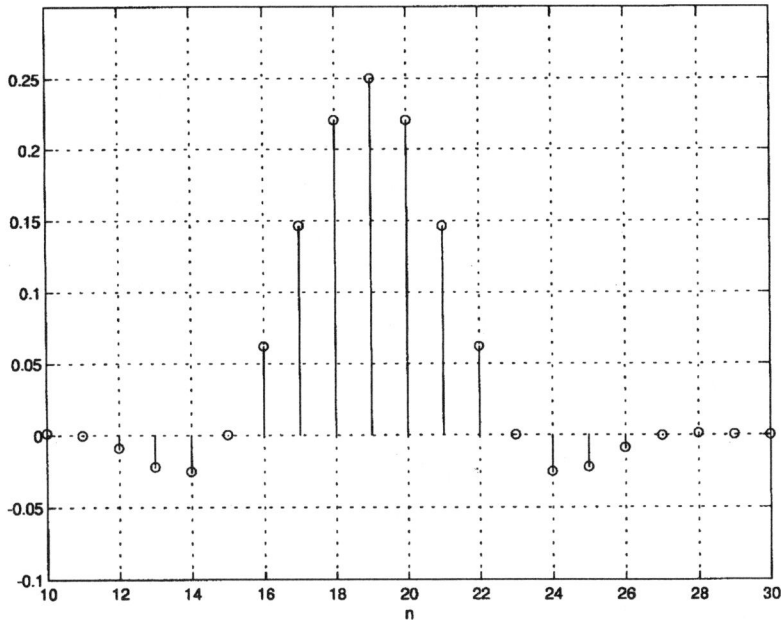

Fig. 2.13. Impulse response of pulse shaping filter in Example 2.3

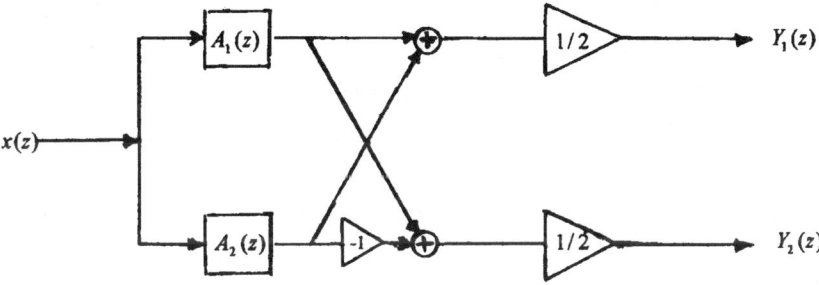

Fig. 2.14. Sum and difference of two allpass filters

$$H(z) = \frac{Y_2(z)}{X(z)} = \frac{1}{2}[A_1(z) - A_2(z)] \qquad (2.108)$$

Since $A_1(z)$ and $A_2(z)$ are assumed to be allpass functions with real coefficients, they can be written as

$$A_1(z) = \frac{z^{-(N-r)}D_1(z^{-1})}{D_1(z)} \qquad (2.109)$$

and
$$A_2(z) = \frac{z^{-r} D_2(z^{-1})}{D_2(z)} \tag{2.110}$$

Therefore
$$G(z) = \frac{1}{2} \left[\frac{z^{-(N-r)} D_1(z^{-1}) D_2(z) + z^{-r} D_2(z^{-1}) D_1(z)}{D_1(z) D_2(z)} \right] \tag{2.111}$$

and
$$H(z) = \frac{1}{2} \left[\frac{z^{-(N-r)} D_1(z^{-1}) D_2(z) - z^{-r} D_2(z^{-1}) D_1(z)}{D_1(z) D_2(z)} \right] \tag{2.112}$$

If we represent them as
$$\begin{aligned} G(z) &= \frac{P(z)}{D(z)} = \frac{\sum_{n=0}^{N} p_n z^{-n}}{D(z)} \\ H(z) &= \frac{Q(z)}{D(z)} = \frac{\sum_{n=0}^{N} q_n z^{-n}}{D(z)} \end{aligned} \tag{2.113}$$

then we can show that the following conditions are satisfied.

Property (1) $P(z^{-1}) = z^N P(z)$. Hence $p_n = p_{N-n}$ i.e. the coefficients of $P(z)$ are symmetric.

Property (2) $Q(z^{-1}) = -z^N Q(z)$. Hence $q_n = -q_{N-n}$ i.e. the coefficients of $Q(\ddagger)$ are antisymmetric.

Property (3) $P(z)P(z^{-1}) + Q(z)Q(z^{-1}) = D(z)D(z^{-1})$. Hence $|G(e^{j\omega})|^2 + |H(e^{j\omega})|^2 = 1$ i.e. $G(z)$ and $H(z)$ are said to form a power complementary pair.

Property (4) $|G(e^{j\omega})| = \frac{1}{2} |e^{j\theta_1(\omega)} + e^{j\theta_2(\omega)}| = \frac{1}{2} |1 + e^{j(\theta_1(\omega) - \theta_2(\omega))}| \leq 1$

In the following section, we will assume that the above four conditions are satisfied and derive the result that $G(z)$ and $H(z)$ can be obtained in the form $G(z) = \frac{1}{2}[A_1(z) + A_2(z)]$ and $H(z) = \frac{1}{2}[A_1(z) - A_2(z)]$.

Consider $P(z)P(z^{-1}) + Q(z)Q(z^{-1}) = D(z)D(z^{-1})$. Using Properties (1) and (2), we get

$$\begin{aligned} P(z) z^N P(z) - z^N Q(z) Q(z) &= D(z) D(z^{-1}) & (2.114) \\ P^2(z) - Q^2(z) &= D(z) z^{-N} D(z^{-1}) & (2.115) \\ [P(z) + Q(z)][P(z) - Q(z)] &= z^{-N} D(z) D(z^{-1}) & (2.116) \end{aligned}$$

Since $[P(z^{-1}) + Q(z^{-1})] = z^N [P(z) - Q(z)]$, we get

$$[P(z) + Q(z)] z^{-N} [P(z^{-1}) + Q(z^{-1})] = D(z) z^{-N} D(z^{-1}) \tag{2.117}$$

2.5. DESIGN OF PULSE SHAPING FILTERS

and also the result that the zeros of $[P(z) - Q(z)]$ are reciprocals

We shall assume that $G(z)$ is asymptotically stable. Hence from Property (4) we infer that $G(z)$ has no poles on the unit circle. In other words, the zeros of $D(z)$ are within the unit circle and the zeros of $D(z^{-1})$ are outside the unit circle. So also the zeros of $[P(z) + Q(z)]$ are not on the unit circle. Let us assume that $D(z)$ has r zeros z_k ($k = 1, 2, \ldots, r$) that are inside the unit circle and $(N - r)$ zeros z_k ($k = r+1, r+2, \ldots, N$) that are outside the unit circle.

Thus we will assume the polynomial $D(z)$ in the form

$$D(z) = \prod_{k=1}^{r}(1 - z^{-1}z_k) \prod_{k=r+1}^{N}(1 - z^{-1}z_k^{-1}) \tag{2.118}$$

Then we can also derive

$$[P(z) + Q(z)][P(z) - Q(z)] = z^{-N}D(z)D(z^{-1}) =$$

$$\prod_{k=1}^{r}(1 - z^{-1}z_k) \prod_{k=r+1}^{N}(1 - z^{-1}z_k^{-1}) \prod_{k=1}^{r}(z^{-1} - z_k) \prod_{k=r+1}^{N}(z^{-1} - z_k^{-1}) \tag{2.119}$$

Thus we identify

$$P(z) + Q(z) = \alpha \prod_{k=1}^{r}(1 - z^{-1}z_k) \prod_{k=r+1}^{N}(1 - z^{-1}z_k^{-1}) \tag{2.120}$$

and

$$P(z) - Q(z) = \frac{1}{\alpha} \prod_{k=1}^{r}(1 - z^{-1}z_k) \prod_{k=r+1}^{N}(z^{-1} - z_k^{-1}) \tag{2.121}$$

Then

$$G(z) + H(z) = \frac{P(z) + Q(z)}{D(z)} = \alpha \prod_{k=r+1}^{N}\left(\frac{z^{-1} - z_k^{-1}}{1 - z^{-1}z_k^{-1}}\right) = \alpha A_1(z) \tag{2.122}$$

and

$$G(z) - H(z) = \frac{P(z) - Q(z)}{D(z)} = \frac{1}{\alpha}\left(\frac{z^{-1} - z_k}{1 - z^{-1}z_k}\right) = \frac{1}{\alpha}A_2(z) \tag{2.123}$$

But from the power complementary property, we must have $\alpha^2 = 1$. Therefore, $\alpha = 1$ so that

$$G(z) = \frac{1}{2}[A_1(z) + A_2(z)] \tag{2.124}$$

$$H(z) = \frac{1}{2}[A_1(z) - A_2(z)] \tag{2.125}$$

So we have proved that when the four Properties listed above are satisfied, we can synthesize $G(z)$ as the sum of two allpass functions. Indeed, it has been shown that the four Properties are both necessary and sufficient conditions [42].

Using the above results, one can show how to synthesize a digital transfer function $G(z)$ as the sum of two allpass filters. We know that the magnitude response of allpass structures is not sensitive to changes in the word length of the multiplier coefficients and therefore the transfer function $G(z)$ has low sensitivity to changes in the multiplier coefficients. But the real purpose of the above discussion is not to offer a method for *synthesis* of a transfer function. It serves only as the theoretical background for the *approximation* of both the magnitude of a lowpass filter and a constant group delay. This subject is treated next.

Renfors and Saramaki [38] proposed that one of the allpass network be chosen as a pure delay network, for example, $A_2(z) = z^{-M}$. Then the coefficients of the other function $A_1(z)$ are considered as the unknown variables which we wish to optimize such that the magnitude and the group delay of the lowpass filter $G(e^{j\omega})$ approximates a constant value in the passband. Thus we force the magnitude $|A_1(e^{j\omega})|$ to follow the magnitude $|e^{-j\omega M}| = 1$. We constrain their phase to be equal in the passband but differ by π radians in the stopband, so that $G(e^{j\omega})$ is equal to zero in the stopband. This approach enables us to approximate the magnitude of a lowpass (bandpass) filter with constant group delay and at the same time, to get a highpass(bandstop) filter $H(e^{j\omega})$ which has its passband and stopband that are complement of those of the lowpass(bandpass) filter $G(e^{j\omega})$. We describe this method in greater detail below.

When $A_2(z)$ is chosen as a pure delay network function, the parallel structure shown in Fig. 2.14 gives

$$G(z) = \frac{1}{2}[A_1(z) + z^{-M}] \qquad (2.126)$$

$$H(z) = \frac{1}{2}[A_1(z) - z^{-M}] \qquad (2.127)$$

where we assume $A_1(z)$ is an Nth order, stable allpass function, in which $a_0 = 1$, i.e.

$$A_1(z) = z^{-N} \frac{\sum_{n=0}^{N} a_n z^n}{\sum_{n=0}^{N} a_n z^{-n}} \qquad (2.128)$$

It was pointed out that the phase of a stable N^{th} order allpass section is 0 at $\omega = 0$ and decreases monotonically to the value $-N\pi$ at $\omega = \pi$. (Here we have normalized the frequency by T.) We see that the phase of $A_1(z)$ and z^{-M} are 0 at $\omega = 0$. Hence at $\omega = 0$, we have $G(1) = 1$ and $H(1) = 0$ so that $G(z)$ is a lowpass (LP) filter or a bandstop(BS) filter, while $H(z)$ is a highpass (HP) filter or a bandpass (BP) filter. We also recollect that $A_1(z)$ given by (2.128) and $A_2(z) = z^{-M}$ are allpass functions that satisfy the four Properties listed above. If $\theta_1(\omega)$ is the desired (ideal) phase of $A_1(e^{j\omega})$, and since the phase of $A_2(e^{j\omega}) = e^{-j\omega M}$ is obviously equal to $-M\omega$, we must approximate $\theta_1(\omega)$ as follows:

$$\theta_1(\omega) = \begin{cases} -M\omega & \text{in the passband of } G(e^{j\omega}) \\ -M\omega \pm \pi & \text{in the stopband of } G(e^{j\omega}) \end{cases} \qquad (2.129)$$

2.5. DESIGN OF PULSE SHAPING FILTERS

when $G(e^{j\omega})$ and $H(e^{j\omega})$ are required to be LP-HP filter pair, and

$$\theta_1(\omega) = \begin{cases} -M\omega & \text{in the first passband of } G(e^{j\omega}) \\ -M\omega \pm \pi & \text{in the stopband of } G(e^{j\omega}) \\ -M\omega \pm 2\pi & \text{in the second passband of } G(e^{j\omega}) \end{cases} \quad (2.130)$$

when $G(e^{j\omega})$ and $H(e^{j\omega})$ are required to be the BP-BS filter pairs. We pointed out that the phase of $A_1(z)$ is $-N\pi$ at $\omega = \pi$; so we must choose $N = M \pm 1$ for the LP-HP pair and $N = M \pm 2$ for the BP-BS pair to assure the stability of these filters. It has been found that the choice of the plus sign gives better results. Instead of approximating $\theta_1(\omega)$ as specified in (2.130), we can formulate an equivalent problem as described below.

Consider

$$F(z) = \frac{A_1(z)}{z^{-M}} = z^{M-N} \frac{\sum_{n=0}^{N} a_n z^n}{\sum_{n=0}^{N} a_n z^{-n}} \quad (2.131)$$

Its frequency response is given by

$$F(e^{j\omega}) = e^{j(M-N)\omega} \frac{\sum_{n=0}^{N} a_n e^{j\omega n}}{\sum_{n=0}^{N} a_n e^{-j\omega n}} = e^{j2\varphi} \quad (2.132)$$

where

$$\varphi = \tan^{-1} \frac{\sum_{n=0}^{N} a_n \sin(n + \frac{M-N}{2})\omega}{\sum_{n=0}^{N} a_n \cos(n + \frac{M-N}{2})\omega} = \tan^{-1} \Phi(\omega) \quad (2.133)$$

From (2.131) and (2.132), we note that φ is one half of the difference between the phases of $A_1(z)$ and z^{-M}. Therefore we can consider the desired value for φ to be 0 in the passband and $-\frac{\pi}{2}$ in the stopband of the lowpass filter $G(e^{j\omega})$. Let $\varphi_p = \tan^{-1} \delta_p$ and $\varphi_s = \tan^{-1} \delta_s$ be the deviations of φ from the desired values in the passband and stopband of the lowpass filter respectively. (The design of the BP-BS pair follows a similar procedure.) Hence the design problem is to find the coefficients a_n such that the following conditions are satisfied.

$$\begin{cases} -\delta_p \leqslant \Phi(\omega) \leqslant \delta_p & \text{in the passband of } G(e^{j\omega}) \\ -\delta_s \leqslant \Phi(\omega) \leqslant \delta_s & \text{in the stopband of } G(e^{j\omega}) \end{cases} \quad (2.134)$$

This problem can be solved using the Remez Exchange Algorithm. We choose $M - N = -1$ in (2.133) and evaluate $\Phi(\omega)$ at a number of discrete frequencies $\omega_i (i = 1, 2, \ldots, k)$ in the passband $[0, \omega_p]$ and ω_j ($j = 1, 2, \ldots, l$) in the stopband $[\omega_s, \pi]$. We derive the conditions (2.135), (2.136) from (2.133), as the equations to be satisfied, when the Remez Exchange Algorithm is used successfully. The final values of the frequencies at which these equalities are satisfied will be different from the initial values chosen in the iteration procedure. When the algorithm is terminated, the filter approximates a linear phase over the frequency band $\{\omega : [0, \omega_p] \bigcup [\omega_s, \pi]\}$ in an equiripple sense, while simultaneously approximating the magnitude to a constant value in the passband $[0, \omega_p]$ and zero value in the stopband $[\omega_s, \pi]$. By increasing the order N of the filter, we can increase the attenuation in the stopband to reach the prescribed value of attenuation.

$$\Phi(\omega_i) = \frac{\sum_{n=0}^{N} a_n \sin(n - \frac{1}{2})\omega_i}{\sum_{n=0}^{N} a_n \cos(n - \frac{1}{2})\omega_i} = (-1)^i \delta_p \qquad (2.135)$$

and

$$\frac{1}{\Phi(\omega_j)} = \frac{\sum_{n=0}^{N} a_n \cos(n - \frac{1}{2})\omega_j}{\sum_{n=0}^{N} a_n \sin(n - \frac{1}{2})\omega_j} = (-1)^i \delta \qquad (2.136)$$

An example of a lowpass filter designed according to the above design theory is given in [38]. It shows a very good lowpass magnitude and group delay response.

2.5.3 Method 2.5: Use of Singular Value Decomposition

A method based on the singular value decomposition was reported by [6] and [49], for the design of two-dimensional recursive digital filters that approximate magnitude only. But this method was restricted to separable filters only and hence can be used to design filters with a passband or stopband region that is only rectangular in shape (though some of the examples chosen in these references specify circular lowpass filters and show their amplitude response as a 3-D plot, a plot of their contours would obviously exhibit rectangular regions). Also the method is restricted to filters of low order because it chooses only one dominant eigen value. In a new approach [19], we use singular value decomposition for the design of both one-dimensional and two-dimensional recursive digital filters that approximate both magnitude and constant group delay. Here we describe the method for the one-dimensional case, leaving the case of 2-D, IIR filter design for Chap. 4.

Let $G(e^{j\omega})$ be the prescribed frequency response of an IIR digital filter. Let us find the Fourier series coefficients of $G(e^{j\omega})$ and truncate them by a rectangular window of length $2N + 1$. Then we obtain an FIR filter

$$F(z) = \sum_{k=-N}^{N} f_k z^{-k} \qquad (2.137)$$

where $f_k = f_{-k}$, for $k = 1, 2, \ldots, N$ such that $F(e^{j\omega})$ represents a very good approximation of $G(e^{j\omega})$, in the least squares sense. We shift this noncausal FIR filter by m samples where m is an integer $0 < m < N$ to obtain

$$\mathbf{G}(z) = z^{-m} F(z) \qquad (2.138)$$

so $\mathbf{G}(z)$ is not necessarily causal. It has a causal part $C_a(z) = \sum_{k=0}^{N+m} C_k z^{-k}$ and also an anticausal part $A_c(z) = \sum_{k=1}^{N-m} A_k z^k$. But $\mathbf{G}(e^{j\omega})$ has exactly the same magnitude as $F(e^{j\omega})$ and the same linear phase with a group delay of m. It is easy to show that the following relations hold.

$$\begin{array}{ll} C_k = f_{-m+k} & \text{for } k = 0, 1, 2, \ldots, N+m \\ A_k = f_{-m-k} & \text{for } k = 1, 2, \ldots, N-m \end{array} \qquad (2.139)$$

2.5. DESIGN OF PULSE SHAPING FILTERS

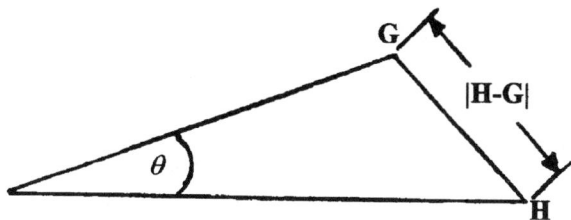

Fig. 2.15. Difference of complex valued functions

We seek to find a causal IIR filter $H(z)$ of fairly low order as an approximation of $\mathbf{G}(z)$ such that the error in Chebyshev norm ϵ defined in (2.140) is minimized. This definition of the Chebyshev norm is similar to that given in (1.21).

$$\epsilon = \left\| H(e^{j\omega}) - G(e^{j\omega}) \right\|_c = \sup \left\{ \left| H(e^{j\omega}) - G(e^{j\omega}) \right| : \omega \in [-\pi, \pi] \right\} \quad (2.140)$$

Notice that the Chebyshev norm to be minimized is the difference of the complex valued functions and not the difference in just their magnitude only. So, we argue that when the error ϵ is minimized, both the magnitude and the phase of $H(z)$ approximate the magnitude and phase of $\mathbf{G}(z)$.

From Fig. 2.15, we see that if ϵ is very small, the error between the phase of $H(e^{j\omega})$ and $G(e^{j\omega})$ satisfies (2.141)

$$\theta_\epsilon = \left| \angle H(e^{j\omega}) - \angle G(e^{j\omega}) \right| \leq \gamma(\omega)\epsilon, ; \quad \omega \in [-\pi, \pi] \quad (2.141)$$

where $\gamma(\omega) > 0$ is a suitable constant, which is nearly equal to one in the passband. So θ_ϵ is small for ω in the passband. The minimization of θ_ϵ in (2.140) no doubt provides us with a new approach for approximating the magnitude and group delay of an IIR filter but it also presents a new problem since efficient numerical techniques for minimization of an error function in the complex domain are not available [4, 17]. Hence instead of minimizing the above error function θ_ϵ, we propose the following two-part procedure. First we design a causal IIR filter $H_1(z)$ such that

$$\epsilon_1 = \|A_c - H_1\|_c = \sup_{\omega \in [-\pi, \pi]} \left\{ \left| A_c(e^{j\omega}) - H_1(e^\omega) \right| \right\} \quad (2.142)$$

is minimized. The second part is to find an IIR filter $H(z)$ of a fairly low order such that

$$\epsilon_2 = \|C_a + H_1 - H\|_c \quad (2.143)$$

$$= \sup_{\omega \in [-\pi, \pi]} \left\{ \left| C_a(e^{j\omega}) + H_1(e^{j\omega}) - H(e^\omega) \right| \right\} \quad (2.144)$$

is minimized. Then the error ϵ in (2.140) becomes bounded by $\epsilon_1 + \epsilon_2$ as shown below:

$$\epsilon = \|G - H\|_c = \|C_a + A_c - H\|_c$$

$$\leq \|A_c - H_1\|_c + \|C_a + H_1 - H\|_c = \epsilon_1 + \epsilon_2 \qquad (2.145)$$

It is true that the causal IIR filter $H(z)$ obtained by these two steps may not be the optimal solution, but it does yield a small value for the Chebyshev norm ϵ, because an optimal solution for finding $H_1(z)$ and a suboptimal solution for finding $H(z)$ as defined above are available in the literature on modern control theory. So the major thrust of this approach is to adapt the results from control theory to the unsolved problem posed in (2.140) and this approach is also prompted by the availability of sophisticated numerical algorithms for finding these solutions.

Now we consider the first part. Note that the anticausal part $A_c(z)$ is a rational polynomial in z and we seek to find a causal $H_1(z)$ which is the ratio of two polynomials such that ϵ_1 is minimized. To do so, let us define the Hankel matrix of $A_c(z)$ by

$$H_{AC} = \begin{bmatrix} A_1 & A_2 & A_3 & \cdots & A_{N-m} \\ A_2 & A_3 & \cdots & A_{N-m} & 0 \\ A_3 & \cdots & A_{N-m} & 0 & 0 \\ \cdots & \cdots & \cdots & \cdots & \cdots \\ A_{N-m} & 0 & \cdots & \cdots & 0 \end{bmatrix} \qquad (2.146)$$

Let us represent the maximum singular value of this Hankel matrix by $\sigma_{\max}(H_{AC})$. It has been proved by Nehari [32] that the minimum value for the error ϵ is given by $\sigma_{\max}(H_{AC})$ i.e.

$$\epsilon_1^* = \inf \{\|A_c - H_1\|_c : H_1(z) \text{ causal}\} \qquad (2.147)$$
$$= \|A_c - H_1^*\|_c = \sigma_{\max}(H_{AC}) \qquad (2.148)$$

The procedure to find the optimal $H_1^*(z)$ that minimizes ϵ_1 in (2.142) is found in the papers [2] and [3]. To find this optimal function, we obtain the singular value decomposition $H_{AC} = USV^T$ where U and V are unitary matrices and $\mathbf{S}$ is a diagonal matrix, i.e. $\mathbf{S} = \text{diag}(\sigma_1, \sigma_2, \ldots, \sigma_n)$ with $\sigma_1 \geq \sigma_2 \geq \sigma_3 \cdots \geq \sigma_n \geq 0$. Thus we have $\sigma_{\max}(H_{AC}) = \sigma_1$.

From the singular value decomposition, we can easily identify the first columns of V and U. We represent them by ξ and η as

$$\xi = [\xi_1, \xi_2, \ldots, \xi_{N-m}]^T$$

and

$$\eta = [\eta_1, \eta_2, \ldots, \eta_{N-m}]^T \qquad (2.149)$$

Next define a transfer function

$$H_t(z) = \sigma_{\max}(H_{AC}) \frac{\sum_{k=1}^{N-m} \eta_k z^k}{\sum_{k=1}^{N-m} \xi_k z^{-k+1}} \qquad (2.150)$$

Then the optimal solution of (2.147) is given by

$$H_1^*(z) = A_c - H_t(z) = \frac{P(z)}{z^{N-m-1} \sum_{k=1}^{N-m} \xi_k z^{-k+1}} \qquad (2.151)$$

2.5. DESIGN OF PULSE SHAPING FILTERS

where the numerator polynomial $P(z)$ is a polynomial of order less than $N-m$. The authors in [2] and [3] have shown that (2.150) is an allpass function and $H_1^*(z)$ given by (2.151) is the unique optimal solution of (2.147). But it must be pointed out that ϵ_1^* depends on the value of the pure delay m, which is not surprising. And the impact of choosing $A_c(z) = z^{-N}$ so that $H_1^*(z)$ is of the same form as (2.127), on the resulting $H_1^*(z)$ is an interesting problem that needs to be further investigated.

Next we consider the second part. The causal recursive digital filter $H(z)$ can be obtained from the approximation of the causal filter $C_a(z) + H_1^*(z)$ as the error ϵ_2 approaches zero. But the order of such a filter will be very high and hence this procedure calls for the suboptimal methods of model reduction that are well known. There are several methods that are available; some of them like the balanced realization method are discussed in [19]. Those who are interested in these methods may refer to the original papers and textbooks on modern control theory [18, 28]. But one part of the model reduction procedure requires that an approximation of the causal $C_a(z)$ by a transfer function $H_2(z)$ is obtained. The balanced model reduction method takes a simpler form in this case, because $C_a(z)$ is a polynomial and not a rational function. A classical method is the Prony's method that can also be used for this purpose. The balanced model reduction method for this case will be explained in Chap. 4 since the same procedure will be used there for the design of 2-D filters.

2.5.4 Example 2.4

A numerical example is chosen from [19] to illustrate the efficiency of the method proposed in that paper and compare the results with those from [12]. The specification of a lowpass filter $G(z)$ is prescribed as follows:

$$|H(e^{j\omega})| = \begin{cases} 1.0 & \text{for } 0.0 \le \omega \le 0.5\pi \\ 0.0 & \text{for } 0.6\pi \le \omega \le \pi \end{cases} \quad (2.152)$$

and a delay $m = 22$.

An 18th order IIR causal digital filter is obtained, using the method outlined above. A lowpass filter of the same order was designed by [12] using Linear Programming techniques, to obtain the same magnitude response in the passband $[0 \le \omega \le 0.5\pi]$ but with a delay of 15. The magnitude and group delay responses of these two filters are plotted in Fig. 2.16 and Fig. 2.17 for comparison.

2.5.5 Method 2.6: Application of Linear Programming

In the previous section, we have already referred to the design example given in [12], and compared its magnitude and group delay responses with those obtained by [19]. Now we will describe this method presented in [12] which uses Linear Programming Technique for approximating both the prescribed magnitude and group delay of an IIR digital filter.

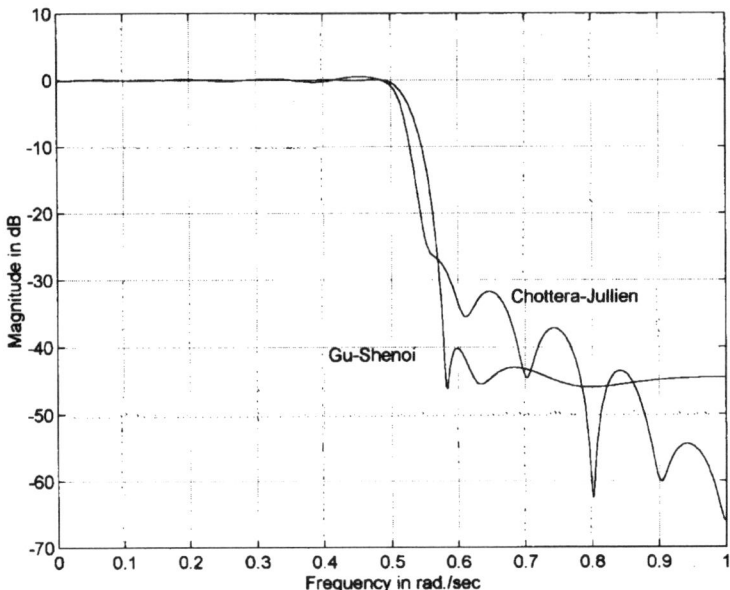

Fig. 2.16. Magnitude of Gu-Shenoi and Chottera-Jullien filters

Let us consider the design of a digital filter given by

$$H(z) = \frac{P(z)}{Q(z)} = \frac{a_0 + a_1 z^{-1} + a_2 z^{-2} + \cdots + a_N z^{-N}}{b_0 + b_1 z^{-1} + b_2 z^{-2} + \cdots + b_M z^{-M}} \quad (2.153)$$

so that its frequency response can be written in the form

$$H(e^{j\omega_i}) = \frac{P(e^{j\omega_i})}{Q(e^{j\omega_i})} = \frac{\sum_{n=0}^{N} a_n \cos(n\omega_i) - j \sum_{n=0}^{N} a_n \sin(n\omega_i)}{\sum_{m=0}^{M} b_m \cos(m\omega_i) - j \sum_{m=0}^{M} b_m \sin(m\omega_i)}$$

Let us assume that the desired magnitude $G(\omega_i)$ and the linear phase response corresponding to a constant group delay τ_d are prescribed at k discrete frequencies i.e.

$$G(\omega_i) e^{-j\tau_d \omega_i} = G(\omega_i) \cos(\omega_i \tau_d) - j G(\omega_i) \sin(\omega_i \tau_d);$$
$$i = 1, 2, 3, \ldots k \quad (2.154)$$

We define an error

$$\mathbf{J}(\mathbf{a}, \mathbf{b}, \omega_i) = G(\omega_i) e^{-j\tau_d \omega_i} - H(e^{j\omega_i}) = G(\omega_i) e^{-j\tau_d \omega_i} - \frac{P(e^{j\omega_i})}{Q(e^{j\omega_i})} \quad (2.155)$$

Multiplying both sides by $Q(e^{j\omega_i})$, and separating the real and imaginary part $e_R(\omega_i)$ and $e_I(\omega_i)$ respectively, we get

$$Q(e^{j\omega_i}) \mathbf{J}(\mathbf{a}, \mathbf{b}, \omega_i) = G(\omega_i) e^{-j\tau_d \omega_i} Q(e^{j\omega_i}) - P(e^{j\omega_i})$$
$$= e_R(\omega_i) + j e_I(\omega_i) \quad (2.156)$$

2.5. DESIGN OF PULSE SHAPING FILTERS

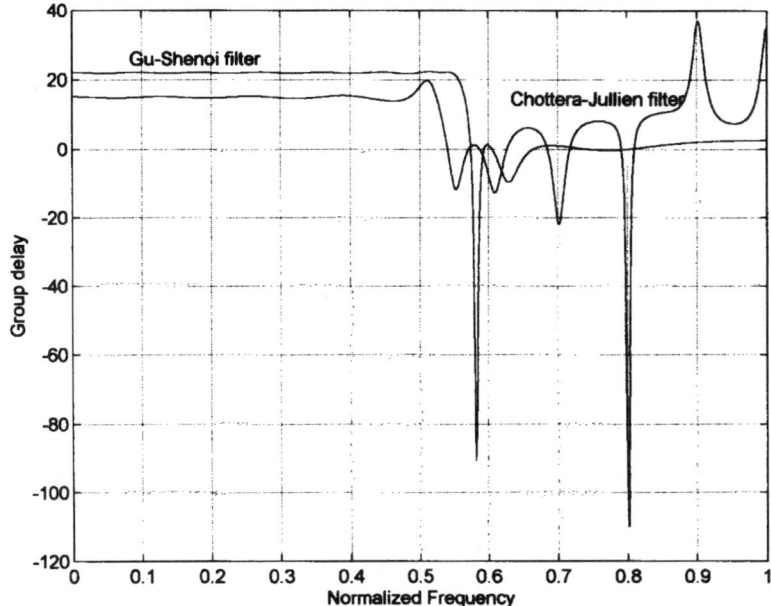

Fig. 2.17. Group Delay of Gu-Shenoi and Chottera-Jullien filters

where

$$e_R(\omega_i) = \sum_{m=0}^{M} b_m \left[G(\omega_i) \cos\{\omega_i(m + \tau_d)\} \right] - \sum_{n=0}^{N} a_n \cos(n\omega_i) \tag{2.157}$$

and

$$e_I(\omega_i) = \sum_{m=0}^{M} b_m \left[-G(\omega_i) \cos\{\omega_i(m + \tau_d)\} \right] + \sum_{n=0}^{N} a_n \sin(n\omega_i)$$
$$i = 1, 2, \ldots, k \tag{2.158}$$

Notice that these two error components are linear functions with respect to the unknown coefficients a_n and b_m. It is obvious from (2.156) that when both $e_R(\omega_i)$ and $e_I(\omega_i)$ are zero, the error function $\mathbf{J}(\mathbf{a}, \mathbf{b}, \omega_i)$ is zero. Therefore we seek to find the unknown coefficients in order to simultaneously minimize the error functions $e_R(\omega_i)$ and $e_I(\omega_i)$— if not possible to make them exactly equal to zero at all the frequencies. This means we like to force $|e_R(\omega_i)| \leqslant \epsilon$ and $|e_I(\omega_i)| \leqslant \epsilon$ where ϵ is a small positive value, i.e.

$$\begin{aligned} e_R(\omega_i) - \epsilon &\leqslant 0 \\ -e_R(\omega_i) - \epsilon &\leqslant 0 \\ e_I(\omega_i) - \epsilon &\leqslant 0 \\ -e_I(\omega_i) - \epsilon &\leqslant 0 \qquad i = 1, 2, \ldots, k \end{aligned} \tag{2.159}$$

If we now define $\zeta = -\epsilon$ and also let $b_0 = 1$ without loss of generality, the above inequalities can be reformulated in the framework of a Linear Programming problem as follows.

Maximize

$$g = [b_1, b_2, b_3, \ldots, b_M, a_0, a_1, a_2, \ldots, a_N, \zeta] \begin{bmatrix} 0 \\ 0 \\ 0 \\ \cdot \\ \cdot \\ 0 \\ 1 \end{bmatrix} \quad (2.160)$$

subject to the following constraints:

$$\sum_{m=0}^{M} b_m \left[G(\omega_i) \cos\{\omega_i(m + \tau_d)\} \right] - \sum_{n=0}^{N} a_n \cos(n\omega_i) + \zeta \leqslant 0 \quad (2.161)$$

$$\sum_{m=0}^{M} b_m \left[-G(\omega_i) \cos\{\omega_i(m + \tau_d)\} \right] + \sum_{n=0}^{N} a_n \cos(n\omega_i) + \zeta \leqslant 0 \quad (2.162)$$

$$\sum_{m=0}^{M} b_m \left[-G(\omega_i) \cos\{\omega_i(m + \tau_d)\} \right] + \sum_{n=0}^{N} a_n \sin(n\omega_i) + \zeta \leqslant 0 \quad (2.163)$$

$$\sum_{m=0}^{M} b_m \left[G(\omega_i) \cos\{\omega_i(m + \tau_d)\} \right] - \sum_{n=0}^{N} a_n \sin(n\omega_i) + \zeta \leqslant 0 \quad (2.164)$$

Minimizing the value of ϵ is equivalent to maximizing the value of g since $g = \zeta = -\epsilon$ and the upper bound on g is zero, because ϵ is a nonnegative real number. As in many cases of computer-aided design of circuits and systems, when a maximum value of g is found by Linear Programming method and the coefficients of $H(z)$ are thereby known, there is no guarantee that the transfer function $H(z)$ is a stable function. The authors prove that the condition $Re[Q(e^{j\omega})] \geqslant 0$ is a sufficient condition for stability. This condition can be included as additional linear constraints in the Linear Programming Problem. When $b_0 = 1$, this condition incorporated in the algorithm, reduces to the following form:

$$-\sum_{m=1}^{M} b_m \cos(m\omega_i) \leqslant 1 - \delta, \qquad \text{for } 0 \leqslant \omega_i \leqslant \pi \quad (2.165)$$

It is added to those inequalities given by (2.161) - (2.164) in defining the Linear Programming (LP) problem. The solution to this problem gives the values for the coefficients of $H(z)$. We notice that for each value of ω_i, there are five constraining inequalities and hence the Linear Programming method maximizes the value of g, subject to $5k$ constraints. For each value of τ_d, another set of $5k$ constraints are defined and the new maximum value for g is obtained. The

2.5. DESIGN OF PULSE SHAPING FILTERS

authors have found that the maximum value of g does not increase endlessly as the value of τ_d is increased but attains a peak value and then decreases as τ_d is further increased. They suggest that the algorithm choose an initial value for $\tau_d = N - 1$ where N is the order of the filter, and solve the Linear Programming problem to get the maximum value for g, as well as the corresponding set of coefficients. Then the value of the group delay is decreased by one in the next iteration, and the LP problem is again solved to get a new set of coefficients. The iteration is continued if the new value of g increases, till it attains a peak value and when it begins to decrease, the iteration is stopped. The authors state from their experience that one needs to continue the iteration till the value of the group delay is about $N/2$. In this method, the total number of equations to be solved in each stage is large and because of the large number of iterations, a satisfactory solution requires a very large amount of computation time. However, in contrast to nonlinear programming procedures, this one does not require the choice of starting values for the unknown coefficients that are very close to the optimum values. The Linear Programming method gives the optimum values, whenever a solution to the problem exists. The method can be used for the design of filters with arbitrary magnitude and phase specifications. For example, the authors have designed a five band filter, with three stopbands and two passbands. They have also shown in another paper [13] how the Linear Programming method can be used for the design of 2-D IIR filters with arbitrary magnitude and group delay responses. This is explained in Chap. 4.

2.5.6 Method 2.7: Theory of Commensurate Distributed Networks

In Chap. 1, we denoted the transfer function of a continuous-time filter network by $H(s)$. When the unit impulse response $h(t)$ of this analog filter is sampled and a discrete-time sequence $h(nT)$ was obtained, we denoted the impulse invariant z-transform of the sequence as the transfer function $H(z)$ of the digital filter. The mapping relationship between the frequency response $H(j\omega)$ of the analog filter and $H(e^{j\omega})$ of the impulse invariant digital filter was derived and is similar to equation (1.101). In the second half of that chapter, we applied the bilinear transformation $\mathbf{s} = \frac{z-1}{z+1}$ to map the frequency response of the digital filter $H(e^{j\omega T})$ to $H(j\lambda)$, the mapping relationship being defined by $j\lambda = j\tanh(\frac{\omega T}{2})$. By scaling the frequency response of the transfer function $H(\mathbf{s})$ or by application of the proper analog frequency transformation, the analog lowpass prototype filter $H(\mathbf{p})$ was obtained. This was designed by using the same approximation theory as was used for designing the filter $H(p)$. In that context, it was pointed out that the frequency response of $H(\mathbf{p})$ and $H(\mathbf{s})$ and their unit impulse responses were not the same as those of $H(p)$ and $H(s)$ from which $H(z)$ was derived.

In this section, we will give another definition of the function $H(s)$ derived from $H(z)$, which will be identified as the transfer function of commensurate distributed networks. The theory of commensurate distributed networks has

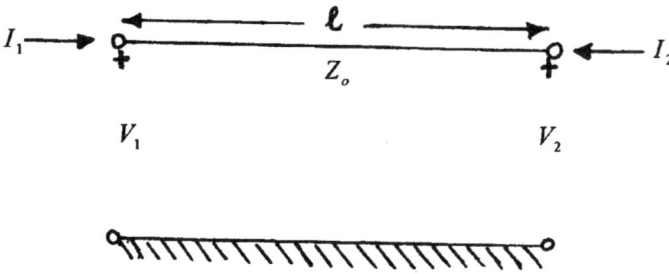

Fig. 2.18. Distributed parameter 2-port network

been used to develop a method for approximating the magnitude and linear phase response of a digital filter. For a brief introduction of this theory, let us start with the model of a uniform, lossless, transmission line with a length $= l$ and characteristic impedance Z_o as shown in Fig. 2.18.

The chain matrix parameters (A, B, C, D) or the transmission matrix parameters for this distributed parameter 2-port network, which relate the voltage and current variables at the two ports of the network are given below.

$$\begin{bmatrix} V_1 \\ I_1 \end{bmatrix} = \begin{bmatrix} \cosh s\bar{\tau} & Z_o \sinh s\bar{\tau} \\ Y_o \sinh s\bar{\tau} & \cosh s\bar{\tau} \end{bmatrix} \begin{bmatrix} V_2 \\ -I_2 \end{bmatrix} \qquad (2.166)$$

Under sinusoidal steady state conditions, we let $s = j\omega$ and in order to explicitly show the dependence of the transmission matrix parameters on the unit length l, we introduce a new parameter β called the propagation constant so that $\beta l = \omega \bar{\tau}$. Denoting the velocity of propagation of the electromagnetic waves (TEM only) by c, we get $l = c\bar{\tau}$, from which we derive $\beta = \frac{\omega}{c}$ as an expression for the propagation constant.

The transmission matrix under sinusoidal steady state conditions is now given by

$$\begin{bmatrix} \cosh \beta l & jZ_o \sinh \beta l \\ jY_o \sinh \beta l & \cosh \beta l \end{bmatrix} \qquad (2.167)$$

If we define a new complex variable $\lambda = \tanh s\bar{\tau}$, then $\sqrt{1-\lambda^2} = \frac{1}{\cosh s\bar{\tau}}$ and the transmission matrix can be written in a compact form given by

$$\begin{bmatrix} \cosh s\bar{\tau} & Z_o \sinh s\bar{\tau} \\ Y_o \sinh s\bar{\tau} & \cosh s\bar{\tau} \end{bmatrix} = \frac{1}{\sqrt{1-\lambda^2}} \begin{bmatrix} 1 & Z_o\lambda \\ Y_o\lambda & 1 \end{bmatrix} \qquad (2.168)$$

Note that when the unit element (UE) defined by the above 2-port parameters, is terminated at the output terminals 2-2' by a load impedance Z_L, the driving point input impedance $Z_{in}(\lambda)$ is given by

$$Z_{in}(\lambda) = \frac{AZ_L + B}{CZ_L + D} = \frac{Z_o\lambda + Z_L}{Y_oZ_L\lambda + 1} \qquad (2.169)$$

2.5. DESIGN OF PULSE SHAPING FILTERS

When the output terminals are shorted, the input impedance becomes $Z_{sc} = Z_o \lambda$ and the corresponding network element is called a short-circuited stub which therefore behaves in the λ domain just like an inductance in the s domain. Similarly, when the output terminals are left open, the one-port network element called an open circuited stub, has an input impedance $Z_{oc} = \frac{1}{Y_o \lambda}$ which therefore behaves like a capacitor in the λ domain. Networks that contain UE's and ideal transformers as two-port elements, the short-circuited and open-circuited stubs as the one-port elements and resistors connected only at the terminal pairs at the output port are classified as distributed parameter networks. If the length of all the elements have the same value l or are integer multiples of l, then they are said to be commensurate to the same basic length l and networks containing only elements of commensurate length are called as commensurate, distributed parameter networks. Properties of the network functions of this class of networks defined in the λ domain are similar to those of the linear, time-invariant, lumped parameter networks containing inductors, capacitors, resistors and ideal transformers, defined in the s domain. For example, these network functions are rational functions in λ. Their driving point impedances are positive real functions of λ and can be synthesized by a procedure similar to Brune's synthesis. But the commensurate, distributed parameter networks contain the unit elements (UE) for which there exist no analogous lumped parameter elements and consequently some of the properties are different from those of the lumped parameter networks. The properties of commensurate distributed networks can be found in greater detail in [7],[39]. In this chapter, however, we will use only those results which are relevant to the approximation of the frequency response of these networks and the digital filters. It is easy to show the connection between the properties of the network functions for the commensurate, distributed networks and the digital filters. In anticipation of showing this relationship we have already introduced the proper notation for the variables involved in this section.

From the bilinear transformation $\mathbf{s} = \frac{z-1}{z+1}$ used for the design of IIR transfer functions $H(z)$, we derived $\mathbf{s} = \tanh(\frac{sT}{2})$ and from the theory of commensurate distributed networks, we have derived $\lambda = \tanh s\bar{\tau}$. Hence we see that the transfer functions of digital filters when transformed to the $\mathbf{s}$ domain are analogous to the transfer functions of commensurate, distributed parameter networks, if and when we choose the delay $\bar{\tau}$ of the unit element to be equal to half the sampling period T chosen for the digital filter. In other words, we have $\mathbf{s} = \lambda$ when $s\bar{\tau} = \frac{sT}{2}$. The importance of this result lies in the fact that the transfer function of any lowpass commensurate, distributed parameter network has a frequency response $H(j\lambda)$ which is also the frequency response of the digital filter mapped to the same λ domain. Hence if we find a solution $H(\lambda)$ to the approximation problem of a commensurate, distributed network transfer function, we choose it and apply the bilinear transformation $\lambda = \frac{z-1}{z+1} = \frac{1-z^{-1}}{1+z^{-1}}$ to get the transfer function $H(z)$ of the digital filter. It is this approach that has been used by the authors of the two books cited above, for deriving some significant results. Their results are not so widely known among researchers in digital filter design;

those interested in the details should refer to these books. Only a few of the important results that are applicable for the simultaneous approximation of the magnitude and linear phase of a digital filter are extracted and outlined briefly, without proof, in the next section.

First we consider the transfer function of a commensurate, distributed network, which approximates a constant group delay in the maximally flat sense. This has been derived by Rhodes [39] and it is in the form

$$T_{21}(\lambda) = \frac{1}{Q_n(\lambda)} \qquad (2.170)$$

where the first (2n-1) derivatives of its group delay at $\lambda = 0$ must be zero. The polynomials $Q_n(\lambda)$ are generated by the recursive relationship

$$Q_{n+1}(\lambda) = Q_n(\lambda) + \frac{\alpha^2 - n^2}{4n^2 - 1}\lambda^2 Q_{n-1} \qquad (2.171)$$

where $Q_o = 1$ and $Q_1 = 1 + \alpha\lambda$.

When we substitute $\lambda = \frac{z-1}{z+1}$ in (2.170), we get a maximally flat group delay for the resulting digital filter. The polynomial $Q_n(\lambda)$ has been shown by Rhodes to be a strictly Hurwitz polynomial when $\alpha \geq n - 1$ where α is called the *bandwidth scaling factor*. Now, we recollect that Thiran derived the formula for the coefficients of the denominator of an all-pole digital filter function which approximates a constant group delay in the maximally flat sense. Usually the nominal group delay τ is prescribed and Thiran's formula generates the coefficients explicitly in terms of that group delay. In comparison with Thiran's formula, we find that (2.171) does not provide a direct method for generating the transfer function in terms of the desired or prescribed group delay. Also the impact of the bandwidth scaling factor α on the actual value of the group delay is not explicitly known. We also note that when we substitute $\lambda = \frac{z-1}{z+1}$ in $Q_n(\lambda)$, we get an nth order zero at $z = -1$ in the transfer function (2.170), whereas Thiran's transfer function is an all-pole function.

Rhodes derived another transfer function, using the polynomial $Q_n(\lambda)$, such that it simultaneously approximates a constant magnitude and a constant group delay in the maximally flat sense. This transfer function is of the form

$$H(\lambda) = \frac{Q_{n*} + k\lambda Q_{n-1*}}{Q_n + k\lambda Q_{n-1}} \qquad (2.172)$$

where k is chosen to produce a zero of transmission at $\lambda = \infty$ i.e.

$$k = \left.\frac{Q_n}{kQ_{n-1}}\right|_{\lambda \to \infty} \qquad (2.173)$$

In (2.172), $Q_{n*} = Q_n(-\lambda)$, where $Q_n = Q_n(\lambda)$. When we substitute $\lambda = \frac{z-1}{z+1}$ in (2.172) we get the transfer function for the IIR filters. Extensive simulation of the frequency response characteristics of this class of filters were carried out by the author. It is found that the passband over which the group delay is

2.5. DESIGN OF PULSE SHAPING FILTERS

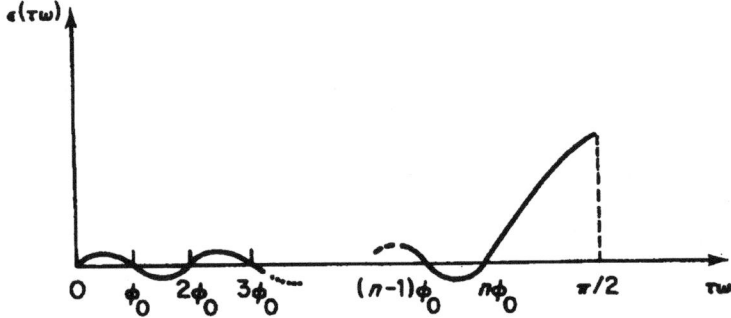

Fig. 2.19. Phase error of the equidistant linear phase polynomial (From [7], by permission from John Wiley & Sons)

constant within a small percentage of error from the nominal constant value is small, though the group delay exhibits a maximally flat response at $\omega = 0$. However, the magnitude exhibits a wide, maximally flat characteristics.

2.5.7 Equidistant Linear Phase Approximation

Instead of obtaining a maximally flat group delay approximation, Rhodes [39] has also shown how to approximate the corresponding linear phase response of the digital filter function $H(e^{j\omega})$. Usually we choose an error criteria like the maximally flat error, equiripple or Chebyshev type of error, or the least squares error to approximate the given frequency response characteristics. But he has chosen a 'new' type of error criteria to approximate the linear phase characteristic of the transfer function of the form

$$T_{21}(\lambda) = \frac{K}{A_n(\lambda \,|\phi_o)} \qquad (2.174)$$

in which the polynomial $A_n(\lambda|\phi_o)$ satisfies the condition

$$\arg A_n(\pm j \tan r\phi_o|\phi_o) = \pm br\phi_o, \qquad r = 0, 1, 2, \ldots, n \qquad (2.175)$$

with an associated error function $\varepsilon(\tau\omega) = b\omega - \psi(\tan \tau\omega)$. This error function is plotted in Fig. 2.19 which shows that the phase is exactly linear at a set of equidistant points $0, \phi_o, 2\phi_o, 3\phi_o, \ldots, n\phi_o$ in the prescribed passband, i.e., (2.175) is an interpolation formula that matches the values of the linear phase exactly at these $n+1$ equidistant points. The polynomial $A_n(\lambda \,|\phi_o)$ is obtained by the recursive formula

$$A_{n+1}(\lambda \,|\phi_o) = A_n(\lambda \,|\phi_o) + \gamma_n(\lambda^2 + \tan^2 n\phi_o)A_{n-1}(\lambda \,|\phi_o) \qquad (2.176)$$

where $A_o(\lambda \,|\phi_o) = 1$ and $A_1(\lambda \,|\phi_o) = 1 + \frac{\tan b\phi_o}{\tan \phi_o}\lambda$ and

$$\gamma_n = \frac{\cos(n-1)\phi_o \cos(n+1)\phi_o \sin(b+n)\phi_o \sin(b-n)\phi_o \cos^2 n\phi_o}{\sin(2n-1)\phi_o \sin(2n+1)\phi_o \cos^2 b\phi_o} \qquad (2.177)$$

subject to the constraints $b \geqslant n-1$, $n\phi_o \leqslant \frac{\pi}{2}$ and $(b+n-1)\phi_o \leqslant \pi$.

Let us consider a transfer function of the form

$$H(\lambda) = \frac{N(\lambda)}{D(\lambda)} \qquad (2.178)$$

such that it satisfies the following conditions:

$$H(j \tan r\phi_o) = e^{-j2br\phi_o} \qquad r = 0, 1, 2, \ldots, (n-1) \qquad (2.179)$$

and

$$H(\infty) = 0 \qquad (2.180)$$

This transfer function therefore interpolates to a magnitude of unity and linear phase response at equidistant points in the passband as defined by (2.175).

The transfer function that provides such an interpolation has been derived by Rhodes analytically, in terms of the polynomial $A_n(\lambda \mid \phi_o)$ given in (2.175), i.e.

$$H(\lambda) = \frac{A_{n*} + k\lambda A_{n-1*}}{A_n + k\lambda A_{n-1}} \qquad (2.181)$$

where k is chosen to produce a zero of transmission at $\lambda = \infty$ (or $\omega = \pi$), i.e.

$$k = \left.\frac{A_n}{\lambda A_{n-1}}\right|_{\lambda \to \infty} \qquad (2.182)$$

Now we substitute $\lambda = \frac{z-1}{z+1}$ in these equations to get the transfer functions of the digital filter that approximates both the magnitude and linear phase. As an example, the group delay response obtained from the equidistant approximation to linear phase by a 10th order transfer function was shown in Fig. 2.12 under the label 'Baher filter'.

The equations (2.176), (2.181) and (2.182) show that the algorithm provides only the values of b and n to control the magnitude and group delay response. Therefore, generating several transfer functions and plotting their magnitude and delay characteristics, with different values of b and a is required if specified bandwidth and nominal group delay value have to be met. Computer simulation has shown that sometimes the transfer functions become unstable and/or the magnitude and/or group delay responses are not quite satisfactory.

2.5.8 Method 2.8: Theory of Eigen Filters

A well-known result from matrix theory that has been used in the optimization of an error function is known as the Rayleigh's Principle. If an error function can be expressed in the form

$$J(\mathbf{b}) = \mathbf{v}^t \mathbf{P} \mathbf{v} \qquad (2.183)$$

where $\mathbf{P}$ is a real, symmetric and positive-definite matrix and $\mathbf{v}$ is a real vector of the unknown variables, subject to the constraint $\mathbf{v}^t \mathbf{v} = 1$, then, according to

2.5. DESIGN OF PULSE SHAPING FILTERS

Rayleigh's Principle [43], the solution vector **b** that minimizes the error function $J(\mathbf{b})$ is just the eigenvector of **P** corresponding to the smallest eigenvalue of **P**. This result has been used extensively in the optimal design of circuits, control and communication systems where the error function can be formulated as a quadratic function of the unknown variables. The authors of [54] have used this result for the design of FIR filters which meet an error criteria of the general form

$$J(\mathbf{b}) = \frac{1}{\pi} \int_{\omega_1}^{\omega_2} W(e^{j\omega}) \left[D(e^{j\omega}) - H(e^{j\omega}) \right]^2 d\omega \qquad (2.184)$$

As a simple case, let the lower limit ω_1 be chosen as the lower value of the stopband and the upper limit ω_2 as the Nyquist frequency $(= \pi)$. In the above expression, $D(e^{j\omega})$ is the desired or prescribed response and $H(e^{j\omega})$ is the frequency response of the filter to be designed and hence is a function of the unknown coefficients h_n of the FIR filter. The weighting function $W(e^{j\omega})$ is often used to adjust the relative magnitude of the error over different range of the frequency and/or to improve convergence of the optimization algorithm. In the stopband, we note that $D(e^{j\omega}) = 0$ so that it is easy to express the error function in the form of (2.183), and minimize it simply by use of the Rayleigh's Principle. But when the error function has to be minimized over the passband or the entire frequency range $[0, \pi]$, (over which there may be several passbands and stopbands), the authors have proposed a novel modification to formulate the error function in the form (2.183). This modified error function is easily minimized thereafter. The authors of [33] have shown how the theory of eigen filters developed for the design of FIR filters in [54] can be extended for the design of allpass filters that approximate a constant group delay (or a linear phase response), i.e. allpass filters that compensate for the nonlinear phase response of an IIR filter to obtain a nearly linear phase response, and IIR filters that simultaneously approximate a constant group delay as well as a lowpass magnitude response. A review of [54] serves as an essential background for understanding the work reported by the authors of [33]. The formulation of an error function in a quadratic form for designing an IIR filter is not as straightforward as an FIR filter. We will closely follow the discussion given in [33] in the sequel.

Let us first consider the design of an allpass filter that approximates a prescribed constant group delay. Let its transfer function be denoted as

$$H_a(z) = \frac{a_N + \cdots + a_1 z^{-(N-1)} + a_0 z^{-N}}{a_0 + a_1 z^{-1} + \cdots + a_N z^{-N}} = z^{-N} \frac{D(z^{-1})}{D(z)} \qquad (2.185)$$

with its phase response $\Theta_a(\omega)$ given by

$$\begin{aligned}
\Theta_a(\omega) &= -N\omega + 2\tan^{-1}\left\{ \frac{\sum_{k=0}^{N} a_k \sin(k\omega)}{\sum_{k=0}^{N} a_k \cos(k\omega)} \right\} \\
&= -N\omega + 2\tan^{-1}\left\{ \frac{\mathbf{a}^t \mathbf{s}(\omega)}{\mathbf{a}^t \mathbf{c}(\omega)} \right\} \qquad (2.186)
\end{aligned}$$

where $\mathbf{a} = [a_0 \ a_1 \ \cdots a_N]^t$ is the vector of the coefficients in the allpass filter

and

$$\begin{aligned}\mathbf{s}(\omega) &= [0\ \sin(\omega)\ \sin(2\omega)\cdots\sin(N\omega)]^t \\ \mathbf{c}(\omega) &= [1\ \cos(\omega)\ \cos(2\omega)\cdots\cos(N\omega)]^t\end{aligned} \quad (2.187)$$

We use $\Delta\Theta(\omega)$ to represent the error between the prescribed phase $\Theta_{pre}(\omega)$ and the phase response $\Theta_a(\omega)$ of the allpass filter $H_a(z)$. The error function $E_1(\mathbf{a})$ to be minimized in the least squared sense is defined as

$$E(\mathbf{a}) = \int_R W(\omega)\,|\Theta_{pre}(\omega) - \Theta_a(\omega)|^2\,d\omega \quad (2.188)$$

where R is the frequency range of interest and $W(\omega)$ is a nonnegative weighting function. Defining $\beta(\omega) = \frac{1}{2}\{\Theta_{pre}(\omega) + N\omega\}$, we obtain

$$\begin{aligned}\frac{\Delta\Theta(\omega)}{2} &= \frac{\Theta_{pre}(\omega) - \Theta_a(\omega)}{2} \\ &= \frac{\Theta_{pre}(\omega) + N\omega}{2} - \tan^{-1}\left\{\frac{\sum_{k=0}^N a_k \sin(k\omega)}{\sum_{k=0}^N a_k \cos(k\omega)}\right\} \quad (2.189) \\ &= \beta(\omega) - \tan^{-1}\left\{\frac{\sum_{k=0}^N a_k \sin(k\omega)}{\sum_{k=0}^N a_k \cos(k\omega)}\right\} \quad (2.190)\end{aligned}$$

Using the identity $\tan(x-y) = \frac{\tan x - \tan y}{1+\tan x \tan y}$ on the right side of (2.190) we can derive (2.191).

$$E(\mathbf{a}) = \int_R W(\omega)\left|2\tan^{-1}\left\{\frac{\mathbf{a}^t\mathbf{s}_\beta(\omega)}{\mathbf{a}^t\mathbf{c}_\beta(\omega)}\right\}\right|^2 d\omega \quad (2.191)$$

where

$$\begin{aligned}\mathbf{s}_\beta(\omega) &= [\sin(\beta(\omega))\ \sin((\beta(\omega)-\omega))\ \sin((\beta(\omega)-2\omega))\cdots\sin((\beta(\omega)-N\omega))]^t \\ \mathbf{c}_\beta(\omega) &= [\cos(\beta(\omega))\ \cos((\beta(\omega)-\omega))\ \cos((\beta(\omega)-2\omega))\cdots\cos((\beta(\omega)-N\omega))]^t\end{aligned}$$

Yet, (2.191) is not in the quadratic form $\mathbf{a}^t\mathbf{P}\mathbf{a}$ that is required for using the eigen filter method, because it contains an arctangent function. So the authors make an approximation that $\arctan(x)$ is nearly equal to x, when x is small. Even with this approximation, it is not in the suitable quadratic form because the coefficient vector $\mathbf{a}$ appears both in the numerator and the denominator and they resort to another modification to the error function- which we will discuss a later. With the approximation $arctan(x) \cong x$, when x is small, the error function is chosen as

$$E_1(\mathbf{a}) \cong 4\int_R W(\omega)\left|\left\{\frac{\mathbf{a}^t\mathbf{S}_\beta(\omega)\mathbf{a}}{\mathbf{a}^t\mathbf{C}_\beta(\omega)\mathbf{a}}\right\}\right|d\omega \quad (2.192)$$

where $\mathbf{S}_\beta(\omega)$ and $\mathbf{C}_\beta(\omega)$ are matrices defined as

$$\begin{aligned}\mathbf{S}_\beta(\omega) &= [\mathbf{s}_\beta(\omega)]\,[\mathbf{s}_\beta(\omega)]^t \\ \mathbf{C}_\beta(\omega) &= [\mathbf{c}_\beta(\omega)]\,[\mathbf{c}_\beta(\omega)]^t\end{aligned} \quad (2.193)$$

2.5. DESIGN OF PULSE SHAPING FILTERS

respectively. The authors propose two approaches to resolve the problem that this error function has a quadratic form both in the numerator and denominator. The first approach is to neglect the denominator in (2.192) or to assume that it is nearly a constant. The error function then becomes

$$E_1(\mathbf{a}) \cong 4 \int_R W(\omega) \{\mathbf{a}^t \mathbf{S}_\beta(\omega) \mathbf{a}\} \, d\omega = \mathbf{a}^t \mathbf{P}_1 \mathbf{a} \qquad (2.194)$$

where

$$\mathbf{P}_1 = 4 \int_R W(\omega) \mathbf{S}_\beta(\omega) d\omega \qquad (2.195)$$

is a real valued, symmetric and positive-definite matrix. It is easy to satisfy the constraint $\mathbf{a}^t \mathbf{a} = 1$ so that the Rayleigh's Principle can now be invoked to find the eigen vector that corresponds to the smallest eigenvalue of $\mathbf{P}_1$. The (n, m) element of $\mathbf{P}_1$ is given by $[\mathbf{P}_1]_{n,m} =$

$$\int_R W(\omega) \{\cos[\omega(n-m)] - \cos[\Theta_{pre}(\omega) + N\omega - \omega(n+m)]\} \, d\omega \qquad (2.196)$$

If the frequency range of interest R consists of several bands over which the phase response $\Theta_{pre}(\omega)$ is prescribed, then the integration is carried out as the sum of integration over each of the bands.

Instead of assuming that the denominator term in (2.192) is a constant over the range R, a cost function $E_c(\mathbf{a})$ is added to the error function $E_1(\mathbf{a})$.

$$E_c(\mathbf{a}) = \int_R \left| \frac{d_c(\omega)}{d_c(\omega_0)} \mathbf{a}^t \mathbf{c}_\beta(\omega_0) - \mathbf{a}^t \mathbf{c}_\beta(\omega) \right|^2 d\omega =$$

$$\int_R \left[\frac{d_c(\omega)}{d_c(\omega_0)} \mathbf{a}^t \mathbf{c}_\beta(\omega_0) - \mathbf{a}^t \mathbf{c}_\beta(\omega) \right] \left[\frac{d_c(\omega)}{d_c(\omega_0)} \mathbf{a}^t \mathbf{c}_\beta(\omega_0) - \mathbf{a}^t \mathbf{c}_\beta(\omega) \right]^t d\omega$$

$$= \mathbf{a}^t \mathbf{P}_c \mathbf{a} \qquad (2.197)$$

where $d_c(\omega)$ is the desired response of $\mathbf{a}^t \mathbf{c}_\beta(\omega)$ and $\omega_0 \in R$ is a reference frequency properly chosen as explained in [54]. Therefore it is seen that the term $\frac{\mathbf{a}^t \mathbf{c}_\beta(\omega)}{d_c(\omega)}$ is forced to approximate the constant value $\frac{\mathbf{a}^t \mathbf{c}_\beta(\omega_0)}{d_c(\omega_0)}$ and the error function becomes zero when it is equal to the constant value. So, a new error function to be minimized is defined as the weighted sum of $E_1(\mathbf{a})$ and $E_c(\mathbf{a})$.

$$\begin{aligned} E_2(\mathbf{a}) &= \alpha E_1(\mathbf{a}) + (1-\alpha) E_c(\mathbf{a}) \\ &= \mathbf{a}^t [\alpha \mathbf{P}_1 + (1-\alpha) \mathbf{P}_c] \mathbf{a} = \mathbf{a}^t \mathbf{P}_2 \mathbf{a} \end{aligned} \qquad (2.198)$$

where α is a weighting parameter that can be chosen by trial and error or whatever past experience in programming for a similar design problem suggests. The matrix $\mathbf{P}_2$ is expressed by

$$\alpha \int_R W(\omega) \mathbf{S}_\beta(\omega) d\omega + (1-\alpha) \int_R [\mathbf{c}_\beta(\omega_0) - \mathbf{c}_\beta(\omega)][\mathbf{c}_\beta(\omega_0) - \mathbf{c}_\beta(\omega)]^t d\omega \qquad (2.199)$$

Again, if the frequency range R consists of several bands, then the integration is the sum of integrations over these bands; the reference frequency is properly chosen for each band. The eigenvector corresponding to the minimum eigenvalue of $\mathbf{a}^t\mathbf{P}_2\mathbf{a}$, (subject to the constraint $\mathbf{a}^t\mathbf{a}=1$) gives the coefficients of the allpass filter $H_a(z)$.

Notice that if the denominator term $\mathbf{a}^t\mathbf{C}_\beta(\omega)\mathbf{a}$ in (2.192) were known, it can be used as a weighting function. It is proposed by the authors that an iterative algorithm be used to minimize the following error function $E_3(\mathbf{a})$ wherein the weighting function in the $(q-1)^{th}$ step is used to obtain the value of the error function in the qth step and continue the iterative process till it converges e.g., till $|\Delta\mathbf{a}|^2 = \|\mathbf{a}_{(q)} - \mathbf{a}_{(q-1)}\|^2 \leq 10^{-4}$. The error function $E_3(\mathbf{a})$ at the qth step is given by

$$E_3(\mathbf{a}) = 4\int_R W(\omega)\left\{\frac{\mathbf{a}^t_{(q)}\mathbf{S}_\beta(\omega)\mathbf{a}_{(q)}}{\mathbf{a}^t_{(q-1)}\mathbf{C}_\beta(\omega)\mathbf{a}_{(q-1)}}\right\}d\omega = \mathbf{a}^t_{(q)}\mathbf{P}_3\mathbf{a}_{(q)} \qquad (2.200)$$

When and if the process converges, and when the linear approximation of the arctangent function is justified, we can expect that the final value for the coefficient vector a will yield the transfer function $H_a(z)$ for the allpass filter.

But minimization of the above two error functions $E_2^{(\mathbf{a})}$ and $E_3^{(\mathbf{a})}$ only leads to an approximation to the prescribed phase in the least squared sense. Instead, one could consider minimization of the least p^{th} error where p is a sufficiently large positive even integer, in order to obtain an equiripple approximation. However, Rayleigh's Principle can not be used for that formulation of the error function, though other optimization techniques can still be used. In order to minimize the maximum value of the phase error function, the authors have adapted the iterative weighting method that was used in [54], as an alternative approach. In this method, the weighting function $W(\omega)$ is chosen to be the envelope of the error curve at the previous step. Thus the new error function at the qth step is defined as

$$E_4(\mathbf{a}) = 4\int_R W_e(\omega)\left\{\frac{\mathbf{a}^t_{(q)}\mathbf{S}_\beta(\omega)\mathbf{a}_{(q)}}{\mathbf{a}^t_{(q-1)}\mathbf{C}_\beta(\omega)\mathbf{a}_{(q-1)}}\right\}d\omega = \mathbf{a}^t_{(q)}\mathbf{P}_4\mathbf{a}_{(q)} \qquad (2.201)$$

where the envelope of the phase error in the previous step as defined below, is the curve connecting the local maxima of $\Delta\Theta(\omega)$ in the $(q-1)^{th}$ step.

$$W_e(\omega) = \left[env(|\Delta\Theta_{(q-1)}|)\right]^p \qquad (2.202)$$

Though there is no proof that as p approaches ∞, the error function (2.201) approaches a equiripple or L_∞-norm approximation, the authors claim that when the iterative process converges for a large value of p, the solution obtained as the eigenvector corresponding to the minimum eigen value of $\mathbf{a}^t_{(q)}\mathbf{P}_4\mathbf{a}_{(q)}$ approaches the L_∞-norm approximation.

The authors have designed an allpass filter to equalize the nonlinear phase response of a lowpass filter. They have also designed a lowpass filter with a

2.5. DESIGN OF PULSE SHAPING FILTERS

constant group delay by choosing the structure shown in Fig. 2.14 and the prescribed phase repose as given in (2.129). Recollect that in [38], Remez Exchange Algorithm was used to minimize the error function. The same design specifications were used by the authors of [33] but the eigen filter method was used to minimize the least squares error functions $E_1(\mathbf{a})$, $E_2(\mathbf{a})$ and $E_3(\mathbf{a})$ and also the equiripple error function $E_4(\mathbf{a})$.

Given the order N of the allpass filter, and the desired phase response $\Theta_{pre}(\omega)$ which may be specified over several frequency bands in the range $(0, \pi)$, the design requires us to choose the initial value for the coefficient vector $\mathbf{a}$, the weighting function(s) $W(\omega)$, the weighting parameter α, reference frequency ω_0 (or reference frequencies over the different bands) and the value of p used in the envelope function. Based on their experience in extensive simulation of the above method, the authors have provided useful guidelines for choosing the values for the above items. They also point out that the phase response (or the corresponding group delay) must be prescribed carefully; otherwise the allpass filter may not be stable. The authors state the obvious fact that stability of IIR filters designed by least squares error minimization algorithms can not be guaranteed in general and the above design procedure is no exception. In contrast to the above method of equalizing the phase response of a lowpass filter to get a nearly linear phase characteristic, the method 2.3 described in sections 2.4.3 and 2.4.6 using a least squares approximation guarantees the stability of the filter, when we design lowpass filters with a maximally flat group delay. They attempt the values for $p = 2, 3, 4, 6, 8, 10, 15, 20, 25, 30, 35, 40, 45, 50$ and find that $p = 50$ yields a very good approximation for an equiripple phase response. But this approach increases the computation time, the computation of the envelope function becomes more complex, the number of iterations increases and the rate of convergence decreases as p is increased. However, by choosing a weighted average of the new and old value for the coefficient vector as indicated below, the number of iterations required was found to be considerably reduced, for each value of p.

$$\mathbf{a}_{(q)} = \frac{1}{p}[\mathbf{a}_{(q)} + (p-1)\mathbf{a}_{(q-1)}] \qquad (2.203)$$

2.5.9 Example 2.5

We consider an example from [33] to illustrate the theory of eigen filters. First assume that we have designed a 6th order, lowpass, Chebyshev II filter $H(z)$ to achieve an attenuation of 40 dB at the stopband cutoff frequency 0.3π. The magnitude response of this IIR filter is shown in Fig. 2.20.

Next a 6th order allpass filter is designed by choosing the third and fourth error functions $E_3(\mathbf{a})$ and $E_4(\mathbf{a})$, respectively yielding a least squares and equiripple approximation. The phase response of the Chebyshev II filter, denoted by $\Theta_H(\omega)$ is highly nonlinear and when it is cascaded by the allpass filter, it does not change the magnitude response of $H(z)$ but adds a phase so that the overall phase approximates a linear phase response given by $-\tau_0\omega$. This phase response corresponds to a group delay of τ_0. An empirical formula for an optimum value

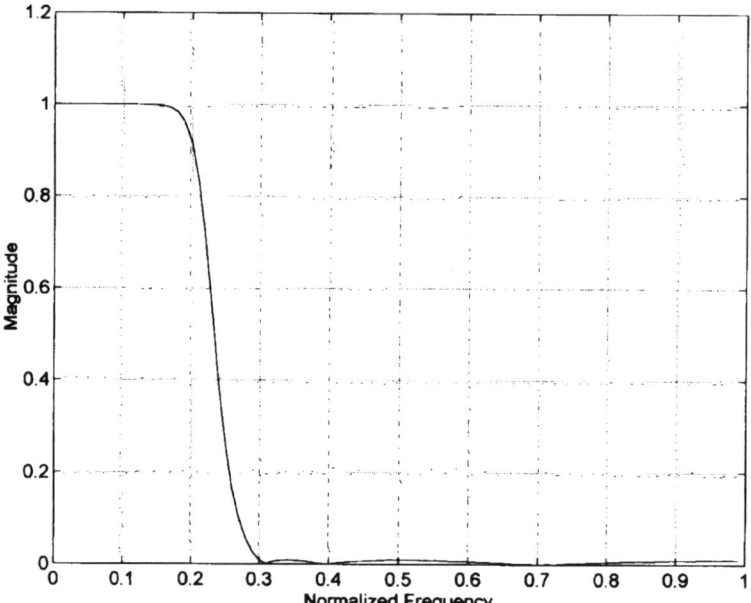

Fig. 2.20. Magnitude response of a lowpass filter in Example 2.5

of τ_0, based on extensive numerical experiments, has been proposed by the authors. It is given as

$$\tau_0 = 0.8 \left(\frac{N\pi - \Theta_H(\omega_U)}{\omega_U} \right) \qquad (2.204)$$

where ω_U is the upper cutoff frequency in the passband of $H(z)$ and $\Theta_H(\omega_U)$ is the phase of $H(z)$ at that frequency.

The group delay of the IIR filter $H(z)$ and the overall group delay obtained by minimizing the third and fourth error functions which approximate the constant group delay in the least squares and equiripple sense are shown respectively in Fig. 2.21. An enlarged view of the group delay over the passband is shown in Fig. 2.22. (The least squares approximation took only four iterations but the equiripple approximation took 26 iterations in solving the equalization problem.) The error resulting from the least squares solution is smaller than that from the equiripple solution as seen in Fig. 2.22.

Since the publication of [33], another paper [11] has reported some research work that uses the theory of eigen filters and the Remez Multiple Exchange Algorithm for the design of allpass filters. The formulation of the error function is nearly the same as in [33] but the algorithm makes an approximation of the linear phase of the filter instead of a constant group delay and uses a least squares approximation to get initial values for the filter coefficients and the equiripple approximation by using the Remez algorithm to complete the design.

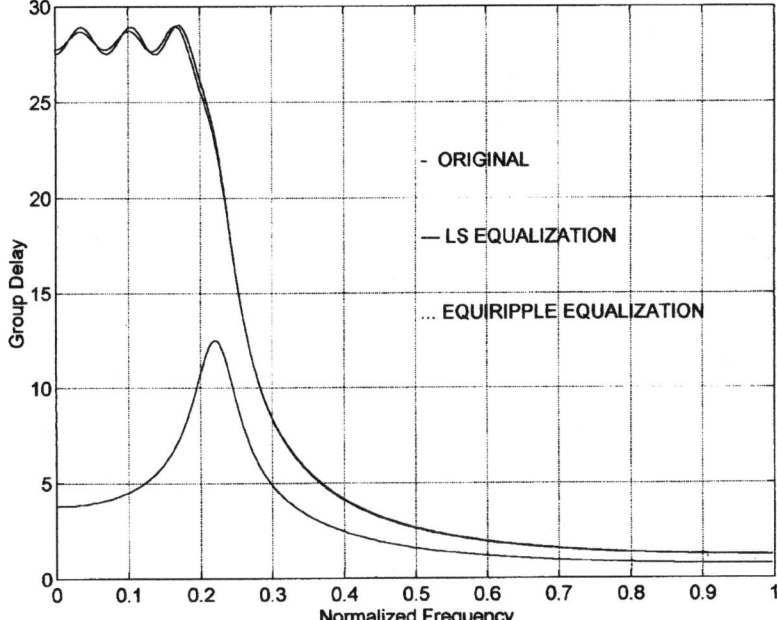

Fig. 2.21. Original and equalized group delays

2.6 Conclusion

In this chapter, we have described a few methods for the approximation of constant group delay or linear phase and several methods for the approximation of both constant group delay and magnitude response of IIR filters. Some of the methods develop elegant theory but practical implementation by computer programming shows that they do not present satisfactory results. Some methods are based mainly on linear and nonlinear programming techniques and they require a lot of computation. For example, both Deczky's method and the eigen filter method formulate the design problem of phase equalizers by connecting allpass filters in cascade with filters that approximate given magnitude response. But the error function to be minimized is a highly nonlinear function of the unknown coefficients and hence the nonlinear programming effort is tedious and complex. There exists no study making a comparison of the computational efficiency and complexity of these two methods, though the MATLAB$^{(R)}$ code is easily available for implementing the eigenfilter method [33]. From these points of view, it is claimed that the method proposed by [19, 22, 42] and [33] are very attractive. For this reason, they have been described in greater detail than the other methods. But the first two methods are suitable for approximating

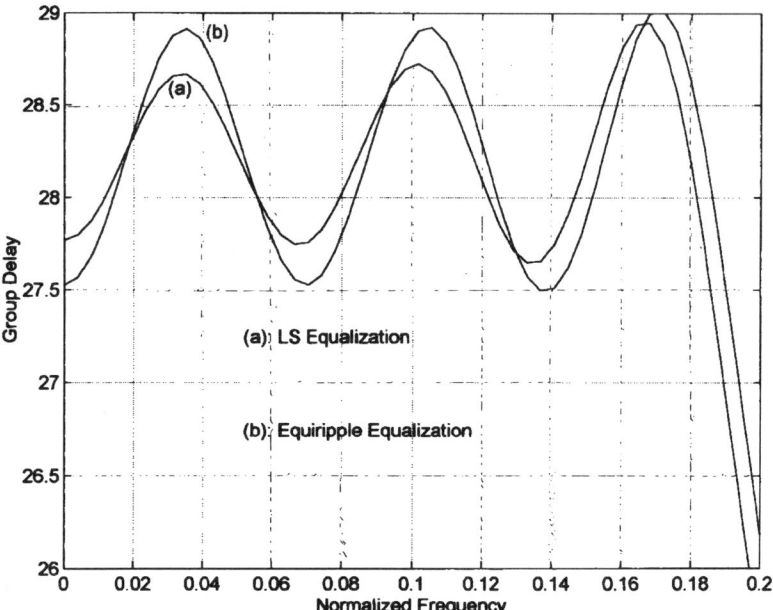

Fig. 2.22. Group delays after equalization

the magnitude that is constant over a finite frequency range, whereas the fourth method can be used to design phase equalizers as well. It should be remembered that there are other types of filter responses which are equally important in digital signal processing, for example, filters that are used to minimize intersymbol interference in digital data transmission [8, 10]. In describing the method 2.3, we carried out the design of a maximally flat delay filter with the required raised cosine magnitude characteristics for the data transmission filters. They can also be designed by using method 2.8. However, the method presented by [42] is not suitable for designing such filters. This leaves the methods of [22] and [33] as the two preferred tools for the design of IIR filters approximating both magnitude and group delay specifications.

Bibliography

[1] M. Abramowitz and I.A. Stegun, *Handbook of Mathematical Functions with Formulae, Graphs and Mathematical Tables*, National Bureau of Standards, 1964.

[2] V.M. Adamjan, D.Z. Arove and M.G. Krein, "Analytic properties of Schmidt pairs for Hankel operator and the generalized Schur-Takagi problem," Math.USSR Sbornik, vol. 15, pp. 15-73, 1971.

[3] ——ibid———, "Infinite block Hankel matrices and related extension problems," AMS Trans., vol. 111, pp. 133-156, 1978.

[4] S. Alliney and F. Sgallari, "Chebyshev approximation of recursive digital filters," Signal Processing, vol. 2, pp. 317-321, 1980.

[5] A. Antoniou, *Digital Filters: Analysis, Design, and Applications*, McGraw-Hill, 1993.

[6] A. Antoniou and W-S. Lu, "Design of two-dimensional digital filters by using the singular value decomposition," IEEE Trans. on Circuits and Systems, CAS-34, pp. 1191-1198, October 1987.

[7] H. Baher, *Synthesis of Electrical Networks*, John Wiley & Sons, 1984.

[8] H. Baher and J. Beneat, "Design of analog and digital data transmission filters," IEEE Trans. on Circuits and Systems, Part II. vol. 40, pp. 449-460, July 1993.

[9] A. Chaisawadi, T. Takebe, T. Matsumoto and K. Nishimura, "IIR partial response digital filter design with equiripple stopband attenuation, (Class I)," IEEE Trans. on Circuits and Systems, vol. 37, pp. 1209-1216, October 1990.

[10] A. Chaisawadi, T. Takebe, T. Matsumoto and K. Nishikawa, "IIR partial response digital filter design with equiripple stopband attenuation (Class I)," IEEE Trans. on Circuits and Systems, vol. 37, pp. 1200-1216, October 1990.

[11] C-K. Chen and C-C. Tseng, "Minimax design of digital allpass filters using multiple exchange eigen decomposition algorithms," Signal Processing, vol. 57, pp. 93-102, 1997.

[12] A.T. Chottera and G.A. Jullien, " A linear programming approach to recursive digital filter design with linear phase," IEEE Trans. on Circuits and Systems, CAS-29, pp. 139-149, March 1982.

[13] A.T. Chottera and G.A. Jullien, "Design of Two-dimensional recursive digital filters using Linear Programming," IEEE Trans. on Circuits and Systems, CAS-29, pp. 817-826, December 1982.

[14] User's Guide for SWITCHCAP, version 5, Columbia University.

[15] A.G. Deczky, "Synthesis of recursive digital filters using the minimum p error criterion," IEEE Trans. on Audio and Electroacoustics. AU-20, pp. 257-263, 1972.

[16] ——ibid——, "Recursive digital filters having equiripple group delay," IEEE Trans. on Circuits and Systems, CAS-21, pp. 131-134, Jan 1974.

[17] S. Ellacott and J. Williams, "Rational Chebyshev approximation in the complex plane," SIAM J. Numerical Analysis, vol. 13, no. 3, pp. 310-323, 1976.

[18] L. Fortuna, G. Nunnari and A. Gallo, *Model Order Reduction Techniques with Applications in Electrical Engineering*, Springer-Verlag, London, 1992.

[19] G. Gu and B.A. Shenoi, "A novel approach to the synthesis of recursive digital filters with linear phase," IEEE Trans. on Circuits and Systems, vol. 38, pp. 602-612, June 1991.

[20] Rajamohana Hegde, Design of Digital Filters Using the Maximally Flat Criterion, M.S.thesis, Wright State University, 1996

[21] Rajamohana Hegde and B.A. Shenoi, "Magnitude approximation of IIR digital filters with constant group delay response," Proc. IEEE Int'l Sympo. on Circuits and Systems, vol. IV, pp. 2200-2203, 1997.

[22] Rajamohana Hegde and B.A. Shenoi, "Magnitude approximation of digital filters with specified degrees of flatness and constant group delay characteristics," IEEE Trans. on Circuits and Systems, Part II, CAS-45 no. 11, pp. 1476-1486, Nov 1998.

[23] Simon Haykin, *Communication Systems*, John Wiley & Sons, 1983.

[24] T. Hinamoto and S. Maekawa , "Design of two-dimensional recursive digital filters using mirror image polynomials," IEEE Trans. on Circuits and Systems, CAS-33, pp. 750-758, 1986.

[25] D.S. Humpherys, *The Analysis Design and Synthesis of Electrical Filters*, Prentice-Hall, 1970.

[26] T. Inukai, "A unified approach to optimal recursive digital filter design," IEEE Trans. on Circuits and Systems, CAS-27, pp. 644-649, July 1980.

[27] A.T. Johnson, "Simultaneous magnitude and phase equalization using digital filters," IEEE Trans. on Circuits and Systems, CAS-25, pp. 319-321, 1978.

[28] P.T. Kabamba, "Balanced gains and their significance in L_2 model reduction," IEEE Trans. on Automatic Control, AC-30, pp. 690-693, 1985.

[29] W.S. Kafri, "Phase and delay approximation for 1-D digital IIR filter in the L_∞ norm," Signal Processing, vol. 57, pp. 163-175, 1997.

[30] G.A. Maria and M.M. Fahmy, "l_p Approximation of the group delay response of one- and two-dimensional filters," IEEE Trans. on Circuits and Systems. vol. CAS-21, pp. 431-436, 1974.

[31] ——ibid——, "A new design technique for recursive digital filters," IEEE Trans. on Circuits and Systems, CAS-23, pp. 323-325, May 1976.

[32] Z. Nehari, "On bounded linear forms," Ann. Math, vol. 65, pp. 153-162.

[33] T.Q. Nguyen, T.I. Laakso and R.D. Koilpillai, "Eigenfilter approach for the design of allpass filters approximating a given phase response," IEEE Trans. on Signal Processing, vol. 42, pp. 2257-2263, September 1994.

[34] K. Nishikawa and T. Takebe, "Maximally flat group delay low-pass IIR digital filters with flat passband and quasi-equiripple stopband attenuation characteristics," Trans. IECE, vol. J64-A, no. 10, pp. 819-826, 1981(in Japanese).

[35] T.W. Parks and J.H. McClellan, "Chebyshev approximation for non recursive digital filters with linear phase," IEEE Trans. on Circuit Theory, CT-19, pp. 189-194, March 1972.

[36] S-C. Pei and C-C. Tseng, "Design of equiripple log FIR and IIR filters using multiple exchange algorithm," Signal Processing, vol. 59, pp. 291-303, 1997.

[37] Staff of REA, *Handbook of Mathematical, Scientific and Engineering Formulas*, REA, 1994.

[38] M. Renfors and T. Saramaki, " A class of approximately linear phase filters composed of allpass sub filters," Proc. IEEE Int'l Sympo. on Circuits and Systems, pp. 678-681, 1986.

[39] J.D. Rhodes, *Theory of Electrical Filters*, John Wiley & Sons, 1976.

[40] T. Saramaki, Y. Neuvo and T. Saarinen, "Equal ripple amplitude and maximally flat group delay digital filters," Proc. IEEE Int'l Symposium on Circuits and Systems, pp. 236-239, 1981.

[41] T. Saramaki, "Design of optimum recursive digital filters with maximally flat passband and equiripple stopband magnitude," Int'l J. Circuit Theory and Applications, vol. 13, pp. 269-286, April 1985.

[42] T. Saramaki, "On the design of digital filters as a sum of two all-pass filters," IEEE Trans. on Circuits and Systems, CAS-32, pp. 1191-1193, Nov 1985.

[43] G. Strang, *Linear Algebra and its Applications*, Harcourt, Brace and Javanovich Inc. 1988.

[44] J.L. Sullivan and J.W. Adams, "PCLS IIR Digital Filters with simultaneous frequency response magnitude and group delay specifications," IEEE Trans. on Signal Processing, vol. 46, no. 11, pp. 2853-2861, November 1998.

[45] G. Szentirmai, "FILSYN-A general purpose filter synthesis program", Proc. IEEE, vol. 65, pp. 1443-1458, October 1977.

[46] G. Szentirmai, "Computer-aided design methods in filter design: S/FILSYN and other packages, *Handbook of Electrical Filters* (Chapter 3), CRC Press, Boca Raton, 1997.

[47] J.P. Thiran, "Recursive digital filters with maximally flat group delay," IEEE Trans. on Circuit Theory, CT-18, pp. 659-664, Nov. 1971.

[48] –ibid——, "Equal ripple delay recursive digital filters," ibid, . CT-18, pp. 664-669, Nov. 1971.

[49] R.E. Twogood and S.K. Mitra, "Computer-aided design of separable Two-dimensional digital filters," IEEE Trans. ASSP-25, pp. 165-169, April 1977.

[50] E. Ulbrich and H. Piloty, "Uber den Entwurf von Allpässen,Tiefpässen und Bandpässen mit einer im Tchebyscheffschen Sinne approximierten konstanten Gruppenlaufzeit," Arch. Elek.Übertragung, vol. 14, pp. 451-467, 1960.

[51] R. Unbehauen, "Recursive digital low-pass filters with maximally flat group delay and Chebyshev stop-band attenuation," Proc.IEEE Int'l Sympo. on Circuits and Systems, pp. 593-596, 1980.

[52] R. Unbehauen, "On the design of recursive digital lowpass filters with maximally flat passband and Chebyshev stopband attenuation," Proc. of IEEE Int'l Sympo. on Circuits and Systems, pp. 528-531, 1981.

[53] P.P. Vaidyanathan, S.K. Mitra and Y. Neuvo, "A new approach to the realization of low sensitivity IIR digital filters," IEEE Trans. ASSP-34, pp. 350-361, April 1986.

[54] P.P. Vaidyanathan and T.Q. Nguyen, "Eigenfilters: A new approach to least squares FIR filter design and applications including Nyquist filters," IEEE Trans. on Circuits and Systems, CAS-34, pp. 11-23, January 1987.

[55] J. Vlach, *Computerized Approximation and Synthesis of Linear Networks*, John Wiley & Sons, 1969.

Chapter 3

Magnitude Approximation of 2-D IIR Filters

3.1 Introduction

In this chapter, we describe several methods for the design of two-dimensional (2-D) IIR digital filters, which have passbands and stopbands defined by regions in the $\omega_1 - \omega_2$ plane. For the design of 1-D IIR filters, typically the given specifications are (1) the type of filters (e.g. lowpass,bandpass) (2) the type of response in the passband and stopbands (e.g. maximally flat or equiripple) (3) the cutoff frequency ω_c defining the passband and the maximum attenuation A_p (or ripple in dB) in the passband (4) the stopband cutoff frequency ω_s and the minimum attenuation A_s in the stopband and so on. For the design of 2-D IIR filters, an additional specification is the shape of the passband and stopband regions in the $\omega_1 - \omega_2$ plane instead of the cutoff frequencies ω_c and ω_s. This additional specification makes a significant difference in the methods for designing 2-D filters in the sense that when all the other specifications are the same, the design methods will be different if the passband and stopband regions are different. Some of the common types of regions are those bounded by straight lines and the methods for the design of such filters are discussed in sections 3.2 and 3.3 of this chapter. Another type of passband and stopband are circular or elliptical regions in the $\omega_1-\omega_2$ plane. In section 3.4, a few well-known methods for the design of filters with circular passband or stopband regions in the $\omega_1 - \omega_2$ plane are discussed. Once such filters are designed, it is easy to design filters with new shapes and in different part of the frequency plane. For example, using some spectral transformations [6, 25] digital filters with an elliptical passband or stopband can be obtained, from the design of circularly symmetric lowpass filters. These spectral transformations will be considered at the end of the chapter.

3.2 Filters with Rectangular Passbands

In this section, we consider the design of 2-D IIR filters which have their passband and stopband regions enclosed by straight lines. One class of such filters have rectangular regions. The second class of filters have regions defined by straight lines inclined to the ω_1 and ω_2 axes, for example, fan filters and filters with diamond shaped regions for their passbands.

Let us first consider the simple case of filters which have passband and stopband regions bounded by straight lines parallel to the ω_1 and ω_2 axes. Let $H_1(z_1)$ be the transfer function of a one- dimensional IIR digital filter, as given by

$$H_1(z_1) = \frac{\sum_{i=0}^{N} a_{1i} z_1^{-i}}{\sum_{i=0}^{M} b_{1i} z_1^{-i}} \tag{3.1}$$

Assume that it has an ideal lowpass magnitude response i.e.

$$\left|H_1(e^{j\omega_1})\right| = \begin{cases} 1 & \text{for} \quad |\omega_1| \leq \omega_{1p} \\ 0 & \text{for} \quad \omega_{1p} < |\omega_1| \leq \pi \end{cases} \tag{3.2}$$

Instead of considering this as a function of the frequency variable ω_1 only, we can consider it as a function of the two frequency variables ω_1 and ω_2 but varying in the direction of ω_1 only, i.e.

$$\left|H(e^{j\omega_1}, e^{j\omega_2})\right| = \left|H_1(e^{j\omega_1})\right| \tag{3.3}$$

Thus, the function (3.3) can be considered as the magnitude of the two-dimensional transfer function which is independent of ω_2.

It can be plotted as a vertical strip of width $2\omega_{1p}$ in the the $\omega_1 - \omega_2$ plane, as shown in Fig. 3.1a. It should be noted that in the following figures, crossed areas represent passbands with a magnitude of one and the remaining areas are stopbands with a zero magnitude.

In a similar manner, the functions (3.4) and (3.5)

$$H_2(z_2) = \frac{\sum_{i=0}^{N} a_{2i} z_2^{-i}}{\sum_{i=0}^{M} b_{2i} z_2^{-i}} \tag{3.4}$$

$$\left|H_2(e^{j\omega_2})\right| = \begin{cases} 1 & \text{for} \quad |\omega_2| \leq \omega_{2p} \\ 0 & \text{for} \quad \omega_{2p} < |\omega_2| \leq \pi \end{cases} \tag{3.5}$$

represent a 2-D filter with a passband in the $\omega_1 - \omega_2$ plane, which is a horizontal strip of width $2\omega_{2p}$ as shown in Fig. 3.1b.

Now if these two filters are cascaded, we get a 2-D filter with the magnitude given by

$$\begin{aligned} \left|H_l(e^{j\omega_1}, e^{j\omega_2})\right| &= \left|H_1(e^{j\omega_1})\right| \left|H_2(e^{j\omega_2})\right| \\ &= \begin{cases} 1 & \text{for} \quad |\omega_1| \leq \omega_{1p} \text{ and } |\omega_2| \leq \omega_{2p} \\ 0 & \text{for} \quad \omega_{1p} < |\omega_1| \leq \pi \text{ and } \omega_{2p} < |\omega_2| \leq \pi \end{cases} \end{aligned} \tag{3.6}$$

3.2. FILTERS WITH RECTANGULAR PASSBANDS

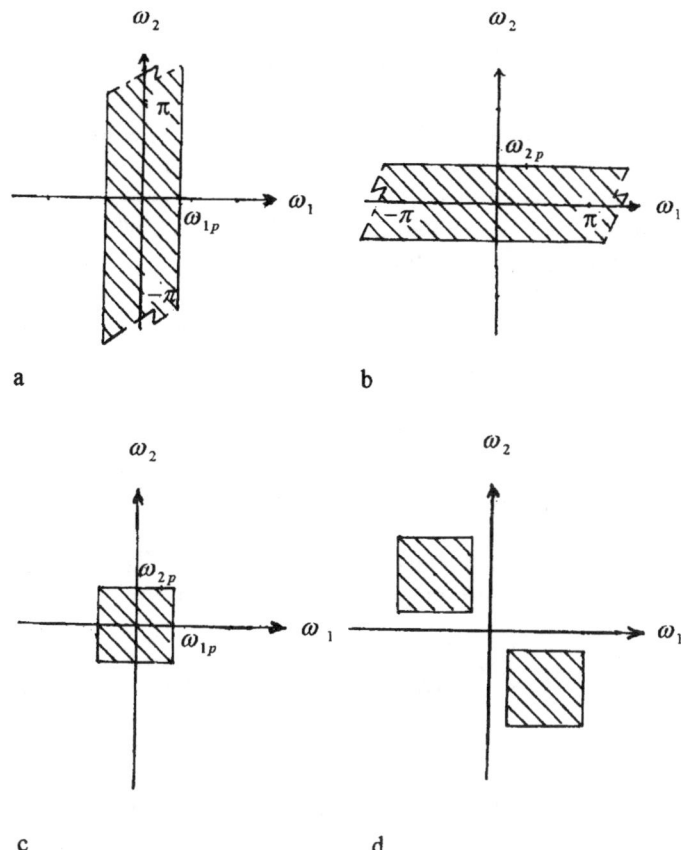

Fig. 3.1. Magnitude response of different 2-D filters. **a** A 2-D lowpass filter independent of ω_2 **b** A 2-D lowpass filter independent of ω_1 **c** A 2-D lowpass filter **d** A half-plane 2-D filter

As shown in Fig. 3.1c, this represents a lowpass filter having a passband which is a rectangular region of width $2\omega_{1p}$ and $2\omega_{2p}$, centered at the origin.

If the prescribed passband of the 2-D lowpass filter is a rectangle centered at other points $(\pm\omega_{1c}, \pm\omega_{2c})$ in one of the four quadrants, we can obtain it by changing the frequency variable ω_1 to $(\omega_1 \pm \omega_{1c})$ and ω_2 to $(\omega_2 \pm \omega_{2c})$. For example, if we transform the variable z_1 to $(z_1 e^{-j\omega_{1c}})$ in $H_1(z_1)$ and z_2 to $(z_2 e^{j\omega_{2c}})$ in $H_2(z_2)$, the corresponding shift in the frequencies, as shown by (3.7) yields a passband which is a rectangle with the same bandwidth but centered at $\omega_1 = \omega_{1c}, \omega_2 = -\omega_{2c}$ in the fourth quadrant. Thus the resulting filter

$$H_4(z_1 e^{-j\omega_{1c}}, z_2 e^{j\omega_{2c}}) = H_1(z_1 e^{-j\omega_{1c}}) H_2(z_2 e^{j\omega_{2c}}) \qquad (3.7)$$

has a magnitude given by

$$\left|H_4(e^{j(\omega_1-\omega_{1c})}, e^{j(\omega_2+\omega_{2c})})\right| = \left|H_1(e^{j(\omega_1-\omega_{1c})})\right|\left|H_2(e^{j(\omega_2+\omega_{2c})})\right|$$

The coefficients of the transfer function $H_4(z_1 e^{-j\omega_{1c}}, z_2 e^{j\omega_{2c}})$ are complex valued. One common method to obtain a transfer function with real coefficients is to connect another transfer function $H_4(z_1^{-1} e^{j\omega_{1c}}, z_2^{-1} e^{-j\omega_{2c}})$ in cascade or in parallel. This function can also be denoted as $H_4^*(z_1^{-1} e^{-j\omega_{1c}}, z_2^{-1} e^{j\omega_{2c}})$, where the asterisk is used to imply that the coefficients are chosen to be the complex conjugates of those in $H_4(z_1 e^{-j\omega_{1c}}, z_2 e^{j\omega_{2c}})$.

The filter function obtained by connecting the above two functions in parallel or in cascade results in functions (3.8) and (3.9) with real coefficients.

$$H_4(z_1 e^{-j\omega_{1c}}, z_2 e^{j\omega_{2c}}) H_4(z_1^{-1} e^{j\omega_{1c}}, z_2^{-1} e^{-j\omega_{2c}}) \tag{3.8}$$

or

$$H_4(z_1 e^{-j\omega_{1c}}, z_2 e^{j\omega_{2c}}) + H_4(z_1^{-1} e^{j\omega_{1c}}, z_2^{-1} e^{-j\omega_{2c}}) \tag{3.9}$$

If the filters are denoted as $\widehat{H}_4(z_1, z_2) = H_4(z_1 e^{-j\omega_{1c}}, z_2 e^{j\omega_{2c}})$ and $\widehat{H}_4^*(z_1^{-1}, z_2^{-1}) = H_4^*(z_1^{-1} e^{-j\omega_{1c}}, z_2^{-1} e^{j\omega_{2c}})$, it is easy to see that they have the same magnitude but opposite phase.

$$\left|H_4(e^{j(\omega_1-\omega_{1c})}, e^{j(\omega_2+\omega_{2c})})\right| = \left|H_4(e^{-j(\omega_1-\omega_{1c})}, e^{-j(\omega_2+\omega_{2c})})\right|$$

But

$$Arg\left\{H_4(e^{j(\omega_1-\omega_{1c})}, e^{j(\omega_2+\omega_{2c})})\right\} = -Arg\left\{H_4(e^{-j(\omega_1-\omega_{1c})}, e^{-j(\omega_2+\omega_{2c})})\right\} \tag{3.10}$$

Then the function $H_{24}(z_1, z_2) = \widehat{H}_4(z_1, z_2) \widehat{H}_4^*(z_1^{-1}, z_2^{-1})$ will have a magnitude of one over the passband in the fourth quadrant and it also has a zero phase.

Similarly let

$$\widehat{H}_3(z_1, z_2) = H_4(z_1 e^{j\omega_{1c}}, z_2 e^{-j\omega_{2c}})$$
$$\widehat{H}_3^*(z_1^{-1}, z_2^{-1}) = H_4^*(z_1^{-1} e^{j\omega_{1c}}, z_2^{-1} e^{-j\omega_{2c}})$$

Then $H_{13}(z_1, z_2) = \widehat{H}_3(z_1, z_2)\widehat{H}_3^*(z_1^{-1}, z_2^{-1})$ has a magnitude of one in the second quadrant and has zero phase. Now we can connect them in parallel - since they are zero phase functions, and we get a filter (3.11).

$$H_{13}(z_1, z_2) + H_{24}(z_1 z_2) \tag{3.11}$$

which has a magnitude of one in the second and fourth quadrant as shown in Fig. 3.1d.

It is pointed here that if we cascade $H_{13}(z_1, z_2)$ and $H_{24}(z_1 z_2)$, the magnitude of $H_{13}(z_1, z_2) H_{24}(z_1 z_2)$ will be zero everywhere in the $\omega_1 - \omega_2$ plane, because the passband and stopband regions of the two filters intersect with each other.

3.2. FILTERS WITH RECTANGULAR PASSBANDS

Whereas the ideal magnitude response has a sharp boundary between the passband and the stopband, the above transfer functions can only approximate the ideal magnitude response, with a passband defined by the maximum attenuation, a stopband defined by a minimum attenuation and a transition band between them. Though we will assume, in the following paragraphs, that the magnitude responses are ideal with sharp boundaries between the passband and stopband regions, and discuss several new design procedures, we have to remember the effect of the non-ideal nature of the frequency responses in the actual design of these 2-D filters when the ideal passband regions are close to each other. One way of taking care of this overlapping problem is to choose a high order function for the 1-D digital filter with narrow transition band and to choose a passband smaller than the ideal region so that over the stopband regions, the attenuation is very high. A detailed discussion of these effects is found in [20] and [22].

In [17], Hirano and Aggarwal used a different and sometimes simpler method to design some of the above filters with quadrantal and half plane symmetries in their magnitude response. To describe their method, let us consider two 1-D transfer functions which have the ideal bandpass responses as given by

$$\left|H_1(e^{j\omega_1})\right|_{bp} = \begin{cases} 0 & \text{for } 0 \le |\omega_1| < \omega_{11} \\ 1 & \text{for } \omega_{11} < |\omega_1| < \omega_{12} \\ 0 & \text{for } \omega_{12} < |\omega_1| \le \pi \end{cases} \quad (3.12)$$

and

$$\left|H_2(e^{j\omega_2})\right|_{bp} = \begin{cases} 0 & \text{for } 0 \le |\omega_2| < \omega_{21} \\ 1 & \text{for } \omega_{21} < |\omega_2| < \omega_{22} \\ 0 & \text{for } \omega_{22} < |\omega_2| \le \pi \end{cases} \quad (3.13)$$

Now when we cascade these two 1-D filters, we obtain an intersection of the passband regions of the two bandpass filters as indicated in Fig. 3.2a. The magnitude is one over the intersecting region and zero elsewhere. The resulting 2-D transfer function has a passband in the four quadrants as shown in Fig. 3.2b. Its magnitude is given by $\left|H(e^{j\omega_1}, e^{j\omega_2})\right|$

$$\begin{aligned} &= \left|H_1(e^{j\omega_1})\right|_{bp} \left|H_2(e^{j\omega_2})\right|_{bp} \\ &= \begin{cases} 1 & \text{for } \omega_{11} \le |\omega_1| \le \omega_{12} \text{ and } \omega_{21} \le |\omega_2| \le \omega_{22} \\ 0 & \text{otherwise} \end{cases} \end{aligned} \quad (3.14)$$

Suppose we are required to design a 2-D bandpass filter which has a passband in the region as shown in Fig. 3.3a or a bandstop filter with a passband as shown in Fig. 3.3b. These filters can not be easily designed by cascading 1-D bandpass or bandstop filters.

Hirano and Aggarwal have proposed a modification to the above method which still uses a cascade connection of 1-D IIR filters as building blocks and which can be used to design the 2-D bandpass and bandstop filters shown in Fig. 3.3a and Fig. 3.3b.

Consider the cascade connection of two 1-D IIR filters as $H_1(z_1)H_2(z_2) = H(z_1, z_2)$, where each of the filters $H_1(z_1)$ and $H_2(z_2)$ can be lowpass, highpass,

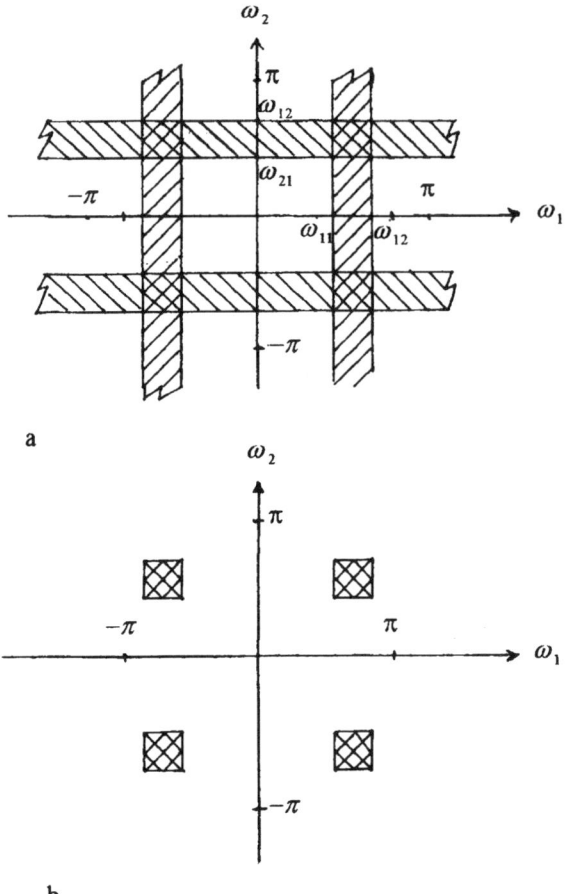

Fig. 3.2. Cascade connection of two bandpass filters

bandpass or bandstop filters. Hence we can obtain 16 digital filters which have their passbands as shown in Figs 3.4-3.7. In these figures, as before, the 1-D IIR filters are assumed to be ideal with a magnitude of one in the passband and zero in the stopband. The widths of these passband and stopband are determined by the given specifications, though in these figures, the same width is assumed for their passband and stopbands.

Let us assume the transfer functions for the 16 filters to be in the form

$$H_1(z_1)H_2(z_2) = \frac{N_1(z_1)N_2(z_2)}{D_1(z_1)D_2(z_2)} \qquad (3.15)$$

Now we choose two 1-D allpass digital transfer functions $H_{A1}(z_1)$ and $H_{A2}(z_2)$

3.2. FILTERS WITH RECTANGULAR PASSBANDS

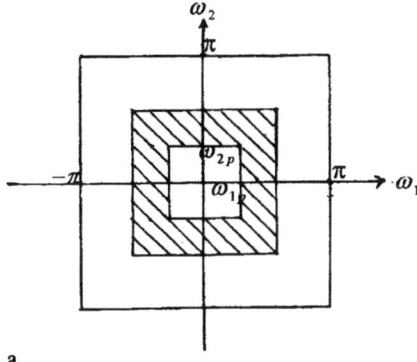

a

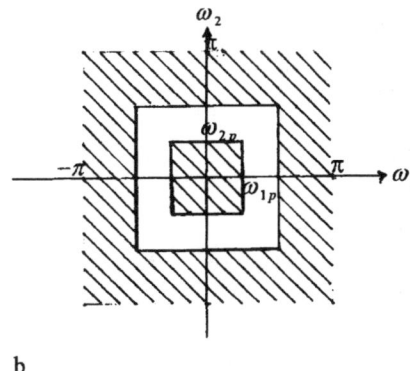

b

Fig. 3.3. Magnitude response of ideal 2-D filters. **a** Bandpass filter **b** Bandstop filter

which have the same poles as (3.15). Therefore they are of the form

$$H_{A1}(z_1) = \frac{z_1^K D_1(z_1^{-1})}{D_1(z_1)} \tag{3.16}$$

and

$$H_{A2}(z_2) = \frac{z_2^R D_2(z_2^{-1})}{D_2(z_2)} \tag{3.17}$$

where K and R are the degrees of the polynomials $D_1(z_1)$ and $D_2(z_2)$ respectively. Their magnitude response is one for all values of the frequency and their phase response is $K\omega_1 - 2\theta_{D_1}$ and $R\omega_2 - 2\theta_{D_2}$ respectively. Hence

$$H_{A1}(e^{j\omega_1}) = e^{j(K\omega_1 - 2\theta_{D_1})} \tag{3.18}$$
$$H_{A2}(e^{j\omega_2}) = e^{j(R\omega_2 - 2\theta_{D_2})} \tag{3.19}$$

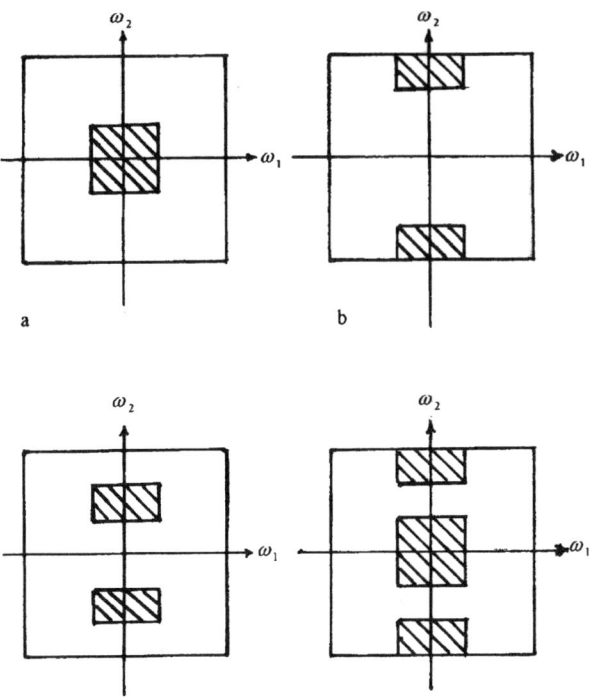

Fig. 3.4. Magnitude response of ideal 2-D filters. **a** $H_{lp}(z_1)H_{lp}(z_2)$ **b** $H_{lp}(z_1)H_{hp}(z_2)$ **c** $H_{lp}(z_1)H_{bp}(z_2)$ **d** $H_{lp}(z_1)H_{bs}(z_2)$

Let

$$\frac{N_1(e^{j\omega_1})}{D_1(e^{j\omega_1})} = \frac{M_{N1}(\omega_1)e^{j\theta_{N1}(\omega_1)}}{M_{D1}(\omega_1)e^{j\theta_{D1}(\omega_1)}} \qquad (3.20)$$

$$\frac{N_2(e^{j\omega_2})}{D_2(e^{j\omega_2})} = \frac{M_{N2}(\omega_2)e^{j\theta_{N2}(\omega_2)}}{M_{D2}(\omega_2)e^{j\theta_{D2}(\omega_2)}} \qquad (3.21)$$

Now if we construct a 2-D digital filter that is given by

$$H(z_1, z_2) = H_{A1}(z_1)H_{A2}(z_2) - e^{-jm\pi}\left[\frac{N_1(z_1)N_2(z_2)}{D_1(z_1)D_2(z_2)}\right]^2 \qquad (3.22)$$

where m is an independent integer variable, we see that its frequency response is given by $H(e^{j\omega_1}, e^{j\omega_2}) =$

$$e^{j(K\omega_1 + R\omega_2 - 2\theta_{D1} - 2\theta_{D2})} - \left[\frac{M_{N1}(\omega_1)M_{N2}(\omega_2)}{M_{D1}(\omega_1)M_{D2}(\omega_2)}\right]^2 e^{j\psi(\omega_1,\omega_2) - jm\pi} \qquad (3.23)$$

where $\psi(\omega_1, \omega_2) = 2\theta_{N1} + 2\theta_{N2} - 2\theta_{D1} - 2\theta_{D2}$. So if the following condition is satisfied,

$$(K\omega_1 + R\omega_2) = 2[\theta_{N1}(\omega_1) + \theta_{N2}(\omega_2)] - m\pi \qquad (3.24)$$

3.2. FILTERS WITH RECTANGULAR PASSBANDS

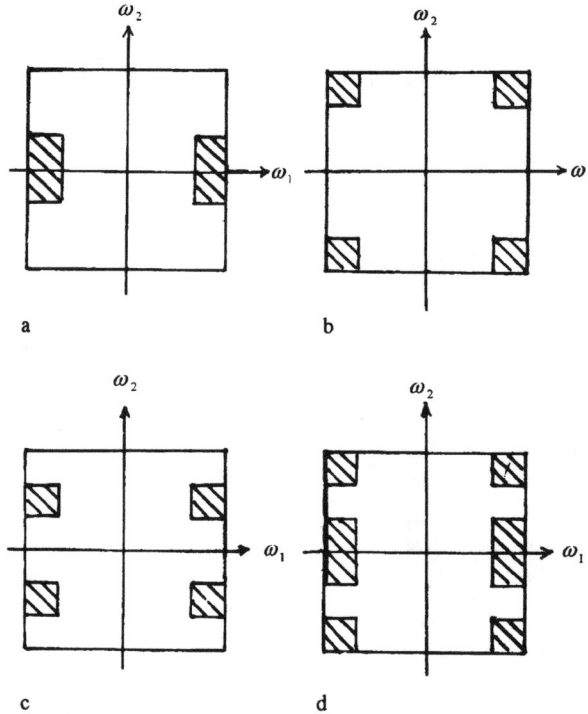

Fig. 3.5. Magnitude response of ideal 2-D filters. **a** $H_{hp}(z_1)H_{lp}(z_2)$ **b** $H_{hp}(z_1)H_{hp}(z_2)$ **c** $H_{hp}(z_1)H_{bp}(z_2)$ **d** $H_{hp}(z_1)H_{bs}(z_2)$

the phase of the two terms on the right side of (3.23) are equal, so we have the final result ($H(e^{j\omega_1}, e^{j\omega_2})$ given by

$$\left\{1 - \left[\frac{M_{N1}(\omega_1)M_{N2}(\omega_2)}{M_{D1}(\omega_1)M_{D2}(\omega_2)}\right]^2\right\} e^{j[K\omega_1+R\omega_2-2\theta_{D1}(\omega_1)-2\theta_{D2}(\omega_2)]} \qquad (3.25)$$

Since we have chosen $H_1(e^{j\omega_1})H_2(e^{j\omega_2})$ with the magnitude of one in the passband and zero in the stopband, it is seen that the magnitude of the filter (3.25) is complementary to that of $H_1(e^{j\omega_1})H_2(e^{j\omega_2})$; hence we are able to design another set of 16 filters which have passbands and stopbands which are complementary to those of Figs 3.4-3.7.

This operation of using 2-D allpass filters to get the complementary magnitude will be called the complementary operation. It is useful in many ways. For example, suppose we again consider the design of a 2-D filter with the ideal passband and stopband as shown in Fig. 3.3a. First by cascading 1-D lowpass filters, we design a 2-D lowpass filter $H_l(z_1, z_2)$ with a passband of width $2\omega_{1p}$ and $2\omega_{2p}$. See Eq.(3.6). Using the complementary operation, we design a highpass filter $H_h(z_1, z_2)$ with a stopband as shown in Fig. 3.8.

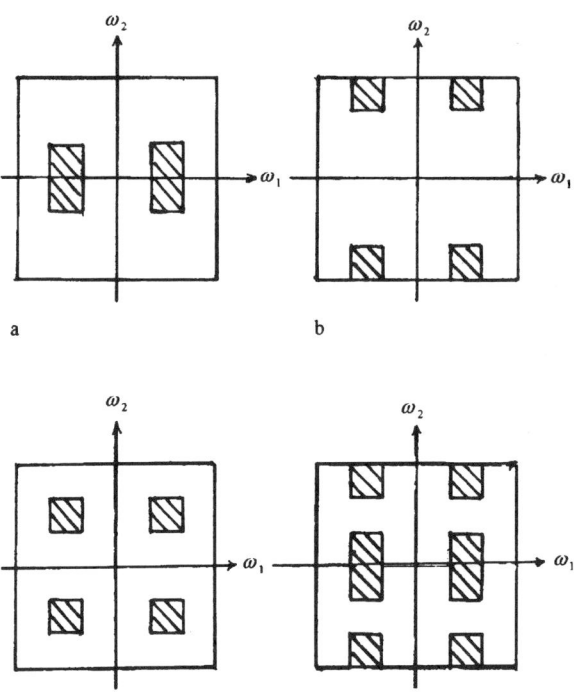

Fig. 3.6. Magnitude response of ideal 2-D filters. **a** $H_{bp}(z_1)H_{lp}(z_2)$ **b** $H_{bp}(z_1)H_{hp}(z_2)$ **c** $H_{bp}(z_1)H_{bp}(z_2)$ **d** $H_{bp}(z_1)H_{bs}(z_2)$

Then by cascading it with another lowpass filter having a width of ω_{1h} and ω_{2h} (see Fig. 3.9c), we get a 2-D bandpass filter with a magnitude response shown in Fig. 3.3a. Though the magnitude response of the bandstop filter in Fig. 3.3b is complementary to that of the bandpass filter of Fig. 3.3a, we can not obtain its transfer function by applying the complementary operation on the transfer function $H_l(z_1, z_2)H_h(z_1, z_2)$, because $H_h(z_1, z_2)$ obtained by the complementary operation, is not separable and hence can not be expressed in the form of (3.15). However, the bandstop filter can be obtained by the parallel connection of a lowpass filter $H_l(z_1, z_2)$ and the highpass filter $H_h(z_1, z_2)$. As another example, Fig. 3.9a shows the magnitude response of a 2-D filter obtained by cascading a highpass filter $H_{hp}(z_1)$ with a highpass filter $H_{hp}(z_2)$. By using the complementary operation described above, we can obtain a filter with its response shown in Fig. 3.9b. When this 2-D filter is cascaded with the lowpass filter of Fig. 3.9c, we get the filter with the response shown in Fig. 3.9d.

For a successful implementation of the complementary operation, we need two conditions. (1) The filter function $H(z_1, z_2)$ which has a passband complementary to what is specified must be a separable function like (3.15) (2) The condition (3.24) must be satisfied. The 1-D IIR filters designed by applying

3.2. FILTERS WITH RECTANGULAR PASSBANDS

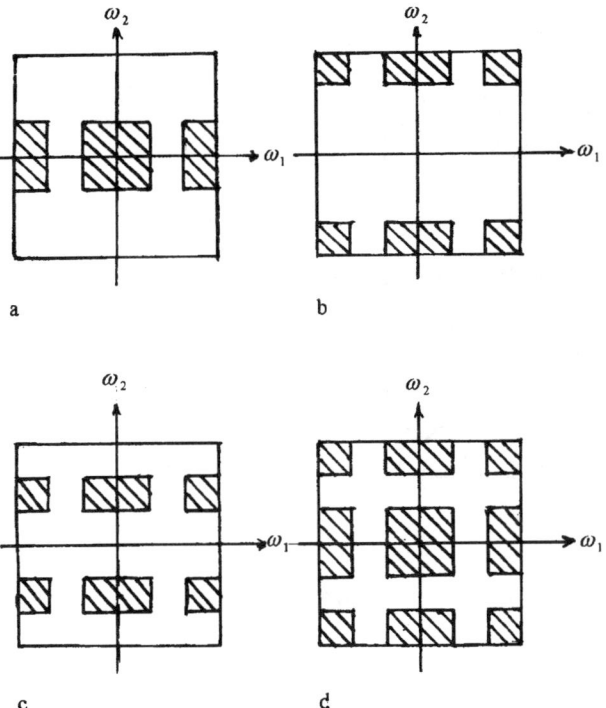

Fig. 3.7. Magnitude response of ideal 2-D filters. **a** $H_{bs}(z_1)H_{lp}(z_2)$ **b** $H_{bs}(z_1)H_{hp}(z_2)$ **c** $H_{bs}(z_1)H_{bp}(z_2)$ **d** $H_{bs}(z_1)H_{bs}(z_2)$

the bilinear transformation on such classical analog filters as the Butterworth or Chebyshev type, have numerators which have zeros at $z = \pm 1$. The phase response of these filter functions is a nonlinear function of ω in general; hence such filter functions will not satisfy (3.24). But in Chap. 2, we described the design of 1-D IIR lowpass filters which have Thiran's polynomial as the denominator and a mirror image polynomial as the numerator in order to approximate both a maximally flat delay and a maximally flat magnitude response. The numerator of this filter has a pure delay term multiplied by the mirror image polynomial. Hence it has almost a linear phase in the passband. This class of filters may be suited to satisfy the condition (3.24). We showed that the 1-D highpass filters also can likewise be easily obtained from the lowpass filters such that they approximate a maximally flat delay and a maximally flat magnitude response as well. Therefore, the design of the filter shown in Fig. 3.8 can be completed easily. But it must be pointed out that the method discussed in Chap. 2 does not provide a procedure for obtaining bandpass and bandstop filters with a constant group delay. Because of the limitations of the complementary operation, as described above, a simpler method - as described next - is used more often. A lowpass filter $H(z_1, z_2)$ is designed first and then a filter with zero

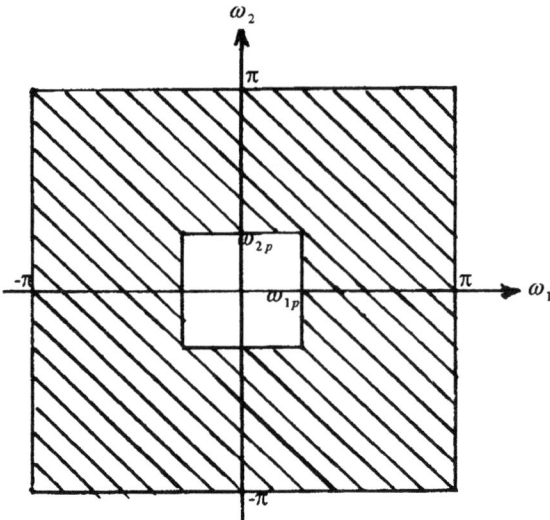

Fig. 3.8. Magnitude response of an ideal, 2-D highpass filter

phase is obtained as the product $H(z_1,z_2)H^*(z_1^{-1},z_2^{-1})$ which has a magnitude $\left|H(e^{j\omega_1},e^{j\omega_2})\right|^2$. Then the transfer function $1-H(z_1,z_2)H^*(z_1^{-1},z_2^{-1})$ will have a highpass response. One could also design the lowpass filter with a magnitude which is the square root of the prescribed magnitude - if it is not one - so the response of $H(z_1,z_2)H^*(z_1^{-1},z_2^{-1})$ in the passband is equal to the prescribed magnitude of the lowpass filter and at the same time has zero phase.

Hirano and Aggarwal proposed another method for designing 2-D filters with half plane symmetry. Consider the design of a 2-D lowpass filter as shown in Fig. 3.10a. This is the special case of the lowpass filter (3.6) in which $\omega_{1p} = \omega_{2p} = \frac{\pi}{2}$, i.e. (3.6) changes to (3.26) given below:

$$\left|H_l(e^{j\omega_1},e^{j\omega_2})\right| = \begin{cases} 1 & \text{for } |\omega_1| \leq \frac{\pi}{2} \text{ and } |\omega_2| \leq \frac{\pi}{2} \\ 0 & \text{for } \frac{\pi}{2} < |\omega_1| \leq \pi \text{ and } \frac{\pi}{2} < |\omega_2| \leq \pi \end{cases} \quad (3.26)$$

It is easy to see that by shifting the frequency response by $\pm\frac{\pi}{2}$ in both ω_1 and ω_2 directions, we get the magnitude response which is equal to one in the complete region of one of the four quadrants. As an example, starting from a 1-D digital filter $H(z)$ with a passband from $-\frac{\pi}{2}$ to $\frac{\pi}{2}$ as shown in Fig. 3.14a, we get $H(z_1,z_2) = H(z_1)H(z_2)$. Then we use (3.27) and (3.28), to get the passband response with a magnitude of one in the second and fourth quadrants respectively[19].

$$H_{22}(z_1,z_2) = H(z_1 e^{j\frac{\pi}{2}})H(z_2 e^{-j\frac{\pi}{2}}) = H(jz_1,-jz_2) \quad (3.27)$$
$$H_{44}(z_1,z_2) = H(z_1 e^{-j\frac{\pi}{2}})H(z_2 e^{j\frac{\pi}{2}}) = H(-jz_1,jz_2) \quad (3.28)$$

3.2. FILTERS WITH RECTANGULAR PASSBANDS

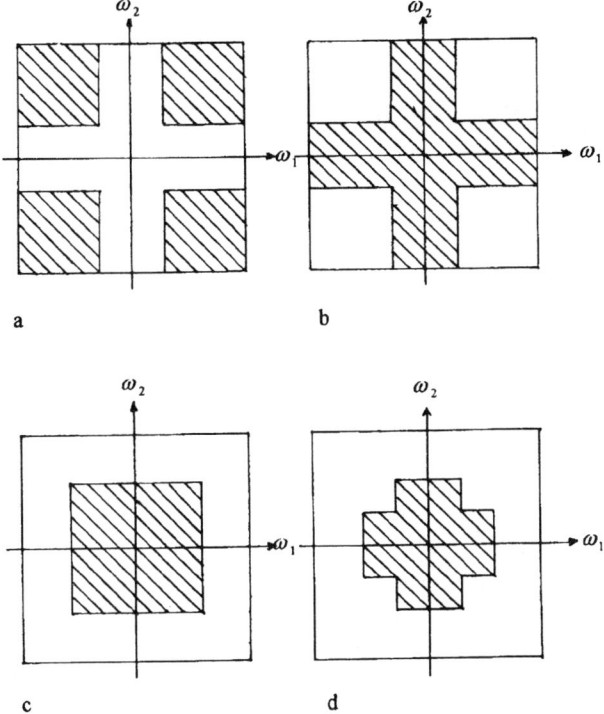

Fig. 3.9. Magnitude response of ideal 2-D filters

Similarly we can obtain filters with a passband in the first and the third quadrants, i.e.

$$H_{11}(z_1, z_2) = H(z_1 e^{-j\frac{\pi}{2}}) H(z_2 e^{-j\frac{\pi}{2}}) = H(-jz_1, -jz_2) \quad (3.29)$$
$$H_{33}(z_1, z_2) = H(z_1 e^{j\frac{\pi}{2}}) H(z_2 e^{j\frac{\pi}{2}}) = H(jz_1, jz_2) \quad (3.30)$$

Note again that the coefficients of the above four functions are complex. The coefficients of $H_{22}(z_1, z_2)$ are conjugate to those of $H_{44}(z_1, z_2)$; hence their sum will have real coefficients but not zero phase. Similarly the coefficients of $H_{11}(z_1, z_2)$ are conjugate to those of $H_{33}(z_1, z_2)$. The magnitude of $H_{22}(z_1, z_2)$+ $H_{44}(z_1, z_2)$ has a half-plane symmetry as shown in Fig. 3.10b and the magnitude of $H_{11}(z_1, z_2) + H_{33}(z_1, z_2)$ has a half-plane symmetry as shown in Fig. 3.10c.

If we wish to design a filter with a passband in two quadrants, we can do so by modifying the above functions to have zero phase and add them as explained earlier.

By judicious connection of any of these filters in cascade and parallel, with the many types of filters designed by the methods described above, we can design another large set of filters with half-plane symmetry as well as quadrantal symmetry. Of course we can design the 1-D filters in z_1 and z_2 domain with

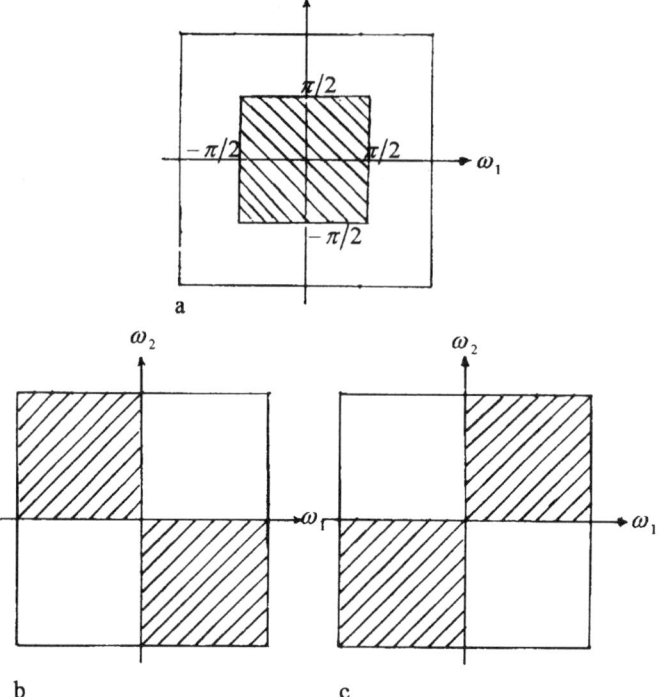

Fig. 3.10. Magnitude response of 2-D filters

multiple passbands in the ω_1 and ω_2 axis; when they are cascaded, we can generate many additional types of passbands and stopbands as illustrated by the example of Fig. 3.11a. So far, we have described the application of several operations to generate such a large set of 2-D digital filters, namely (a) cascading 1-D digital filters $H_1(z_1)$ and $H_2(z_2)$, (b) shifting or translating the frequency variables as shown in (3.7) (c)complementary operation, using all pass filters, (d) use of $H_{11}(z_1, z_2)$ and $H_{22}(z_1, z_2)$ to generate half-plane symmetric filters and so on. An example where almost all of these operations have been used is the 2-D filter with a magnitude response shown in Fig. 3.11b.

But note that the boundaries of the passband and stopband are straight lines parallel to the ω_1 and ω_2 axes in all cases and the denominator of the 2-D filter is always separable. In the next section, we consider the design of filters like the fan filters in which the boundaries of the passband and stopband are straight lines which, however, are not parallel to the ω_1 and ω_2 axes.

3.3. DESIGN OF FAN FILTERS

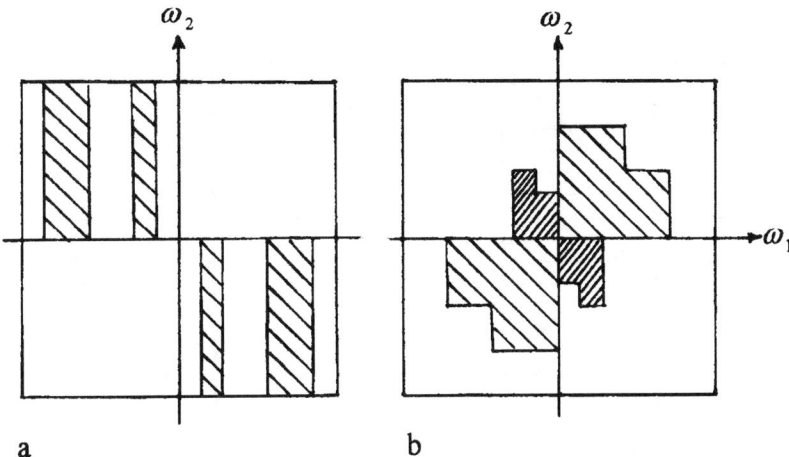

Fig. 3.11. Magnitude response of ideal 2-D filters

3.3 Design of Fan Filters

The most well-known application of fan filters is in the processing of geoseismic signals recorded, as a function of time, by a number of geophones stationed along a line at equal distances. If the signals recorded by each of these geophones are sampled with a period of T seconds, we get a two-dimensional discrete data with space and time forming the two variables. Hence it is called a spatio-temporal signal, with the sampling frequency in time being measured in cycles per second (Hz) and the frequency in space being measured in cycles per feet (or cycles per meters). If f and k are the temporal and spatial frequencies [28], then the frequency response of a typical fan filter is given by

$$Y(f,k) = \begin{cases} 1 & \text{for} \quad -\frac{|f|}{v} \leq k \leq \frac{|f|}{v} \\ 0 & \text{otherwise} \end{cases} \tag{3.31}$$

where $v = \pm \frac{f}{k}$ is the slope of the boundary lines as shown in Fig. 3.12a. In this figure, the slope is ± 1 and the filter is known as a 90° fan filter which has a quadrantal symmetry. Considerable research effort has concentrated on the design of FIR fan filters with 90° angle [4],[21]-[24]. Some effort has also been focussed in recent years, on the design of IIR filters with 90° and other angles. The 90° fan filter, however, is limited in its application for discriminating signals based on speed (not velocity), because of their quadrantal symmetry. When we want to discriminate between the signals coming from opposite directions, we need to measure their velocity and not just the speed. For this purpose, half-plane symmetric fan filters as shown in Fig. 3.12b are required. The design of such fan filters is a real test of the efficiency of any general procedure for

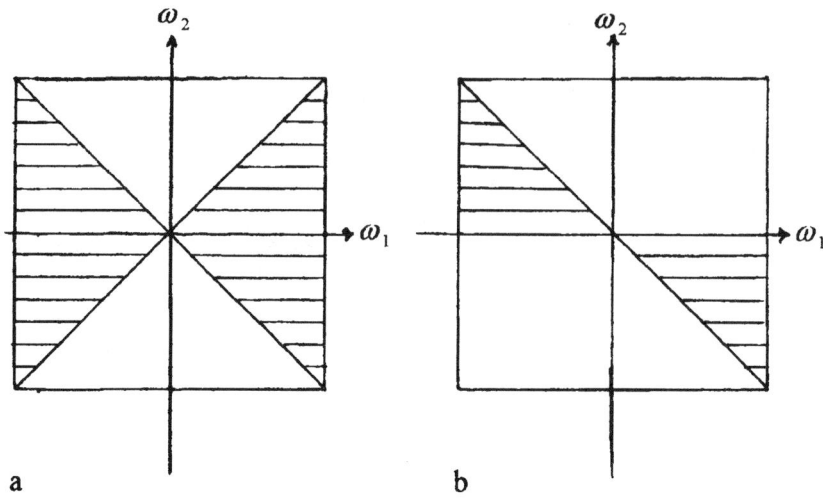

Fig. 3.12. Magnitude response of fan filters. **a** 90° fan filter **b** 45° fan filter

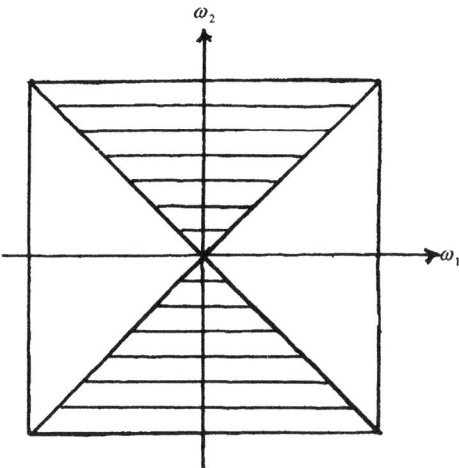

Fig. 3.13. Magnitude response of a 90° fan filter

designing 2-D IIR digital filters.

We discuss three methods for designing 90° fan filters in the following sections.

3.3.1 Method 3.1

The authors of [13] and [19] used a new operation, in addition to some of the operations proposed by [17]. For a description of the method developed in [13],

3.3. DESIGN OF FAN FILTERS

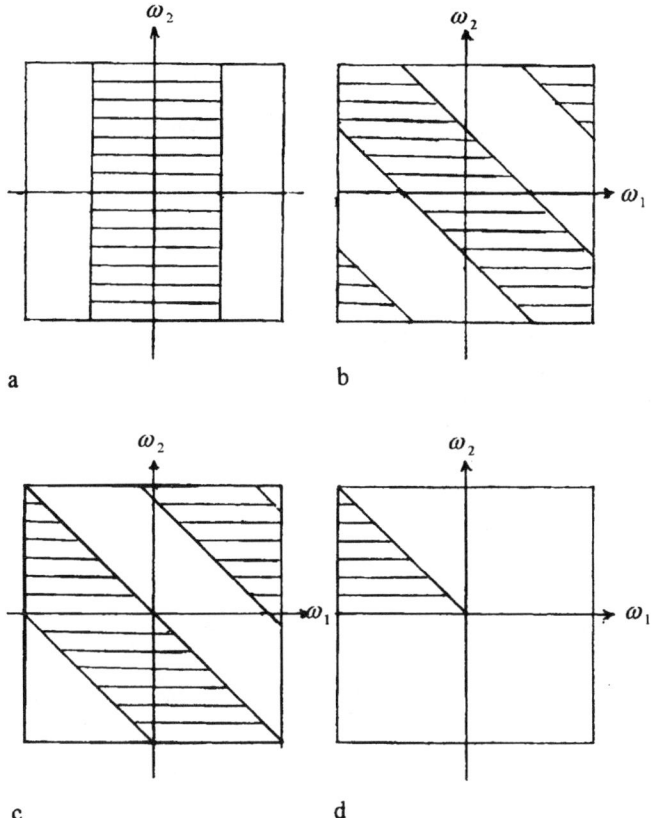

Fig. 3.14. Magnitude response of 2-D filters under different transformations

consider the Digital Spectral Transformation (DST) given by (3.32).

$$z = z_1' z_2' \qquad (3.32)$$

Note that this transformation, when applied to the filter $H(z_1)$ with its magnitude response shown in Fig. 3.14a, rotates its magnitude response counterclockwise by 45° and we get the magnitude response shown in Fig. 3.14b. The transformation $z_1' = z_1 e^{j\frac{\pi}{2}}$, (with $z_2' = z_2$) then shifts this frequency response by $-\frac{\pi}{2}$ along the ω_1 axis, giving the result depicted in Fig. 3.14c.

The Digital Spectral Transformation therefore becomes $z = z_1 e^{j\frac{\pi}{2}} z_2$ and the transfer function $H(z_1 e^{j\frac{\pi}{2}} z_2)$ when cascaded with $H_{22}(z_1, z_2)$ given by (3.27) gives $H(z_1 e^{j\frac{\pi}{2}} z_2) H_{22}(z_1, z_2)$ which has the magnitude response in the second quarter only. This is shown in Fig. 3.14d.

To get a zero phase transfer function, we multiply $H(z_1 e^{j\frac{\pi}{2}} z_2) H_{22}(z_1, z_2)$ by

$H^*(z_1^{-1}e^{j\frac{\pi}{2}}z_2^{-1})H_{22}^*(z_1^{-1},z_2^{-1})$ i.e.,

$$H(z_1e^{j\frac{\pi}{2}}z_2)H_{22}(z_1,z_2)H^*(z_1^{-1}e^{j\frac{\pi}{2}}z_2^{-1})H_{22}^*(z_1^{-1},z_2^{-1}) \qquad (3.33)$$

Similarly we get the fourth quadrant filter with a transfer function (3.34) which also has a zero phase i.e.,

$$H(z_1e^{-j\frac{\pi}{2}}z_2)H_{44}(z_1,z_2)H^*(z_1^{-1}e^{-j\frac{\pi}{2}}z_2^{-1})H_{44}^*(z_1^{-1},z_2^{-1}) \qquad (3.34)$$

This is the second term in (3.35). The first two terms in (3.35) represent the half-plane symmetric filter or the 45° fan filter with the response as shown in Fig. 3.12b. In practical design, we use non-ideal filters with stopbands over which the magnitude is non-zero. Hence the third term in (3.35) is used to remove the overlapping regions between the two regions in the second and fourth quadrants. Notice that (3.35) has only real coefficients and zero phase. This transfer function $H_{45°}(z_1,z_2)$ is given by

$$\begin{array}{l} H(z_1e^{j\frac{\pi}{2}}z_2)H_{22}(z_1,z_2)H^*(z_1^{-1}e^{j\frac{\pi}{2}}z_2^{-1})H_{22}^*(z_1^{-1},z_2^{-1}) \\ +H(z_1e^{-j\frac{\pi}{2}}z_2)H_{44}(z_1,z_2)H^*(z_1^{-1}e^{-j\frac{\pi}{2}}z_2^{-1})H_{44}^*(z_1^{-1},z_2^{-1}) \\ -\left\{ \begin{array}{l} H(z_1e^{j\frac{\pi}{2}}z_2)H_{22}(z_1,z_2)H^*(z_1^{-1}e^{j\frac{\pi}{2}}z_2^{-1})H_{22}^*(z_1^{-1},z_2^{-1}) \times \\ H(z_1e^{-j\frac{\pi}{2}}z_2)H_{44}(z_1,z_2)H^*(z_1^{-1}e^{-j\frac{\pi}{2}}z_2^{-1})H_{44}^*(z_1^{-1},z_2^{-1}) \end{array} \right\} \end{array}$$

(3.35)

Now it is easy to see that the quadrantally symmetric 90° fan filter $H_{90°}(z_1,z_2)$ can be obtained from the above transfer function for the 45° fan filter, as given by

$$H_{90°}(z_1,z_2) = H_{45°}(z_1,z_2) + H_{45°}(z_1,z_2^{-1}) \qquad (3.36)$$

Its magnitude response is shown in Fig. 3.12a.

The authors of [19] follow a similar approach for the design of the 90° fan filter and the 45° fan filter shown in Fig. 3.12b and Fig. 3.13, respectively. They list the following as the advantages of this method:

1. The procedure does not introduce any stability problem. The resulting IIR filters are inherently stable.

2. The original one-dimensional characteristics are preserved and no optimization is needed.

3. The speed of the design is fast.

4. The designed zero-phase two-dimensional filter functions can be implemented by one-dimensional filtering process.

3.3.2 Example 3.1

But there are two important drawbacks in the practical implementation of the method described above. When the frequency response of the 2-D lowpass

3.3. DESIGN OF FAN FILTERS

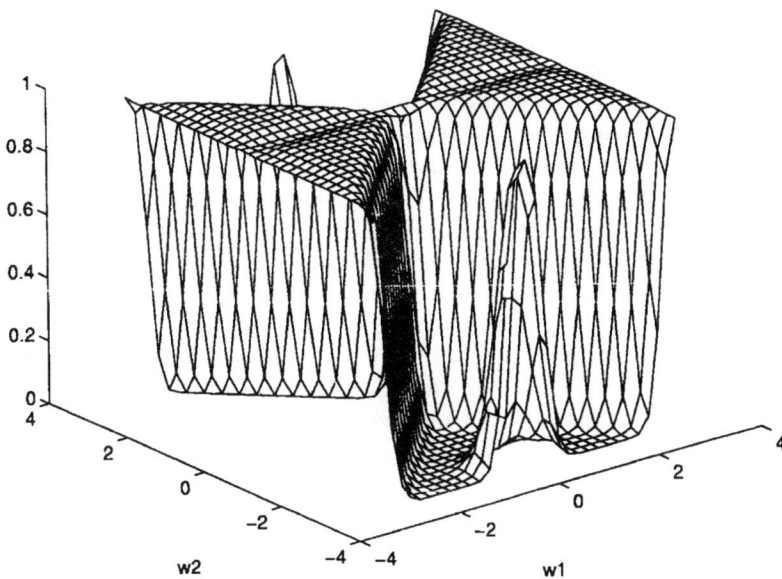

Fig. 3.15. Magnitude response of a 90° fan filter in Example 3.1

filter shown in Fig. 3.14a is rotated under the transformation (3.32), we get the response shown in Fig. 3.14b and Fig. 3.14c. We see that another passband which is a periodic repetition of the passband in Fig. 3.14a spills over mainly into the first quadrant. Hence the half-symmetric filter $H_{45°}(z_1, z_2)$ has a magnitude response with spill over into parts of both the first and the third quadrants, because the third term in (3.35) does not remove the spill overs completely. The effect of the spill over is not shown in the magnitude response of the ideal fan filter, shown in Fig. 3.12a. The magnitude response of a filter obtained by actual simulation starting from a lowpass Butterworth filter is shown in Fig. 3.15. This is included as an example of the design procedure described in the above section. But notice that the passband region consists of four distinct triangular regions joined together to form the 90° fan filter with their edges showing a slow roll off in their magnitude. This is the effect of choosing a 1-D filter that is not an ideal filter with a zero width transition band. The actual result of designing practical digital filter functions and the deviations from the ideal frequency response must be kept in mind when the design procedures discussed above are implemented.

The spill over effect is clearly seen in the magnitude response of the fan filter simulated by using the above design procedure. Any attempt to eliminate this spill over in the stopband region requires excessive computation and is not a very attractive proposition. The second drawback of this method is that the order of the final 2-D filter is also very high. Starting from a numerical example of a 1-D digital filter of fourth order, experience in simulating the performance of the 90° fan filter has shown that the coefficient matrices of the numerator and denominator of the final filter are of order 33 × 33. But a method for the design of circularly symmetric filters will be described later in section 3.4 and its authors [22] also point out the same two drawbacks based on their extensive experience of simulating their design method.

Now if the 90° fan filter is shifted to the right or left along the ω_1 axis by π, we get a diamond shaped passband. Some of the other transformations and operations described above, can be applied to the fan filter to get other interesting magnitude responses. Again, we can rotate the magnitude response of the filters obtained in the previous section by ±45° and generate a few more interesting variations for the passband of the 2-D filters. For example, when the frequency response of Fig. 3.1c is rotated by 45°, we get a diamond shaped passband that is smaller than the one obtained by shifting the 90° fan filter. In other words, the many tools we have introduced in the above two sections are very simple and useful for designing a wide variety of 2-D filters.

Method 3.2

This method is based on (1) some unique properties of the analog Butterworth and elliptical (Cauer) lowpass filters, and (2) some simple digital transformations applied on the corresponding 1-D digital filters obtained from these analog filters. We will discuss these properties in the following two sections and show how the design of fan filters is implemented.

Decomposition of Analog Prototype filters

In this section we show how the transfer function of analog Butterworth and elliptical lowpass filters can be decomposed as the sum (and difference) of two allpass functions. To derive this result, let us recollect the procedure discussed in Chap. 1 to obtain the transfer function of the analog Butterworth lowpass prototype filter from its magnitude squared function.

$$|H_{lp}(j\Omega)|^2 = \frac{1}{1+\epsilon^2 K^2(\Omega^2)} \qquad (3.37)$$

where $K^2(\Omega^2) = \Omega^{2n}$, n being the order of the filter. From this we obtained

$$H_{lp}(p)H_{lp}(-p) = \frac{1}{1+(-1)^n \epsilon^2 p^{2n}} \qquad (3.38)$$

in which we chose p to denote the complex frequency variable of the prototype filter $H_{lp}(p)$. Now we consider the case when n *is odd* and $\epsilon^2 = 1$ so that

$$H_{lp}(p)H_{lp}(-p) = \frac{1}{1-p^{2n}} = \frac{1}{(1+p^n)(1-p^n)} \qquad (3.39)$$

In Chap. 1, we derived the expression for all of the $2n$ zeros of $1 - p^{2n} = 0$. It is given by

$$p_k = e^{j(\frac{2k+n-1}{n})\frac{\pi}{2}}, k = 1, 2, \ldots, (2n) \qquad (3.40)$$

3.3. DESIGN OF FAN FILTERS

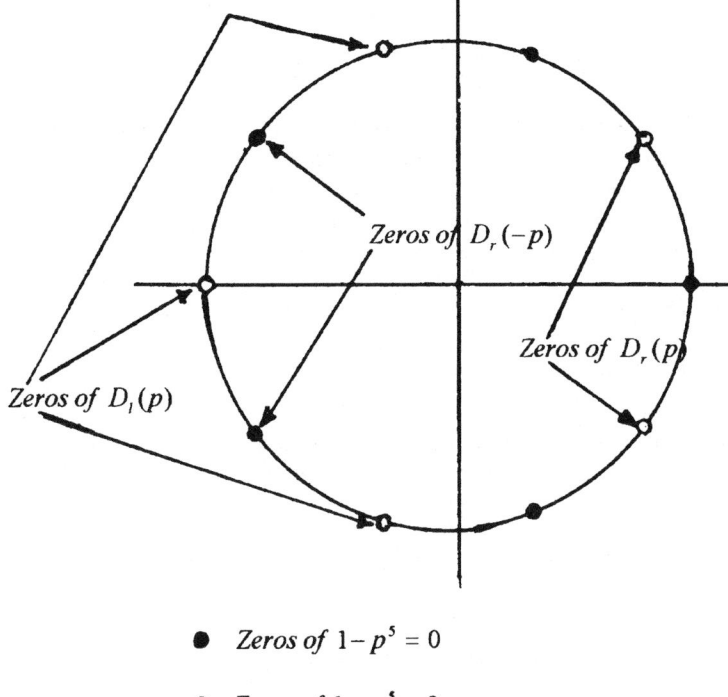

- Zeros of $1 - p^5 = 0$
- Zeros of $1 + p^5 = 0$

Fig. 3.16. Distribution of the zeros of $1 - p^5 = 0$ and $1 + p^5 = 0$

Then we picked the n zeros which lie in the left half of the p-plane as the poles of the filter function $H(p)$. We also observed that when $\epsilon^2 = 1$, these poles lie on a unit circle in the left half plane as shown in Fig. 3.16 for the case $n = 5$. Now let us look at the situation differently by finding the location of the zeros of $D_+(p) = (1 + p^n) = 0$. They are located on the unit circle at

$$p_k = -e^{j(\frac{2k\pi}{n})}, \quad k = 0, \pm 1, \pm 2, \ldots \pm (\frac{n-1}{2}) \qquad (3.41)$$

If we denote the polynomial having the zeros of $1 + p^n = 0$, which lie in the left half p-plane by $D_l(p)$ and the polynomial having the zeros in the right half plane by $D_r(p)$ so that $D_+(p) = D_l(p)D_r(p)$, then the zeros of $D_l(p)$ are given by -1 and $-e^{j(\frac{k\pi}{n})}$ for k even, and $2 \leq k \leq (\frac{n-1}{2})$ whereas the zeros of $D_r(p)$ are given by $e^{j(\frac{k\pi}{n})}$ for k odd, $1 \leq k \leq (\frac{n-1}{2})$. Hence the polynomials $D_l(p)$ and $D_r(p)$ are given by

$$D_l(p) = (1+p) \prod_{k=2,\, k \text{ even}}^{(\frac{n-1}{2})} (1 + 2p \cos \frac{k\pi}{n} + p^2) \qquad (3.42)$$

$$D_r(p) = \prod_{k=1,\,k\text{ odd}}^{(\frac{n-1}{2})} (1 - 2p\cos\frac{k\pi}{n} + p^2) \qquad (3.43)$$

We know that the the zeros of $D_l(p)D_r(p) = 1 + p^n$ and $D_l(-p)D_r(-p) = 1 - p^n$ alternate on the unit circle. We also see from (3.42) and (3.43) another interesting property that the zeros of $D_l(p)$ and $D_r(-p)$ lie in the left half plane and alternate on the unit circle. So we have derived two properties satisfied by the analog Butterworth filters of odd order as listed below.

Property (1): The poles of the filter lie on a unit circle when $\epsilon^2 = 1$.

Property (2): The zeros of $D_l(p)$ and $D_r(-p)$ alternate on the left half of the unit circle.

The magnitude squared function of the highpass filter $H_{hp}(p)$ obtained by the transformation $H_{hp}(p) = H_{lp}(\frac{1}{p})$ is given by

$$|H_{hp}(j\Omega)|^2 = 1 - |H_{lp}(j\Omega)|^2 = \frac{\Omega^{2n}}{1 + \Omega^{2n}} \qquad (3.44)$$

Hence

$$|H_{hp}(j\Omega)|^2 + |H_{lp}(j\Omega)|^2 = 1 \qquad (3.45)$$

Property (3): The filter satisfies the complementary property as defined by (3.45). It is obvious that the poles of the highpass filter $H_{hp}(p)$ are the same as those of $H_{lp}(p)$ and hence lie on the unit circle.

Property (4): The class of analog Butterworth lowpass filters of odd order (and the corresponding highpass filters) discussed above can be expressed as the sum (difference) of allpass filters i.e.

$$H_{lp}(p) = \frac{1}{2}[H_A(p) + H_B(p)] \qquad (3.46)$$

$$H_{hp}(p) = \frac{1}{2}[H_A(p) - H_B(p)] \qquad (3.47)$$

where

$$H_A(p) = \prod_{k=1,\,k\text{ odd}}^{(\frac{n-1}{2})} \frac{(1 - 2p\cos\frac{k\pi}{n} + p^2)}{(1 + 2p\cos\frac{k\pi}{n} + p^2)} = \frac{D_r(p)}{D_r(-p)} \qquad (3.48)$$

$$H_B(p) = \frac{1-p}{1+p} \prod_{k=2,\,k\text{ even}}^{(\frac{n-1}{2})} \frac{(1 - 2p\cos\frac{k\pi}{n} + p^2)}{(1 + 2p\cos\frac{k\pi}{n} + p^2)} = \frac{D_l(-p)}{D_l(p)} \qquad (3.49)$$

Gerheim [11], and Ansari [2] have derived this Property (4) and we will follow the simpler derivation given by [2] as explained below:

3.3. DESIGN OF FAN FILTERS

Let us consider the transfer function of *analog, lowpass, Butterworth filters* of odd order $H_{lp}(p)$ in the form

$$H_{lp}(p) = \begin{cases} \frac{2}{1+p} & \text{for n=1} \\ \frac{2}{(1+p)\prod_{k=1}^{(\frac{n-1}{2})}(p^2+2p\cos(\frac{2k\pi}{n})+1)} & \text{for } n \geq 3 \end{cases} \quad (3.50)$$

where, for convenience, we have chosen the passband magnitude to be 2.0. For $n = 1$ it is obvious that $H_{lp}(p) = \frac{1-p}{1+p} + 1$ and for $n = 3$ we can write $H_{lp}(p) = \frac{1-p}{1+p} + \frac{1-p+p^2}{1+p+p^2}$. For $n \geq 5$, we consider the following transfer function

$$H_n(p) = \frac{2}{1+p^n} \quad (3.51)$$

This function has the same magnitude squared characteristic as $|H_{lp}(j\Omega)|^2 = \frac{4}{1+\Omega^{2n}}$ but the poles of $H_n(p)$ are the zeros of $D_+(p) = D_l(p)D_r(p)$ derived earlier in this section. We observe that

$$H_n(p) = 1 + \frac{1-p^n}{1+p^n} = 1 + \frac{D_l(-p)D_r(-p)}{D_l(p)D_r(p)} \quad (3.52)$$

Next we define a new function

$$H_a(p) = H_n(p)\frac{D_r(p)}{D_r(-p)} = \frac{D_r(p)}{D_r(-p)} + \frac{D_l(-p)}{D_l(p)} \quad (3.53)$$

i.e.

$$H_a(p) = \frac{D_r(p)D_l(p) + D_l(-p)D_r(-p)}{D_r(-p)D_l(p)}$$

$$= \frac{(1+p^{2n}) + (1-p^{2n})}{D_r(-p)D_l(p)} \quad (3.54)$$

$$= \frac{2}{D_r(-p)D_l(p)} \quad (3.55)$$

Notice that (3.50) can also be expressed as

$$H_{lp}(p) = \frac{2}{D_r(-p)D_l(p)} \quad (3.56)$$

By comparing (3.54) and (3.56) we see that $H_a(p) = H_{lp}(p)$ and using the form of (3.54) for $H_a(p)$, we arrive at Property 4 as $H_{lp}(p) =$

$$\prod_{k=1, k\,odd}^{(\frac{n-1}{2})} \frac{(1-2p\cos\frac{k\pi}{n}+p^2)}{(1+2p\cos\frac{k\pi}{n}+p^2)} + \frac{1-p}{1+p} \prod_{k=2, k\,even}^{(\frac{n-1}{2})} \frac{(1-2p\cos\frac{k\pi}{n}+p^2)}{(1+2p\cos\frac{k\pi}{n}+p^2)} \quad (3.57)$$

How do we pick the poles of $H_A(p)$ and $H_B(p)$ when we have found all the poles of $H_{lp}(p)H_{lp}(-p)$? Here we do not pick the poles $H_{lp}(p)H_{lp}(-p)$ which lie in the left half plane. We pick the pole at $p = -1$ as one of the poles of $H_B(p)$. Using the alternation property (Property (2)) we pick every other pole located in the left half of the unit circle as the other poles of $H_B(p)$. The poles left over are identified as those of $H_A(p)$. Since the functions $H_A(p)$ and $H_B(p)$ are allpass functions, their zeros are immediately known. After we have expressed the analog, lowpass, Butterworth filter functions of odd order in the form of (3.57), we apply the bilinear transformation, $p = \frac{z-1}{z+1}$ to get the transfer function of the 1-D digital filter function $H_{lp}(z)$.

If we express (3.57) in the form $H_{lp}(p) =$

$$\prod_{k=1,\,k\,odd}^{(\frac{n-1}{2})} \frac{(1 - a_i p + p^2)}{(1 + a_i p + p^2)} + \frac{1-p}{1+p} \prod_{k=2,\,k\,even}^{(\frac{n-1}{2})} \frac{(1 - b_i p + p^2)}{(1 + b_i p + p^2)} \quad (3.58)$$

then under the bilinear transformation, we get $H_{lp}(z)$ in the form of (3.59).

$$H_{lp}(z) = \frac{1}{2}[H_A(z^2) + z^{-1} H_B(z^2)] \quad (3.59)$$

and the transfer function of the highpass digital filter $H_{hp}(z)$ in the form of (3.60).

$$H_{hp}(z) = \frac{1}{2}[H_A(z^2) - z^{-1} H_B(z^2)] \quad (3.60)$$

where

$$H_A(z^2) = \prod_{i=1}^{N_1} \frac{A_i z^2 + 1}{z^2 + A_i}, \qquad A_i = \frac{2 - a_i}{2 + a_i} \quad (3.61)$$

$$H_B(z^2) = \prod_{i=1}^{N_2} \frac{B_i z^2 + 1}{z^2 + B_i}, \qquad B_i = \frac{2 - b_i}{2 + b_i} \quad (3.62)$$

and $N_1 + N_2 = (n-1)/2$ and $N_1 = N_2$ or $N_1 = N_2 + 1$.

Now let us consider the class of *analog, lowpass, elliptic filters of odd order* n.

$$H_{lp}(p) H_{lp}(-p) = \frac{1}{1 - \epsilon^2 p^2 \prod_{i=1}^{(\frac{n-1}{2})} \left(\frac{p^2 + \Omega_i^2}{p^2 \Omega_i^2 + 1}\right)^2} \quad (3.63)$$

and

$$H_{lp}(j\Omega) H_{lp}(-j\Omega) = \frac{1}{1 + \epsilon^2 \Omega^2 \prod_{i=1}^{(\frac{n-1}{2})} \left(\frac{-\Omega^2 + \Omega_i^2}{-\Omega^2 \Omega_i^2 + 1}\right)^2} \quad (3.64)$$

Obviously, if the highest frequency of the passband or the bandwidth is denoted as Ω_p, then

$$|H_{lp}(j\Omega_p)|^2 = \frac{1}{1 + \epsilon^2 K^2(\Omega_p^2)} \quad (3.65)$$

and the attenuation in dB at this bandwidth frequency is

$$A_p = 10 \log\left[1 + \epsilon^2 K^2(\Omega_p^2)\right] \quad (3.66)$$

3.3. DESIGN OF FAN FILTERS

A unique property of elliptic filters is that the frequency $\Omega_a = \frac{1}{\Omega_p}$ is the lowest frequency of the stopband at which the attenuation is A_s. From the particular form of $K^2(\Omega_p^2)$ for the elliptic filters found in (1.72), it is seen that

$$K^2(\Omega_a^2) = \frac{1}{K^2(\Omega_p^2)} \tag{3.67}$$

and therefore

$$A_s = 10\log\left[1 + \frac{\epsilon^2}{K^2(\Omega_p^2)}\right] \tag{3.68}$$

From (3.67) and (3.68), we get

$$\epsilon^2 K^2(\Omega_p^2) = 10^{0.1 A_p} - 1 \tag{3.69}$$

and

$$\frac{\epsilon^2}{K^2(\Omega_p^2)} = 10^{0.1 A_s} - 1 \tag{3.70}$$

Therefore by multiplying these two expressions, we have the result

$$\epsilon^2 = \left[(10^{0.1 A_p} - 1)(10^{0.1 A_s} - 1)\right]^{\frac{1}{2}} \tag{3.71}$$

(This constraint between A_p and A_s is also satisfied by elliptic lowpass filters of even order.) Because of (3.71), the attenuations A_p and A_s can not be specified independently. The constraint $\epsilon^2 = 1$ usually leads to a very small value for A_p for a given specification A_s. However, we know that this small value for the passband attenuation is highly desirable in many applications.

When we assume $\epsilon^2 = 1$, it can be shown that the class of analog, elliptic lowpass filters of odd order satisfy the same four properties derived for the Butterworth filters. In his paper, Gazsi [10] shows that the Chebyshev, and inverse Chebyshev analog lowpass filters of odd order can also be realized by the parallel connection of allpass filters which are then used to derive the corresponding allpass digital filters.

Next we will show that elliptical lowpass filters of both odd and also even order can be realized by the parallel connection of allpass filters [30]. Let us choose

$$P(p^2) = \prod_{i=1}^{n}(p^2\Omega_i^2 + 1) \quad \text{and} \quad Q(p^2) = \prod_{i=1}^{n}(p^2 + \Omega_i^2) \tag{3.72}$$

where

$$n = \begin{cases} (N-1)/2 \; ; \; N \text{ odd} \\ N/2 \quad\quad\; ; \; N \text{ even} \end{cases} \tag{3.73}$$

Then, we can write

$$H_{lp}(p)H_{lp}(-p) = \frac{[P(p^2)]^2}{[P(p^2)]^2 - [pQ(p^2)]^2} \quad ; \; N \text{ odd} \tag{3.74}$$

and
$$H_{lp}(p)H_{lp}(-p) = \frac{[P(p^2)]^2}{[P(p^2)]^2 + [Q(p^2)]^2} \quad ; N \text{ even} \quad (3.75)$$

Case 1: odd N

The proof is generalized from the earlier case where we considered the Butterworth lowpass filter.

Briefly, let $D_l(p)$ and $D_r(p)$ be the polynomials formed by the zeros of $P(p^2) + pQ(p^2)$ and $P(p^2) - pQ(p^2)$ which are in the left and right half p-plane respectively. The zeros of $D_r(-p)$ are in the left half p-plane and alternate with the zeros of $D_l(p)$. Then, we can write

$$H_{lp}(p) = \frac{P(p^2)}{D_l(p)D_r(-p)} = \frac{1}{2}[H_A(p) + H_B(p)]$$

$$H_{hp}(p) = \frac{pQ(p^2)}{D_l(p)D_r(-p)} = \frac{1}{2}[H_A(p) - H_B(p)] \quad (3.76)$$

where the allpass filters are of the form

$$H_A(p) = \prod_{i=1}^{N_1} \frac{1 - A_i p + p^2}{1 + A_i p + p^2} = \frac{D_r(p)}{D_r(-p)}$$

$$H_B(p) = \frac{1-p}{1+p} \prod_{i=1}^{N_2} \frac{1 - B_i p + p^2}{1 + B_i p + p^2} = \frac{D_l(-p)}{D_l(p)} \quad (3.77)$$

in which $N_1 + N_2 = (N-1)/2$.

They are similar to the two terms in (3.58) and when the bilinear transformation is applied, we get equations of the form (3.59) and (3.60).

Case 2: even N

From (3.75), we see that

$$[P(p^2)]^2 + [Q(p^2)]^2 = [P(p^2) + jQ(p^2)][P(p^2) - jQ(p^2)] \quad (3.78)$$

Now we assume that the polynomial containing all the zeros of $[P(p^2) + jQ(p^2)]$ in the left half s-plane is of the form

$$D_l(p) = \prod_{i=1}^{N/2} (p + e^{j\theta_i}), \quad -\frac{\pi}{2} \le \theta_i \le \frac{\pi}{2} \quad (3.79)$$

and the polynomial containing all the zeros of $[P(p^2) - jQ(p^2)]$ in the right half s-plane is of the form [1]

$$D_r(p) = \prod_{i=1}^{N/2}(-p + e^{-j\theta_i}) = (-1)^{N/2} \prod_{i=1}^{N/2}(p - e^{-j\theta_i}) \quad (3.80)$$

---

[1] The author acknowledges the help received from Dr.W-P Zhu (Concordia University, Montreal) in deriving the following results for Case 2.

3.3. DESIGN OF FAN FILTERS

Notice that the zeros of $D_r(-p)$ are in the left half p-plane which alternate with the zeros of $D_l(p)$.

$$[P(p^2)]^2 + [Q(p^2)]^2 = \left[\prod_{i=1}^{N/2}(p^2\Omega_i^2+1)\right]^2 + \left[\prod_{i=1}^{N/2}(p^2+\Omega_i^2)\right]^2 \quad (3.81)$$

whereas $D_l(p)D_l(-p)D_r(-p)D_r(p) =$

$$\prod_{i=1}^{N/2}(p+e^{j\theta_i})(-p+e^{j\theta_i})\prod_{i=1}^{N/2}(p+e^{-j\theta_i})(-p+e^{-j\theta_i}) \quad (3.82)$$

Comparing the coefficients of the highest degree term p^{2N} in (3.81) and (3.82), we find that

$$\begin{aligned}[P(p^2)+jQ(p^2)] &= (-1)^{N/2}AD_l(p)D_l(-p) \\ [P(p^2)-jQ(p^2)] &= (-1)^{N/2}A^*D_r(-p)D_r(p)\end{aligned} \quad (3.83)$$

where

$$A = \prod_{i=1}^{N/2}\Omega_i^2 + j = |A|e^{j\alpha} \quad (3.84)$$

and therefore

$$\alpha = \tan^{-1}\left(\frac{1}{\prod_{i=1}^{N/2}\Omega_i^2}\right) \quad (3.85)$$

Then

$$\begin{aligned}H_{lp}(p)+jH_{hp}(p) &= e^{j\alpha}\frac{D_l(-p)}{D_r(-p)} = H_A(p) \\ H_{lp}(p)-jH_{hp}(p) &= e^{-j\alpha}\frac{D_r(p)}{D_l(p)} = H_A^*(p)\end{aligned} \quad (3.86)$$

and the allpass filter is

$$H_A(p) = e^{j\alpha}\prod_{i=1}^{N/2}\frac{(-p+e^{j\theta_i})}{(p+e^{-j\theta_i})} = e^{j\alpha}\frac{D_l(-p)}{D_r(-p)} \quad (3.87)$$

$$H_A^*(p) = e^{-j\alpha}\prod_{i=1}^{N/2}\frac{(-p+e^{-j\theta_i})}{(p+e^{j\theta_i})} = e^{-j\alpha}\frac{D_r(p)}{D_l(p)} \quad (3.88)$$

Therefore the lowpass filter can be realized as the parallel connection of two allpass filters:

$$H_{lp}(p) = \frac{1}{2}[H_A(p)+H_A^*(p)] \quad (3.89)$$

The highpass filter is given by

$$H_{hp}(p) = \frac{1}{2j}[H_A(p)-H_A^*(p)] \quad (3.90)$$

When the bilinear transformation is applied on (3.89) and (3.90), we get the allpass filter $H_A(z)$ as

$$H_A(z) = e^{j\beta} \prod_{i=1}^{N/2} \frac{ja_i + z^{-1}}{1 - ja_i z^{-1}} \tag{3.91}$$

where

$$a_i = \tan(\frac{\theta_i}{2}) \tag{3.92}$$

and

$$\beta = \alpha + \sum_{i=1}^{N/2} \theta_i \tag{3.93}$$

The lowpass and highpass digital filters are generated by

$$H_{lp}(z) = \frac{1}{2}[H_A(z) + H_A^*(z)] \tag{3.94}$$

$$H_{hp}(z) = \frac{1}{2}[H_A(z) - H_A^*(z)] \tag{3.95}$$

Now we will briefly discuss the case of decomposing an analog, Butterworth, lowpass prototype filter of even order as the sum of two allpass filters [31]. In this case, we have $H_{lp}(p)$ in the form

$$H_{lp}(p) = \frac{1}{\prod_{k=1}^{N/2}(p^2 + a_k p + 1)}$$

and assume that $H_{lp}(p)$ and $H_{hp}(p)$ can be written in the form

$$H_{lp}(p) = \frac{1}{2j}[H_C(p) - H_C^*(p)] \tag{3.96}$$

$$H_{hp}(p) = \frac{1}{2}[H_C(p) + H_C^*(p)] \tag{3.97}$$

where

$$H_C(p) = \prod_{k=1}^{N/2} \frac{p - e^{-j\theta_k}}{p + e^{j\theta_k}} \tag{3.98}$$

$$\theta_k = \begin{cases} \frac{\pi}{2} - \frac{(4k-3)\pi}{2N} & N = 4n-2; \quad n = 1, 2, \ldots \\ \frac{\pi}{2} - \frac{(4k-1)\pi}{2N} & N = 4n; \quad n = 1, 2, \ldots \end{cases} \tag{3.99}$$

When the bilinear transformation is used, the corresponding allpass digital filters are given by

$$H_{lp}(z) = \frac{1}{2j}[H_C(z) - H_C^*(z)] \tag{3.100}$$

$$H_{hp}(z) = \frac{1}{2}[H_C(z) + H_C^*(z)] \tag{3.101}$$

3.3. DESIGN OF FAN FILTERS

where

$$H_C(z) = e^{j\phi} \prod_{k=1}^{N/2} \frac{jC_k - z^{-1}}{1 + jC_k z^{-1}} \tag{3.102}$$

$$\phi = -\sum_{k=1}^{N/2} \theta_k = \begin{cases} -\frac{\pi}{4} & N = 4n-2; \quad n = 1, 2, \ldots \\ \frac{\pi}{4} & N = 4n; \quad n = 1, 2, \ldots \end{cases} \tag{3.103}$$

and

$$C_k = \tan\left(\frac{\theta_k}{2}\right) \tag{3.104}$$

When N is even, it is seen from the above, that the decomposition of elliptic and Butterworth lowpass filters produces allpass filters with complex coefficients. When the input is real, the outputs from the allpass filter $H_A(z)$ or $H_C(z)$ is a complex valued sequence, and the real part of this output is equivalent to the output of the lowpass filter. In this case, the computational amount is about twice as much as that needed for the odd order filter. Besides, the coefficients of these allpass filters are not even as it is in the case of odd ordered, prototype, lowpass filters. So in this section, we have shown that only when the order of the filters is odd, Butterworth or elliptical lowpass filters (and the corresponding highpass filters which are complementary to them) can be realized by the parallel connection of two allpass filters $H_A(z^2)$ and $H_B(z^2)$. This result has great significance and yields many advantages in the design of 1-D digital filters, e.g. (1) the filters require almost half the number of multipliers (coefficients) required in the conventional methods (2) the sensitivity of the frequency response in the passband (due to finite word length of the coefficients) is lower and (3) the allpass filters run at half the sampling rate. The result is also used in the design of half band filters [3].

3.3.3 Design of Fan Filters using Method 3.2

The above result for Butterworth and elliptical lowpass, prototype filters of odd order is crucial in the design of fan filters using the Method 3.2, because $H_A(z^2)$ and $H_B(z^2)$ have only even degree terms. This method is now described in the following sections. We consider the one-dimensional lowpass filter $H_{lp}(z)$ defined by (3.59), (3.61) and (3.62) and assume that it has been designed to have a bandwidth of $\pi/2$. Then we generate a 2-D lowpass filter $H_{lp}(z_1, z_2) = H_{lp}(z_1)H_{lp}(z_2)$ which has a magnitude of one in its passband region that is a square region of widths equal to π. The magnitude response of such a filter was given by (3.26) and shown in Fig. 3.2a. But remember that the crossed area represents the passband of an elliptical filter response (and not the ideal passband with a magnitude of one). The next step in our design procedure is to shift this passband region to the first and third quadrant as in (3.29) and (3.30). The result is given as (3.105).

$$\begin{aligned} H_{13}(z_1, z_2) &= H_{11}(z_1, z_2) + H_{33}(z_1, z_2) \\ &= H_{lp}(-jz_1)H_{lp}(-jz_2) + H_{lp}(jz_1)H_{lp}(jz_2) \end{aligned} \tag{3.105}$$

It is true that this filter in general has complex coefficients and we have to generate filters with real coefficients and zero phase, while using the earlier procedure. But in this method, that procedure is not necessary as will be evident soon. When the above frequency transformations are applied to the allpass filters of (3.61) and (3.62) we get (3.106).

$$H_{13}(z_1, z_2) = \frac{1}{2}\left[H_A(-z_1^2)H_A(-z_2^2) - z_1^{-1}z_2^{-1}H_B(-z_1^2)H_B(-z_2^2)\right] \quad (3.106)$$

where

$$H_A(-z_k^2) = \prod_{i=1}^{N_1} \left(\frac{1 - A_i z_k^2}{A_i - z_k^2}\right) \quad (3.107)$$

and

$$H_B(-z_k^2) = \prod_{i=1}^{N_2} \left(\frac{1 - B_i z_k^2}{B_i - z_k^2}\right), \quad (k = 1, 2) \quad (3.108)$$

The passband of the filter (3.106) is as shown in Fig. 3.10b. In this context, we also point out that the magnitude of

$$H_{24}(z_1, z_2) = \frac{1}{2}\left[H_A(-z_1^2)H_A(-z_2^2) + z_1^{-1}z_2^{-1}H_B(-z_1^2)H_B(-z_2^2)\right] \quad (3.109)$$

is one in the second and fourth quadrants of the $\omega_1 - \omega_2$ plane as shown in Fig. 3.10c. Our next step in the design procedure is to apply the digital transformation $z_1 \Rightarrow \sqrt{z_1 z_2^{-1}}$ and $z_2 \Rightarrow \sqrt{z_1 z_2}$ (or substitute $z_1 z_2^{-1}$ for z_1^2 and $z_1 z_2$ for z_2^2) in the above three equations. This rotates the frequency response of Fig. 3.10b clockwise by 45° to that shown in Fig. 3.13. This is the magnitude response of the 90° fan filter given as $H_F(z_1, z_2) =$

$$\frac{1}{2}\left[H_A(-z_1 z_2^{-1})H_A(-z_1 z_2) - z_1^{-1}H_B(-z_1 z_2^{-1})H_B(-z_1 z_2)\right] \quad (3.110)$$

It is significant to note that though the transformations $z_1 \Rightarrow \sqrt{z_1 z_2^{-1}}$ and $z_2 \Rightarrow \sqrt{z_1 z_2}$ applied to a general 1-D digital filter would result in the presence of fractional powers of z_1 and z_2, the transfer function (3.110) has terms with only integer powers because the allpass functions (3.107) and (3.108) have only even powers. Similarly, when we apply the above transformation on (3.106) the magnitude is mapped to the fan filter as shown in Fig. 3.12a. Its transfer function is $H_f(z_1, z_2)$

$$= \frac{1}{2}\left[H_A(-z_1 z_2^{-1})H_A(-z_1 z_2) + z_1^{-1}H_B(-z_1 z_2^{-1})H_B(-z_1 z_2)\right] \quad (3.111)$$

3.3.4 Design Procedure

In Chap. 2, as the basis for Method 2.4, we had shown that the transfer function of a 1-D, digital filter could be expressed as the sum of two *digital allpass* transfer

3.3. DESIGN OF FAN FILTERS

functions, when it satisfied some conditions. In the above section, we proved that the transfer function of an analog lowpass filter function can also be expressed as the sum of two *analog allpass* filter functions, when it satisfied a few conditions - as shown for example by (3.76) and (3.77). This fundamental result forms the basis for the design of 90° fan filters, and the design procedure based on it has been developed above. Now let us summarize the design procedure, before an example is worked out.

1. First we transform the specification of a 1-D IIR filter $H_{lp}(z)$ with a bandwidth of 0.5π to the design of a lowpass prototype, Butterworth or elliptic filter, of odd order with an arbitrary low value for A_p and a high value for A_s (when the filter is an elliptic filter, they are not independent as shown by (3.71)).

2. Then the analog filter function is decomposed as the sum of two analog, allpass functions in the form (3.58).

3. Applying the bilinear transformation, we reduce the transfer function to the form of (3.59).

4. A 2-D lowpass digital filter $H_{lp}(z_1, z_2)$ is obtained by cascading two 1-D lowpass digital filters $H_{lp}(z_1)$ and $H_{lp}(z_2)$, such that $H_{lp}(z_1)H_{lp}(z_2)$ has a rectangular passband with a width of π in both ω_1 and ω_2 directions.

5. This filter is shifted to the first and third quadrant to get $H_{13}(z_1, z_2)$ shown as (3.106) and to the second and fourth quadrant to get $H_{24}(z_1, z_2)$ shown as (3.109).

6. Next the transformation $z \Rightarrow \sqrt{z_1 z_2^{-1}}$ and $z_2 \Rightarrow \sqrt{z_1 z_2}$ is applied on the 2-D filter $H_{24}(z_1, z_2)$, to get a 90° fan filter approximating the ideal magnitude response shown in Fig. 3.13. If the same transformation is applied on the filter $H_{13}(z_1, z_2)$, we get the 90° fan filter approximating the ideal magnitude response shown in Fig. 3.12a.

3.3.5 Example 3.2

To design a fan filter using the above design procedure, Gerheim [11] chose a 7th order, lowpass Butterworth filter and Zhu and Nakamura [30] chose a 7th order lowpass elliptic filter with a stopband attenuation $A_s = 40$ dB and hence under the constraint (3.71) we have to chose $A_p = 0.000434316$ dB. (The coefficients used in (3.77) for this filter are $A_1 = 0.189895$, $A_2 = 0.860016$ and $B_1 = 0.551678$). When the design procedure is completed, we get the transfer function for the 90° fan filter which is of the form given below.

$$H_F(z_1, z_2) = \frac{\sum_{i=0}^{7} \sum_{j=0}^{7} b_{ij} z_1^{-i} z_2^{-j}}{\sum_{i=0}^{7} \sum_{j=0}^{7} a_{ij} z_1^{-i} z_2^{-j}} \quad (3.112)$$

168   CHAPTER 3. MAGNITUDE OF 2-D FILTERS

If we choose to write it in the matrix form $H_F(z_1, z_2) = \frac{\mathbf{c}_1 \mathbf{B} \mathbf{c}_2'}{\mathbf{c}_1 \mathbf{A} \mathbf{c}_2'}$ then we get

$$\mathbf{c}_1 = \begin{bmatrix} 1 & z_1^{-1} & z_1^{-2} & z_1^{-3} & z_1^{-4} & z_1^{-5} & z_1^{-6} & z_1^{-7} \end{bmatrix} \quad (3.113)$$

$$\mathbf{c}_2 = \begin{bmatrix} 1 & z_2^{-1} & z_2^{-2} & z_2^{-3} & z_2^{-4} & z_2^{-5} & z_2^{-6} & z_2^{-7} \end{bmatrix} \quad (3.114)$$

and the matrices $\mathbf{B}$ and $\mathbf{A}$ are given by (3.115) and (3.116) respectively.

$$\begin{bmatrix}
0 & 0 & 0 & .0267 & 0 & 0 & 0 & 0 \\
0 & 0 & -.1862 & .3043 & -.1862 & 0 & 0 & 0 \\
0 & .2579 & -.8712 & 1.2996 & -.8712 & .2579 & 0 & 0 \\
-.0901 & .6289 & -1.8003 & 2.4939 & -1.8003 & .6289 & -.0901 & 0 \\
-.0901 & .6289 & -1.8003 & 2.4939 & -1.8003 & .6289 & -.0901 & 0 \\
0 & .2579 & -.8712 & 1.2996 & -.8712 & .2579 & 0 & 0 \\
0 & 0 & -.1862 & .3043 & -.1862 & 0 & 0 & 0 \\
0 & 0 & 0 & .0267 & 0 & 0 & 0 & 0
\end{bmatrix} \quad (3.115)$$

and

$$\begin{bmatrix}
0 & 0 & 0 & 1.0000 & 0 & 0 & 0 & 0 \\
0 & 0 & -1.6016 & 0 & -1.6016 & 0 & 0 & 0 \\
0 & .7425 & 0 & 2.5651 & 0 & .7425 & 0 & 0 \\
-.0901 & 0 & -1.1892 & 0 & -1.1892 & 0 & -.0901 & 0 \\
0 & .1443 & 0 & .5513 & 0 & .1443 & 0 & 0 \\
0 & 0 & -.0669 & 0 & -.0669 & 0 & 0 & 0 \\
0 & 0 & 0 & .0081 & 0 & 0 & 0 & 0 \\
0 & 0 & 0 & 0 & 0 & 0 & 0 & 0
\end{bmatrix} \quad (3.116)$$

The magnitude response of the 1-D elliptic, lowpass digital filter $H_{lp}(z)$ is shown in Fig. 3.17. The magnitude response of the filter $H_{13}(z_1, z_2)$ generated from the 1-D filter is shown in Fig. 3.18. The magnitude response of the 90° fan filter $H_F(z_1, z_2)$ with the numerator and denominator coefficients given in (3.115) and (3.116) is plotted in Fig. 3.19.

As expected the magnitude responses of the 1-D digital filter and that of the 90° fan filter obtained by using the method 3.2 [30] are much better than that in [11]. Also it should be noted that the order of $\mathbf{A}$ used in the above example is only 8×8 which is much lower than the order for the examples 3.1 and 3.3 chosen to illustrate the other two methods, namely the method 3.1 and 3.3.

3.3.6 Method 3.3

A different approach has been used by the authors of [8]. They design 2-D IIR filters $H(z_1, z_2)$ assuming a non-symmetrical half-plane (NSHP) region of support for the coefficients of its impulse response $h(n_1, n_2)$. Because the region of support is a wedge shaped region - as shown in Fig. 3.20 - the filter would be recursively computable but may or may not be stable. When and if the filter is unstable, the authors resort to the spectral factorization method for stabilizing the unstable filter. The impulse response $h_s(n_1, n_2)$ of such a stabilized filter

3.3. DESIGN OF FAN FILTERS

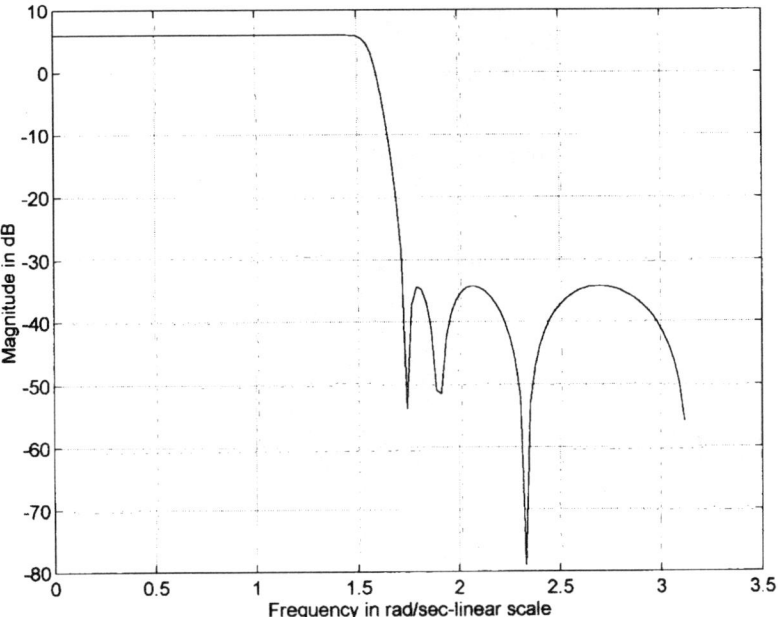

Fig. 3.17. Magnitude response of an elliptic lowpass 1-D filter

$H_s(z_1, z_2)$ would have the same region of support as $H(z_1, z_2)$; it is recursively computable and stable. The magnitude response of $H_s(z_1, z_2)$ is also equal to that of $H(z_1, z_2)$ but its impulse response $h_s(n_1, n_2)$ may not be equal to $h(n_1, n_2)$ - except when $H(z_1, z_2)$ is a stable digital filter. Using this argument, they define a stability error function $e_s(n_1, n_2) = h(n_1, n_2) - h_s(n_1, n_2)$. The error function for designing the NSHP fan filter chosen by them is a weighted sum of the error in the frequency domain and the stability error i.e.,

$$\mathbf{J} = \sum_{i,j \in R} \sum \mathbf{W}(\omega_{1i}, \omega_{2j}) [H_d(\omega_{1i}, \omega_{2j}) - |H(\omega_{1i}, \omega_{2j})|] + \alpha \left[h(n_1, n_2) - h_s^k(n_1, n_2) \right] \quad (3.117)$$

In the above $(\omega_{1i}, \omega_{2j})$ are the discrete values of the frequencies ω_1 and ω_2 in the region of interest R - which needs to be taken as a triangular region in only one quadrant because of the quadrantal symmetry. The magnitude response specified for the fan filter is $H_d(\omega_{1i}, \omega_{2j})$ and $h_s^k(n_1, n_2)$ is the value of $h_s(n_1, n_2)$ at the kth iteration, whereas $\mathbf{W}(\omega_{1i}, \omega_{2j})$ and α are the weighting functions that are arbitrarily chosen during the optimization process. When and if the function $|\mathbf{J}|^2$ is minimized to zero, it is obvious that the magnitude response of the NSHP filter would be equal to the specified magnitude response and the 2-D filter would also be stable. The practical details about the nonlinear programming technique used for minimizing the error in the least squares sense, is given by the authors

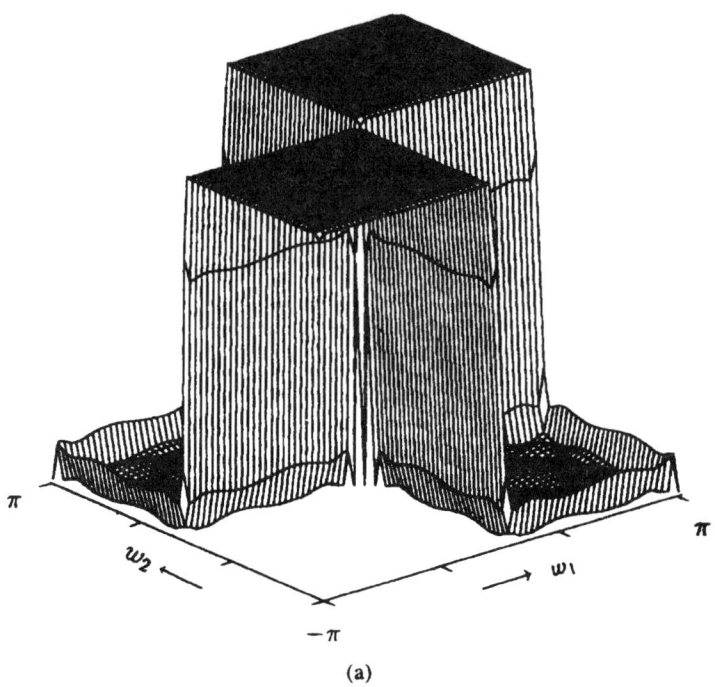

(a)

Fig. 3.18. Magnitude response of the 2-D quadrantal filter in Example 3.2

in [8]. But there is no guarantee that the method will always end with a stable filter. This method is applicable for approximating the magnitude of any 2-D IIR filters like those described in the next section - since it uses nonlinear optimization techniques for minimizing the extended objective function in the least squares sense. Readers interested in a more detailed explanation of this method should refer to the original work cited above.

3.3.7 Example 3.3

One example of the fan filter designed by the authors [8], has a transfer function given by

$$H(z_1^{-1}, z_2^{-1}) = \frac{\mathbf{c}_1 \mathbf{N} \mathbf{c}_2'}{\mathbf{d}_1 \mathbf{D} \mathbf{d}_2'} \quad (3.118)$$

where the row vectors $\mathbf{c}_1, \mathbf{c}_2, \mathbf{d}_1, \mathbf{d}_2$ and the coefficient matrices $\mathbf{N}, \mathbf{D}$ are given by the following:

$$\mathbf{c}_1 = [1 \quad z_1^{-1} \quad z_1^{-2} \quad z_1^{-3}] \quad (3.119)$$
$$\mathbf{c}_2 = [1 \quad z_2^{-1} \quad z_2^{-2} \quad z_2^{-3}] \quad (3.120)$$
$$\mathbf{d}_1 = [z_1^3 \quad z_1^2 \quad z_1 \quad 1 \quad z_1^{-1} \quad z_1^{-2} \quad z_1^{-3}] \quad (3.121)$$

3.3. DESIGN OF FAN FILTERS

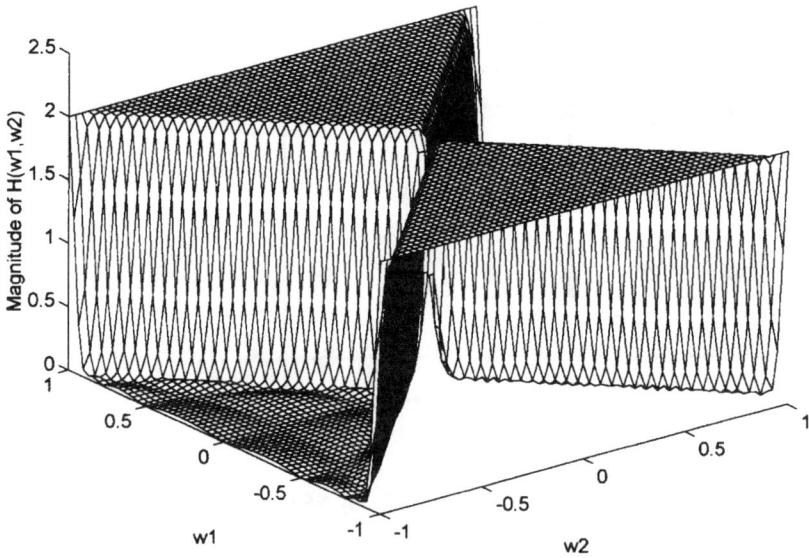

Fig. 3.19. Magnitude response of Zhu-Nakamura fan filter

$$\mathbf{d}_2 = [1 \quad z_2^{-1} \quad z_2^{-2} \quad z_3^{-3}] \qquad (3.122)$$

and

$$\mathbf{N} = \begin{bmatrix} 0.4130 & 0.6390 & 0.2045 & 0.0364 \\ -0.4953 & -0.7232 & -0.4309 & 0.0365 \\ 0.0627 & 0.3232 & 0.0663 & 0.0369 \\ 0.0103 & 0.0546 & -0.0483 & -0.0077 \end{bmatrix} \qquad (3.123)$$

$$\mathbf{D} = \begin{bmatrix} 0.0000 & -0.0083 & 0.0024 & 0.0012 \\ 0.0000 & -0.0721 & 0.1044 & 0.0003 \\ 0.0000 & -0.8787 & 0.0665 & -0.0494 \\ 1.1825 & 0.0103 & 0.6778 & -0.0812 \\ 0.1992 & -1.0150 & 0.1526 & -0.0741 \\ -0.0335 & -0.2063 & 0.2022 & -0.0044 \\ -0.0115 & 0.0024 & 0.0144 & -0.0052 \end{bmatrix} \qquad (3.124)$$

The numerator **N** gives the input mask which lies in the first quadrant. But the output mask given by the denominator **D** has nonsymmetric half plane symmetry; because the first three coefficients in the first column are zero i.e. $d_{ij} = 0$, $i = -3, -2, -1$ and $j = 0$ [and $d_{00} = 1.1825$] in the denominator polynomial $D(z_1^{-1}, z_2^{-1}) = \sum_{i=-3}^{3} \sum_{j=0}^{3} d_{ij} z_1^{-i} z_2^{-j}$. In the matrices **B**, **A**, **N** and **D**, the index i increases from top to bottom and the index j increases from left to right. In the matrix **D**, however, the index i increases from -3 to $+3$ and the index j increases from 0 to $+3$, since it has a non-symmetric half-plane support. The region of support for this NSHP filter is shown in Fig. 3.20.

172 CHAPTER 3. MAGNITUDE OF 2-D FILTERS

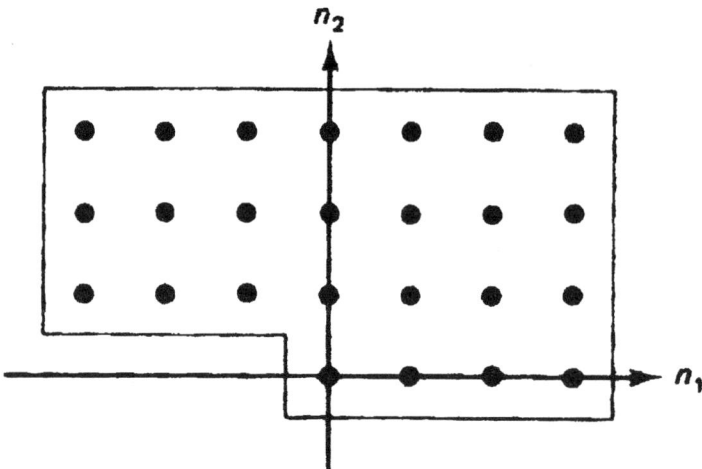

Fig. 3.20. Region of support for $h(n_1, n_2)$ of the NSHP filter

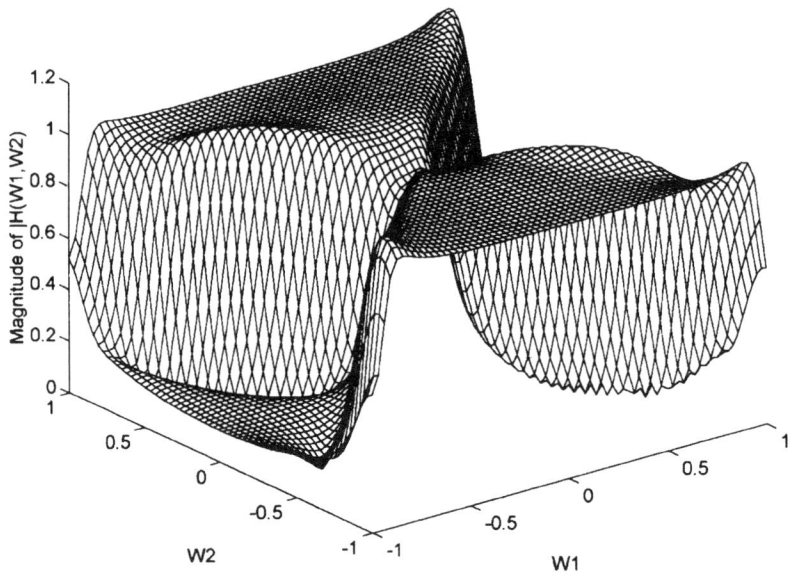

Fig. 3.21. Magnitude response of the fan filter in Example 3.3

The magnitude of the above fan filter is shown in Fig. 3.21. The magnitude response of this fan filter in Fig. 3.21 may be compared with that of Fig. 3.19 and Fig. 3.15. In all three methods, if we assume that the lowpass prototype filter has an ideal magnitude of one in the passband and zero in the stopband, the magnitude characteristic of the fan filter is independent of the method.

3.4. DESIGN OF CIRCULARLY SYMMETRIC FILTERS

In practice, the magnitude responses in the passband regions of the fan filters designed by using method 3.1 and 3.2 are different. However, these two methods are analytical and guarantee stability of the 2-D fan filters, whereas the third method uses a nonlinear optimization algorithm and does not guarantee stability. The first method suffers from two drawbacks that were pointed out earlier. Though the order of the 2-D fan filter designed by method 3.1 is found to be high, it involves only matrix operations and does not take as much computer time (e.g. when MATLAB is used) as the optimization algorithm used in method 3.3. We see that the second method [30] is the best among the three methods described in section 3.3 and the magnitude response of the fan filter is far superior than anything that has been published in the literature. The same method was briefly described earlier by Ansari in [2]. Although we have preferred the analog, lowpass filters of odd order as the prototype filters in the above design procedure, the authors of [30] also discuss the case when the order is even. The resulting fan filters give equally good magnitude response as in the case of odd order. But they require almost twice the computational amount and hence an example for this case is not included here.

3.4 Design of Circularly Symmetric Filters

3.4.1 Survey of the literature

In the extensive literature published on the design of 2-D digital filters, much of the attention is focussed on the design of circularly symmetric lowpass filters. The reason for this emphasis is that once we have designed a circularly symmetric, 2-D IIR, lowpass filter, we can use shifting, rotation, complementary operations, and apply frequency transformations on circularly symmetric filters to generate additional types of filters. A survey of this literature shows that most of the design procedures can be classified into the following four methods.

Method 4.1 Optimization by linear and nonlinear programming.

The next two methods use transformations on a one-dimensional, continuous time (analog), lowpass filter $H(s)$, designed by the well-established theory of filter approximation-which was described in Chap. 1. There are two tracks which use such transformations and they are illustrated below:

Method 4.2

$$H(s)\big|_{s=g(s_1,s_2)} \Rightarrow H(s_1,s_2)\big|_{s_1=\frac{2}{T_1}\left(\frac{z_1-1}{z_1+1}\right), s_2=\frac{2}{T_2}\left(\frac{z_2-1}{z_2+1}\right)}$$
$$\Rightarrow H(z_1,z_2) \tag{3.125}$$

Method 4.3

$$H(s)\big|_{s=\frac{2}{T}\left(\frac{z-1}{z+1}\right)} \Rightarrow H(z)\big|_{z=f(z_1,z_2)} \Rightarrow H(z_1,z_2) \tag{3.126}$$

Method 4.4 This method is a combination of Method 4.1 and either Method 4.2 or 4.3 and has also been used for designing 2-D circularly symmetric lowpass recursive filters.

The first method is very general in scope in the sense that it can be applied for the design of not only lowpass filters with a circular passband but even other types of filters with an arbitrary passband region. Some of the well-known research on the computer-aided design of 2-D recursive (and non-recursive) digital filters has been reviewed by Venetsanopolous in a comprehensive chapter in [29]. It reviews the work published up to 1986 including some methods on the design of 2-D filters that approximate not only the prescribed magnitude but also the group delay $\tau_1 = -\frac{\partial \theta(\omega_1,\omega_2)}{\partial \omega}$ and $\tau_2 = -\frac{\partial \theta(\omega_1,\omega_2)}{\partial \omega_2}$. Significant amount of research on the design of 2-D IIR filters that simultaneously approximate the magnitude and group delay has been reported since 1986 and we will cover this in Chap. 4 in more detail. The methods for the design of 2-D filters that approximate only the magnitude response are therefore special cases of such methods. They will be treated in the remaining part of this chapter. Whenever the design of 2-D IIR digital filters is carried out by linear or nonlinear programming, [Method 4.1] several important issues arise. There is no analytical formula to choose the optimum order for the 2-D filter-only a trial and error method of increasing the order of the filter until the design specifications can be met is the recourse available. Depending on the frequency response specification and the objective function chosen for minimization by nonlinear programming, the amount of computational complexity can be very significant. There is always the vexing problem of checking for stability of the 2-D transfer function at every iteration, and this is a very challenging problem since the testing procedures are neither simple nor reliable.

The problem of guaranteeing stability is of primary concern in Method 4.2 also. Indeed the design procedures are so much concerned with guaranteeing stability of the 2-D transfer function that approximation of the circular contour with a prescribed radius appears to be a secondary objective for the authors who have chosen this method. The transformation $s = g(s_1, s_2)$ must be chosen such that all points on the imaginary axis $s = j\omega$ map into points on the imaginary axes $s_1 = j\omega_1$ and $s_2 = j\omega_2$ i.e., it is globally type-preserving [5]. This means that the magnitude response of a given type (e.g. a lowpass type) is transformed into the magnitude response of the same type for the 2-D transfer function; otherwise the transformation $s = g(s_1, s_2)$ is said to be locally type-preserving.

The transformation $z = f(z_1, z_2)$ proposed in [12] is found to be marginally stable. Hence attention in this method [Method 4.4] is mainly focussed on making it a stable function and then on approximation of a circular contour for the passband by cascading many functions obtained by rotation-just as in the Method 4.2. However, the two transformations $z = f(z_1, z_2)$ proposed by the authors in [14, 15] are á priori chosen to be stable functions and the 2-D IIR transfer functions constructed from the stable 1-D IIR lowpass filter function are guaranteed to be stable. Thus, attention is focussed on choosing optimal values for the coefficients of the digital spectral transformation such that a con-

3.4. DESIGN OF CIRCULARLY SYMMETRIC FILTERS

tour in the $\omega_1 - \omega_2$ plane approximates a circle of specified radius as best as possible. These are the only two transformations in all of published literature that have been shown to preserve not only the magnitude response but also the phase and group delay characteristics [15, 27] in the passband. Hence if the one-dimensional lowpass IIR digital filter $H(z)$ is designed to have a maximally flat magnitude as well as maximally flat group delay characteristics, the 2-D IIR filter obtained through these transformations will not only approximate a maximally flat lowpass magnitude response but also a maximally flat group delay characteristics as well. Because these transformations are able to approximate both the prescribed magnitude as well as group delay response for the 2-D circularly symmetric filters, they will be described in the next chapter. These filters as well as filters designed by some other methods to be described in the next chapter are guaranteed to be stable.

Since the transformations that are used in Method 4.2 are able to approximate only the magnitude response of the circularly symmetric lowpass filters, we will describe them in the remainder of this chapter.

3.4.2 Design Procedures

Shanks, Treitel and Justice [26] were the first authors who proposed the transformations

$$s = g_1(s_1, s_2) = -s_1 \sin \beta + s_2 \cos \beta \qquad (3.127)$$
$$s = g_2(s_1, s_2) = s_1 \cos \beta + s_2 \sin \beta \qquad (3.128)$$

for the design of 2-D IIR digital filters, where $0 < \beta < \frac{\pi}{2}$. So first an analog, 1-D lowpass filter $H_{A1}(s)$ is designed with a desired magnitude response characteristic like the Butterworth or Chebyshev response over a normalized bandwidth of one radian/sec - using classical approximation theory discussed in Chap. 1. When one of the above transformations is applied to $H_{A1}(s)$, a 2-D analog filter function $H_{A2}(s_1, s_2)$ is generated according to (3.129).

$$H_{A2}(s_1, s_2) = H_{A1}(s)\big|_{s=g_i(s_1,s_2)} \quad \text{for} \quad i = 1 \text{ or } 2 \qquad (3.129)$$

The corresponding mapping function of all points $s = j\omega$ into $s_1 = j\omega_1$, $s_2 = j\omega_2$ is described by $\omega_2 = \omega_1 \tan \beta + \frac{\omega}{\cos \beta}$ or $\omega_2 = -\omega_1 \cot \beta + \frac{\omega}{\sin \beta}$ under the transformations g_1 and g_2 respectively. It can be shown that both g_1 and g_2 satisfy the conditions for global type preservation which are given in [5]. We will choose the transformation (3.127) in which β takes many values β_k, to obtain $H_{A2}(s_1, s_2)$. When the bilinear transformations $s_1 = \frac{2}{T_1}\left(\frac{z_1-1}{z_1+1}\right)$, and $s_2 = \frac{2}{T_2}\left(\frac{z_2-1}{z_2+1}\right)$ are applied on $H_{A2}(s_1, s_2)$, it is converted to a 2-D digital function $H_{D2k}(z_1, z_2)$. If $H_{A1}(s)$ is assumed to be

$$H_{A1}(s) = K_0 \frac{\prod_{i=1}^{M_N}(s - z_{ai})}{\prod_{i=1}^{M_D}(s - p_{ai})} \qquad (3.130)$$

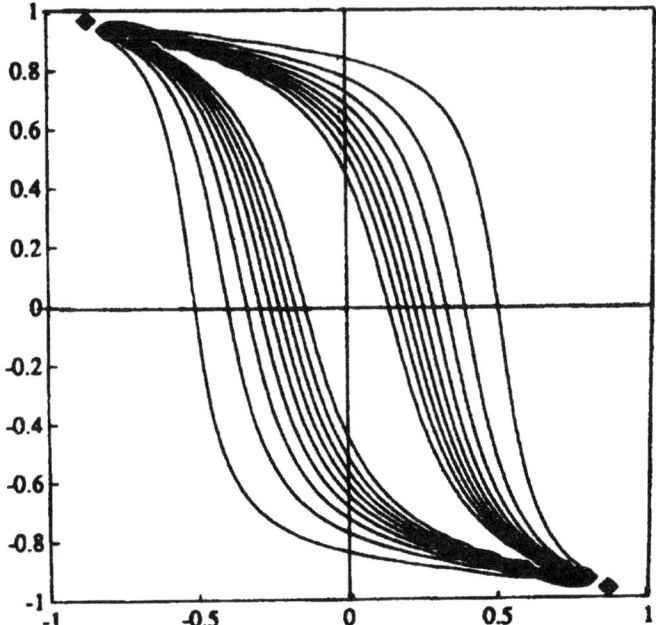

Fig. 3.22. Contours of constant magnitude for the rotated filter

the resulting 2-D digital transfer function under the above transformations is of the form

$$H_{D2k}(z_1, z_2) = K_1 \frac{\prod_{i=1}^{M_N}(a_{11i} + a_{21i}z_1 + a_{12i}z_2 + a_{22i}z_1z_2)}{\prod_{i=1}^{M_D}(b_{11i} + b_{21i}z_1 + b_{12i}z_2 + b_{22i}z_1z_2)} \qquad (3.131)$$

where the coefficients depend on the value chosen for β_k, besides the poles and zeros z_{ai} and p_{ai}. The magnitude response of this transfer function shows closed contours in the $\omega_1 - \omega_2$ plane. As an example, the magnitude plot for an angle $\beta_k = 285°$ is shown in Fig. 3.22.

As the angle is changed, the orientation of the plot changes accordingly. So Costa and Venetsanopolous [7] proposed cascading multiple 2-D transfer functions to approximate a circularly symmetric filter. These transfer functions are generated by choosing different values for $\beta_k, k = 1, 2, 3, \ldots, K$ as given by

$$\beta_k = \begin{cases} [(2k-1)/2K + 1]\pi & \text{for} \quad \text{even} \quad K \\ [(k-1)/K + 1]\pi & \text{for} \quad \text{odd} \quad K \end{cases} \qquad (3.132)$$

so that the resulting 2-D transfer function is given by

$$H(z_1, z_2) = \prod_{k=1}^{K} H_{D2k}(z_1, z_2) \qquad (3.133)$$

3.4. DESIGN OF CIRCULARLY SYMMETRIC FILTERS

The magnitude response of this filter is the product of the magnitude response of the constituent transfer functions $H_{D2k}(z_1, z_2)$ and it approaches a circular contour as K becomes large.

It has been found that for some values of β_k, the corresponding coefficients of the numerator and denominator term in (3.131) may have the same sign in which case, it is easily seen that they are both zero at $z_1 = -1$ and $z_2 = -1$. Such points on $\{|z_1| = 1 \bigcup |z_2| = 1\}$ are known as non-essential singularities of the second kind [26]. The presence of such singularities in the 2-D transfer function negates the well-known tests for stability of 2-D transfer functions of IIR digital filters or may give rise to unstable digital filters without our knowledge. It has been suggested that the non-essential singularities of the second kind can be eliminated by changing the coefficients b_{12i}, b_{22i} by a small amount. Another suggestion is to eliminate the choice of those angles β_k, which give rise to these singularities. However an elimination of transfer functions generated from such angles, would distort the overall closed contour considerably. To avoid these problems, new transformations have been proposed by some authors, e.g. the transformation given by (3.134) was used in [1]. However it is found to be locally type preserving, because for values of $s_1 = j\omega_1$ and $s_2 = j\omega_2$ at which $\omega_1 \omega_2 = 1/c$, the denominator of $g_3(s_1, s_2)$ has a zero value and the mapping $g_3(j\omega_1, j\omega_2)/j$ is not continuous for all values of ω_1 and ω_2.

$$s = g_3(s_1, s_2) = \frac{a_1 s_1 + a_2 s_2}{1 + c s_1 s_2} \tag{3.134}$$

Another transformation obtained by combining the transformations $g_2(s_1, s_2)$ and $g_3(s_1, s_2)$ was proposed in [22], which is given below:

$$s = g_4(s_1, s_2) = \frac{s_1 \cos \beta + s_2 \sin \beta}{1 + c s_1 s_2} \tag{3.135}$$

If we choose the values for β_k such that $\cos \beta_k > 0$, $\sin \beta_k > 0$ and $c > 0$, the resulting 2-D transfer function $H_{D2k}(z_1, z_2)$ is found to be free of non-essential singularities of the second kind. But it is still a locally type preserving transformation - though by choosing a small value for c, a globally type preserving transformation can be approached as closely as desired. So it serves as a compromise between the two transformations proposed in [26] and [1] and the mapping from ω to the $\omega_1 - \omega_2$ plane is not truly a rotation by β, except when $c = 0$. Hence the authors call it a "pseudorotation".

Let us assume that the 1-D digital transfer function is

$$H_{a1}(s) = K_0 \frac{\prod_{i=1}^{M_N}(s - z_{ai})}{\prod_{i=1}^{M_D}(s - p_{ai})} \tag{3.136}$$

where z_{ai} and p_{ai}, $i = 1, 2 \ldots$ are respectively the zeros and poles of the IIR digital transfer function and K_0 is a multiplier constant. When the transformation (3.135) is applied to (3.136), we get a 2-D analog transfer function

$$H_{a2}(s_1, s_2) = K_0 P_{a2}(s_1, s_2)$$

$$\times \left\{ \frac{\prod_{i=1}^{M_N}(s_1\cos(\beta)+s_2\sin(\beta)-z_{ai}(1+cs_1s_2)}{\prod_{i=1}^{M_D}(s_1\cos(\beta)+s_2\sin(\beta)-p_{ai}(1+cs_1s_2)} \right\}$$

where

$$P_{a2}(s_1, s_2) = (1+cs_1s_2)^{M_D-M_N} \qquad (3.137)$$

By using the double bilinear transformation $s_1 = \frac{2}{T_1}\left(\frac{z_1-1}{z_1+1}\right)$, and $s_2 = \frac{2}{T_2}\left(\frac{z_2-1}{z_2+1}\right)$ to (3.137), with $T_1 = T_2 = T$ and we get

$$\begin{aligned} H_{d2}(z_1,z_2) &= K_1 P_{d2}(z_1,z_2) \\ &\times \left\{ \frac{\prod_{i=1}^{M_N}(a_{00i}+a_{10i}z_1+a_{01i}z_2+a_{11i}z_1z_2)}{\prod_{i=1}^{M_D}(b_{00i}+b_{10i}z_1+b_{01i}z_2+b_{11i}z_1z_2)} \right\} \end{aligned} \qquad (3.138)$$

where

$$K_1 = K_0\left(\frac{T}{2}\right)^{M_D-M_N}$$

$$P_{d2}(z_1,z_2) = \left[\begin{array}{c} \left(1+\frac{4c}{T^2}\right)+\left(1-\frac{4c}{T^2}\right)z_1 \\ +\left(1-\frac{4c}{T^2}\right)z_2+\left(1+\frac{4c}{T^2}\right)z_1z_2 \end{array} \right]^{M_D-M_N} \qquad (3.139)$$

and

$$a_{00i} = -\cos(\beta)-\sin(\beta)-z_{ai}\left(\frac{T}{2}+\frac{2c}{T}\right)$$

$$a_{10i} = \cos(\beta)-\sin(\beta)-z_{ai}\left(\frac{T}{2}-\frac{2c}{T}\right)$$

$$a_{01i} = -\cos(\beta)+\sin(\beta)-z_{ai}\left(\frac{T}{2}-\frac{2c}{T}\right)$$

$$a_{11i} = \cos(\beta)+\sin(\beta)-z_{ai}\left(\frac{T}{2}+\frac{2c}{T}\right)$$

Similarly,

$$b_{00i} = -\cos(\beta)-\sin(\beta)-p_{ai}\left(\frac{T}{2}+\frac{2c}{T}\right)$$

$$b_{10i} = \cos(\beta)-\sin(\beta)-p_{ai}\left(\frac{T}{2}-\frac{2c}{T}\right)$$

$$b_{01i} = -\cos(\beta)+\sin(\beta)-p_{ai}\left(\frac{T}{2}-\frac{2c}{T}\right)$$

$$b_{11i} = \cos(\beta)+\sin(\beta)-p_{ai}\left(\frac{T}{2}+\frac{2c}{T}\right) \qquad (3.140)$$

It can be shown that the above transfer function is free from the nonessential singularities of the second kind, by evaluating it at $z_1 = \pm, z_2 = \pm 1$. It can also be shown that when $\cos(\beta) > 0$, $\sin(\beta) > 0$ and $c>0$, the transfer function

3.4. DESIGN OF CIRCULARLY SYMMETRIC FILTERS

is BIBO stable. So the angle β is restricted to $(0°, 90°)$. But the magnitude response of such a BIBO stable function has a shape similar to that shown in Fig. 3.22 which is observed to have a poor approximation to a circular contour in the first and third quadrant and have only a half plane symmetry. Just as in the method of [7], the authors cascade transfer functions rotated by several angles $\beta_r = r\beta_N$, $r = 1, 2, \ldots, N$ where $\beta_N = \frac{90}{N+1}$; these angles are all in the range $(0°, 90°)$. In order to achieve a good approximation to a circular contour, the authors find it is necessary to choose the angles in the range $(90°, 360°)$ but the resulting transfer functions are found to be unstable. They suggest that this problem can be avoided by transforming the input data of the unstable function and filtering through a corresponding stable filter. Since N is the number of rotated sections recursing in the ++ direction, we have $0° < \beta_r < 90°$, and the total number of cascaded sections used to get a circularly symmetric contour is $4N$.

The authors have designed a large number of circularly symmetric lowpass filters with a passband response that is Butterworth, Chebyshev or elliptic in nature, with different values for the radius of the circular region and for the number of rotations N. The variance for the error in approximating the specified circular contour that defines the passband is defined by them to be

$$\sigma^2 = \frac{1}{J-1} \sum_{j=1}^{J} (\overline{\omega}_p - \omega_{pj})^2 \qquad (3.141)$$

where ω_{pj} are points on the actual passband boundary at which the gain is the specified passband gain, and $\overline{\omega}_p$ is the mean of all ω_{pj} chosen for computing the variance. From the results of simulating such a large number of filters, they have derived an empirical formula as a function of the passband radius for a fixed value for N, to compute the variance for the Butterworth, Chebyshev, elliptic filter. This is repeated for values of $N = 1, 2, \ldots, 15$. The empirical formula obtained by using regression analysis, for the variance σ^2 is in the form $\sigma^2 = A_N + B_n x + C_N x^2 + D_N x^3$, where $x = \frac{2\pi(passband\ frequency)}{(sampling\ frequency)}$ is the normalized passband edge. The values for the parameters A_N, B_N, C_N and D_N for the three types of passband response and each value of $N = 1, 2, \ldots, 15$ have been provided by the authors. This data is helpful in deciding on the number of cascaded sections, when the type of passband response is specified or chosen, in order to achieve a value for σ^2 (which may be specified or may have to be chosen on the basis of experience). They state that their method for lowpass filter design, using the above data is unreliable when $x < 0.5$. But as pointed in [14], one can design a filter with twice the specified radius, and then replace z_1, and z_2 in that filter by z_1^2 and z_2^2 respectively, thereby scaling the frequency by 0.5.

When the maximum ripple in the circular passband and the minimum attenuation in the stopband for the 2-D IIR filter is specified, how do we choose the corresponding values for the analog filter (3.136)? Again the authors arrive at some guidelines for choosing the values required for designing the analog filter

$H_{ai}(s)$, after some simple analysis based on the fact that the analog filter is not an ideal lowpass filter.

They also give the formula for the number of multiplications n_m and additions n_a required to implement the zero phase transfer function $G(z_1, z_2) = H(z_1, z_2)H(z_1^{-1}, z_2^{-1})$. Both n_m and n_a are approximately equal to $(32M_D N)$, where M_D is the order of the 1-D transfer function (3.130) and N is the minimum number of rotations. That is a very large number comparable to the number of multiplications and additions required in the method of [14].

The authors in [22] have shown how their method can be used to design highpass, bandpass and bandstop filters also. Sufficient details are given for designing 2-D highpass filters. They also state that to the best of their knowledge, no other technique was available for the design of 2-D, highpass filters satisfying prescribed specifications. It is pointed out that the method presented in [14] has been used to design highpass filters, by choosing the 1-D digital filter $H(z)$ to be a highpass filter, obtaining a zero phase filter and applying the complimentary operation described in section 3.2. This will be illustrated in the next chapter.

Goodman[12] proposed a digital spectral transformation that belongs to the Method 4.3. This transformation (3.142) converts a 1-D digital IIR filter $H_{D1}(z)$ to a 2-D IIR filter $H_{D2k}(z_1, z_2)$.

$$z = f_k(z_1, z_2) = \frac{1 + c_k z_1 + d_k z_2 + e_k z_1 z_2}{e_k + d_k z_1 + c_k z_2 + z_1 z_2} \qquad (3.142)$$

where

$$c_k = \frac{1 + \sin\beta_k + \cos\beta_k}{1 - \sin\beta_k + \cos\beta_k}$$
$$d_k = \frac{1 - \sin\beta_k - \cos\beta_k}{1 - \sin\beta_k + \cos\beta_k} \qquad (3.143)$$
$$e_k = \frac{1 + \sin\beta_k - \cos\beta_k}{1 - \sin\beta_k + \cos\beta_k} \qquad (3.144)$$

The transfer functions $H_{D2k}(z_1, z_2) = H_{D1}(z)|_{z=f_k(z_1,z_2)}$ are generated for β_k, $k = 1, 2, \ldots, K$ and they are cascaded as in the methods of [7] and [22]. The resulting filter has the same stability problem due to the nonessential singularities of the second kind at $z_1 = -1$, $z_2 = -1$ and the author suggests that the denominator coefficients be changed by a small amount to achieve a stable filter. Again the attention is mainly focussed on achieving a stable function and not on approximating the circular contour. However, he suggested that optimization may be used to achieve a close approximation of the circular contour, subject to constraints which assure that the transfer function is stable.

We may recollect that most of the literature on the classical theory of analog filter approximation is aimed at the approximation of the lowpass prototype filter. The main reason for this emphasis is that the design of highpass, bandpass and bandstop filters can be reduced to that of a lowpass prototype filter by use of appropriate analog frequency transformations and once the design of

3.4. DESIGN OF CIRCULARLY SYMMETRIC FILTERS 181

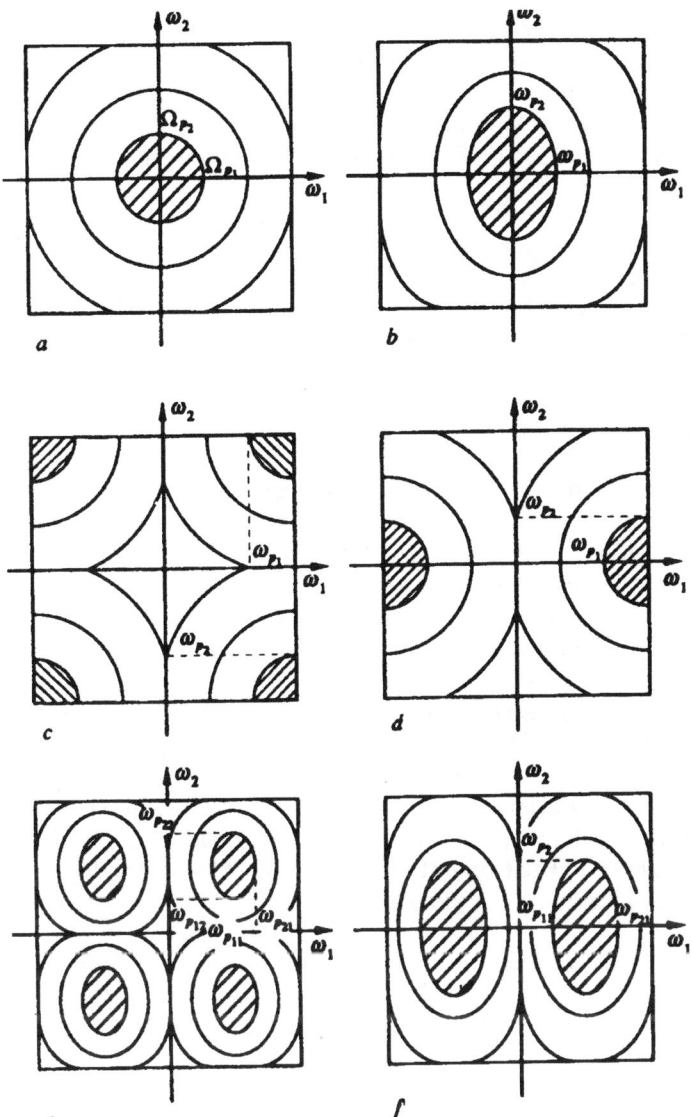

Fig. 3.23. Application of 2-D to 2-D digital spectral transformations (By permission of IEEE). **a** Lowpass prototype filter **b** LP to LP for z_1 and z_2 **c** LP to HP for for z_1 and z_2 **d** LP to HP for z_1 and LP to LP for z_2 **e** LP to BP for z_1 and z_2 **f** LP to BP for z_1 and LP to LP for z_2

such a lowpass prototype filter is completed, the transfer function of the highpass, bandpass or bandstop filter can be easily generated by applying the same transformation. Similarly, the design theory of 1-D, IIR filters uses the digital spectral transformations proposed by Constantinides [6] that convert the transfer function of a lowpass filter with a known passband, to another lowpass filter with a different passband, as well as highpass, bandpass and bandstop filters. These transformations convert 1-D IIR filters to 1-D IIR filters and they were described in Chap. 1. The authors of [25] show how these transformations can be extended to convert a 2-D IIR circularly symmetric lowpass filter to 2-D lowpass filters with a circular or elliptic passband region also to highpass, bandpass and bandstop filters. The mapping of these transformations is shown in Fig. 3.23. One can also use some of the operations such as shifting, and complementary operation that were described in section 3.2, to generate these filters.

3.5 Conclusion

In the first part of this chapter, methods for designing 2-D IIR filters with a passband or stopband region, which is bounded by straight lines parallel to the ω_1 and the ω_2 axes were discussed. Next, three methods for the design of 90° fan filters were described and one of them was found to be superior to the other two methods. Then we considered four different approaches for the design of circularly symmetric filters and concluded that the application of Digital Spectral Transformations (DST) to transform a 1-D IIR filter to a 2-D IIR filter is the more promising approach. The two DSTs proposed in [14] and [15] can be used to design 2-D filters to approximate not only the specified magnitude but also a constant group delay. Use of these two DSTs offers many features and advantages like guaranteed a' priori stability of the 2-D filters and a simpler design procedure and numerical complexity, which make it a generalized approach for the design of 2-D IIR digital filters [15, 16]. This method will be described in greater detail in the next chapter.

Bibliography

[1] M. Ahmadi, A.G. Constantinides and R.A. King, "Design technique for a class of stable two-dimensional recursive digital filters," Proc. IEEE Int'l Conf. on Acoustics, Speech and Signal Processing, pp. 145-147, 1976.

[2] R. Ansari, "A simple derivation for the synthesis procedure for 90° fan filters," IEEE Trans. on Circuits and Systems, vol. CAS-32, pp. 107-108, January 1985.

[3] R. Ansari, "Elliptical filter design for a class of generalized halfband filters," IEEE Trans. on Acoustics, Speech and Signal Processing, Vol. ASSP-33, pp. 1146-1150, October 1985.

[4] A. Antoniou and W-S. Lu, "Design of 2-D nonrecursive filters using window method," IEE Proc. vol. 137, Part G, pp. 247-250, August 1990.

[5] S. Chakrabarti and S.K. Mitra, "Design of two-dimensional digital filters via spectral transformation," Proc.IEEE, pp. 905-914, June 1977.

[6] A.G. Constantinides, "Spectral transformations for digital filters," IEE Proc, vol. 117, pp. 1585-1590, August 1970.

[7] J.M. Costa and A.N. Venetsanopolous, "Design of circularly symmetric two-dimensional recursive filters," IEEE Trans. on Acoustics, Speech and Signal Processing, vol. ASSP-22, pp. 432-443, December 1974.

[8] M.P. Ekstrom, R.E. Twogood and J.W. Woods, "Two-dimensional recursive filter design - A Spectral Factorization Approach," IEEE Trans. ASSP-28, pp. 16-26, February 1980.

[9] G. Garibotto and R. Molpen, "A new approach to half-plane recursive filter design," IEEE Trans. on Acoustics, Speech and Signal Processing, vol. ASSP-29, pp. 111-115, February 1981.

[10] L. Gazsi, "Explicit formulas for lattice wave digital filters," IEEE Trans. on Circuits and Systems, vol. CAS-32, pp. 68-88, January 1985.

[11] A.P. Gerheim, "A synthesis procedure for 90° fan filters," ibid, pp. 858-864, December 1983.

[12] D.M. Goodman, "A design technique for circularly symmetric lowpass filters," IEEE Trans. on Acoustics, Speech and Signal Processing, vol. ASSP-26, pp. 290-304, August 1978.

[13] L. Harn, Design of Two-Dimensional Recursive Digital Filters, Ph.D.thesis, University of Minnesota, 1984.

[14] L. Harn and B.A. Shenoi, "Design of stable two-dimensional IIR filters using digital spectral transformations," IEEE Trans. on Circuits and Systems, vol. CAS-33, pp. 483-490, May 1986.

[15] Rajamohana Hegde and B.A. Shenoi, "Design of 2-D, IIR filters using a new digital spectral transformation", Proc. IEEE Int'l Sympo.on Circuits and Systems, vol. II, pp. 344-347, 1995.

[16] Rajamohana Hegde and B.A. Shenoi, "A unified approach to the design of 2-D IIR digital filters with flat passband magnitude and delay characteristics," Proc. Int'l Symp. on Circuits and Systems, vol. 1, pp. 765-768, 1997

[17] K. Hirano and J.K. Aggarwal, "Design of two-dimensional recursive digital filters," IEEE Trans. on Circuits and Systems, vol. CAS-25, pp. 1066-1076, December 1978.

[18] K. Hirano, H. Sakaguchi and S.K. Mitra, "Explicit design formulae for digital fan filters with low-pass, high-pass, band-pass, and band-stop characteristics," J. Franklin Inst. vol. 307, pp. 263-290, May 1979.

[19] A.H. Kayran and R.A. King, "Design of recursive and nonrecursive fan filters with complex transformation," IEEE Trans. on Circuits and Systems, vol. CAS-30, pp. 849-857, December 1983.

[20] W-S. Lu and A. Antoniou, *Two-Dimensional Digital Filters*, Marcel-Dekker Inc, 1992

[21] J.H. McClellan and T.W. Parks, "Equiripple approximation of fan filters," Geophysics, vol. 7, pp. 573-583, 1972.

[22] G.V. Mendonca. A. Antoniou, and A.N. Venetsanopolous," Design of two-dimensional pseudorotated digital filters satisfying prescribed specifications," IEEE Trans. on Circuits and Systems, vol. CAS-34, pp. 1-10, January 1987.

[23] R.M. Mersereau, W.F.G. Mecklenbrauker and T.F. Quatieri, "McClellan's transformations for two-dimensional digital filtering: I-Design," IEEE Trans. on Circuits and Systems, vol. CAS-23, pp. 405-414, July 1976.

[24] R.M. Mersereau, "The design of arbitrary 2-D zero phase FIR filters using transformations," IEEE Trans. on Circuits and Systems, vol. CAS-27, pp. 142-144, February 1980.

[25] N.A. Pendergras, S.K. Mitra and E.I. Jury, "Spectral transformations for two-dimensional recursive filters," IEEE Trans. on Circuits and Systems, vol. CAS-23, pp. 26-35, January 1976.

[26] J.L. Shanks, S. Treitel and J.H. Justice, "Stability and synthesis of two-dimensional recursive filters," IEEE Trans. on Audio Electroacoustics, vol AU-20, pp. 115-128, June 1972.

[27] B.A. Shenoi and P. Misra, "Design of two-dimensional IIR filters with linear phase," IEEE Trans. on Circuits and Systems II: vol. 42, pp. 124-129, February 1995.

[28] S. Treitel, J.L. Shanks and C.W. Fraisier, "Some Aspects of Fan Filtering," Geophysics, vol. 32, October 1967.

[29] S.G. Tzafestas, (Ed.) *Multi-dimensional Systems: Techniques and Applications*, Marcel Dekker Inc, 1986.

[30] Wei-Ping Zhu and Shogo Nakamura, "An efficient approach for the synthesis of 2-D recursive fan filters using 1-D prototypes," IEEE Trans. on Signal Processing, SP-44, pp. 979-983, April 1996.

[31] Wei-Ping Zhu, "Design of 2-D Recursive Fan Filters via the Butterworth Approximation," Proc. of ICNNS'95, Nanjing, December 1995.

Chapter 4

Magnitude and Delay Approximation of 2-D IIR Filters

4.1 Introduction

In this chapter, we will discuss a few methods that are available for the design of two-dimensional (2-D) IIR filters which approximate not only a specified magnitude response but also a constant group delay. The first three methods assume a certain structure for the desired transfer function that allows the computer-aided optimization procedure to assure stability of the 2-D filter. Optimization of a suitably defined performance index (objective function) is carried out such that the magnitude and group delay are approximated in either maximally flat, or equiripple sense but more often in the least squares sense. Depending on the performance index chosen and the optimization criteria, the complexity of computation can be very high; but irrespective of such choices, there is always the problem of choosing an optimal order for the desired filter. If only the magnitude is to be approximated and the specified response is piece-wise constant over finite regions like circular, or elliptic regions in the $\omega_1 - \omega_2$ plane, then these approximation techniques have to try different order for the filter and choose the 'best' design that meets the magnitude specifications. These three methods can be applied to the design of 1-D IIR filters as a special case of the 2-D IIR filter approximation problem. In turn, these methods can be viewed as extensions of the methods used for approximating the magnitude and group delay response of the 1-D IIR filters.

In the second half of this chapter, we will describe some recent developments in the theory for designing 2-D IIR filters which approximate both the prescribed magnitude and group delay. In contrast to the above methods that use nonlinear programming, these new methods follow a more analytical approach in solving the problem, though they do use computer-aided algorithms, to implement the design procedures. Two of the important applications of 2-D filters are currently found in image processing and seismic signal processing. It has been shown [20, 24] that in image processing the phase response of a 2-D filter is much more important than the magnitude response and usually a zero phase or a linear phase response (corresponding to a constant group delay) is preferable in many cases.

4.2 Design using Nonlinear Programming

4.2.1 Statement of the problem

The direct form for the transfer function of a 2-D IIR digital filter is

$$H(z_1^{-1}, z_2^{-1}) = \frac{\sum_{j=0}^{J_2} \sum_{i=0}^{J_1} b_{ij} z_1^{-i} z_2^{-j}}{\sum_{j=0}^{I_2} \sum_{i=0}^{I_1} d_{ij} z_1^{-i} z_2^{-j}} \quad (4.1)$$

In the design of a 2-D IIR filter using nonlinear programming, however, we assume the transfer function in a particular form as given by (4.2).

$$H_d(z_1^{-1}, z_2^{-1}) = A \prod_{k=1}^{K} \frac{\sum_{j=0}^{J_2^{(k)}} \sum_{i=0}^{J_1^{(k)}} b_{ij}^{(k)} z_1^{-i} z_2^{-j}}{\sum_{j=0}^{I_2^{(k)}} \sum_{i=0}^{I_1^{(k)}} d_{ij}^{(k)} z_1^{-i} z_2^{-j}} = A \frac{N(z_1^{-1}, z_2^{-1})}{D(z_1^{-1}, z_2^{-1})} \quad (4.2)$$

It shows that $H_d(z_1^{-1}, z_2^{-1})$ consists of cascade of K filter sections where $b_{ij}^{(k)}$ and $d_{ij}^{(k)}$; $k = 1, 2, \ldots, K$ are the unknown coefficients of each section, and A is a gain constant that may be left arbitrary during the design procedure.

Further, we restrict $I_1^{(k)}$, $I_2^{(k)}$, $J_1^{(k)}$ and $J_2^{(k)}$ to be *zero, one* or *two* only, in order that we can use a method to assure the stability of the filter.

Two examples of 2-D polynomials are given below, in which we have chosen $d_{00}^{(k)} = 1$.

$$\begin{aligned}
D_1(z_1^{-1}, z_2^{-1}) &= 1 + d_{10}^{(1)} z_1^{-1} + d_{20}^{(1)} z_1^{-2} + d_{01}^{(1)} z_2^{-1} + d_{11}^{(1)} z_1^{-1} z_2^{-1} \\
&\quad + d_{21}^{(1)} z_1^{-2} z_2^{-1} + d_{02}^{(1)} z_2^{-2} + d_{12}^{(1)} z_1^{-1} z_2^{-2} + d_{22}^{(1)} z_1^{-2} z_2^{-2} \quad (4.3) \\
D_2(z_1^{-1}, z_2^{-1}) &= 1 + d_{10}^{(2)} z_1^{-1} + d_{01}^{(2)} z_2^{-1} + d_{11} z_1^{-1} z_2^{-1} \quad (4.4)
\end{aligned}$$

A familiar and useful form of representing the above polynomials is the matrix product form as given below:[1]

$$D_1(z_1^{-1}, z_2^{-1}) = \begin{bmatrix} 1 & z_1^{-1} & z_1^{-2} \end{bmatrix} \begin{bmatrix} 1 & d_{01}^{(1)} & d_{02}^{(1)} \\ d_{10}^{(1)} & d_{11}^{(1)} & d_{12}^{(1)} \\ d_{20}^{(1)} & d_{21}^{(1)} & d_{22}^{(1)} \end{bmatrix} \begin{bmatrix} 1 \\ z_2^{-1} \\ z_2^{-2} \end{bmatrix} \quad (4.5)$$

and

$$D_2(z_1^{-1}, z_2^{-1}) = \begin{bmatrix} 1 & z_1^{-1} \end{bmatrix} \begin{bmatrix} 1 & d_{01}^{(2)} \\ d_{10}^{(2)} & d_{11}^{(2)} \end{bmatrix} \begin{bmatrix} 1 \\ z_2^{-1} \end{bmatrix} \quad (4.6)$$

---

[1] If the coefficient m_{ij} represents the matrix element in the ith row and j^{th} column, then $D_K(z_1^{-1}, z_2^{-1})$ takes the form $\sum_{j=1}^{I_2(k)+1} \sum_{i=1}^{I_1(k)+1} m_{ij} z_1^{-(i-1)} z_2^{-(j-1)}$.

4.2. DESIGN USING NONLINEAR PROGRAMMING

The order of the polynomial is denoted as the order of the matrix used to define it and hence the order of $D_1(z_1^{-1}, z_2^{-1})$ is 3×3 whereas that of $D_2(z_1^{-1}, z_2^{-1})$ is 2×2. Some authors count the order of the transfer function $H_d(z_1^{-1}, z_2^{-1})$ given above as $\sum_{k=1}^{K}(I_1^{(k)} + I_2^{(k)})$ but that can be ambiguous when the denominator has to be expanded in the form $\sum_{j=0}^{I_2} \sum_{i=0}^{I_1} d_{ij} z_1^{-i} z_2^{-j}$ or in the matrix form.

Keeping the notation used in the original papers cited below, we express the frequency response of the filter as

$$H_d(e^{-j\omega_1}, e^{-j\omega_2}) = |H_d(e^{-j\omega_1}, e^{-j\omega_2})| e^{j\phi(\omega_1,\omega_2)} \tag{4.7}$$

The magnitude response is given by

$$H_d(\omega_1, \omega_2) = |H_d(e^{-j\omega_1}, e^{-j\omega_2})| \tag{4.8}$$

and the group delay functions are given by

$$\tau_{di}(\omega_1, \omega_2) = \frac{-\partial \phi(\omega_1, \omega_2)}{\partial \omega_i}, \quad i = 1, 2 \tag{4.9}$$

Eqns (4.8) and (4.9) represent the magnitude and delay functions of the desired 2-D transfer function $H_d(z_1^{-1}, z_2^{-1})$ in which b_{ij}, d_{ij}, A and K are the unknown parameters. Let us define a parameter vector formed by them as
$\mathbf{b} = [(((d_{ij}^{(k)}, i = 0, 1, \ldots, I_1^{(k)}), j = 0, 1, \ldots, I_2^{(k)}), k = 1, 2, \ldots, K),$
$(((b_{ij}^{(k)}, i = 0, 1, \ldots, J_1^{(k)}), j = 0, 1, \ldots, J_2^{(k)}), k = 1, 2, \ldots, K), A]^T$
with the condition $(i, j) \neq (0, 0)$.

Let $(\omega_{1m}, \omega_{2n}), m = 1, 2, \ldots, M, n = 1, 2, \ldots, N$ be the $M \times N$ array of discrete points in the frequency plane at which the magnitude and the delay functions are calculated. Let us denote $H_d(e^{-j\omega_{1m}}, e^{-j\omega_{2n}})$ by $H_d(m, n)$ and $\tau_{di}(\omega_{1m}, \omega_{2n})$ by $\tau_{di}(m, n)$. Similarly, we use the notation $G(m, n)$ and $\tau_i(m, n)$ to denote the specified magnitude and group delays.

$$G(m, n) = |H(e^{-j\omega_{1m}}, e^{-j\omega_{2n}})| \tag{4.10}$$

and

$$\tau_i(m, n) = \tau_i(\omega_{1m}, \omega_{2n}); \quad i = 1, 2 \tag{4.11}$$

With the above notations, we can now state the general problem as follows: Find the values of the parameters in $\mathbf{b}$ such that (1) the magnitude and delay functions $H_d(m, n)$ and $\tau_{di}(m, n)$ approximate the given magnitude and delay functions $G(m, n)$ and $\tau_i(i = 1, 2)$ and (2) the filter is stable.

In order to solve the above problem by nonlinear programming, Aly and Fahmy [1] choose a single performance index or an error function $E(\mathbf{b})$ in the form

$$E(\mathbf{b}) = R_m E_m(\mathbf{b}) + R_{\tau_1} E_{\tau_1}(\mathbf{b}) + R_{\tau_2} E_{\tau_2}(\mathbf{b}) \tag{4.12}$$

where

$$E_m(\mathbf{b}) = \sum_{m=1}^{M} \sum_{n=1}^{N} W(m, n)(G(m, n) - H_d(m, n))^p \tag{4.13}$$

and

$$E_{\tau_i}(\mathbf{b}) = \sum_{m \in M_\tau} \sum_{n \in N_\tau} w(m,n)(\tau_i(m,n) - \tau_d(m,n))^p; \quad i = 1,2 \qquad (4.14)$$

In the above equations, the error function $E(\mathbf{b})$ to be minimized is the weighted sum of three error functions which are respectively the error functions for the magnitude and group delays. Each of these three error functions in turn is again weighted by $W(m,n)$ or $w(m,n)$. Since the values of the three weighted error functions $E_m(\mathbf{b}), E_{\tau_i}(\mathbf{b})$ ($i = 1,2$) have widely different orders of magnitude, they are weighted by the relative weights denoted as R_m, R_{τ_1} and R_{τ_2} respectively. $M_\tau \times N_\tau$ represent the number of discrete frequencies at which the group delays $\tau_i(m,n)$ are required to match $\tau_{d_i}(m,n)$ and often this set of frequencies is confined to the passband region of the filter, whereas $M \times N$ represent the number of frequencies at which the magnitude $H_d(m,n)$ is required to match $G(m,n)$.

4.2.2 Stability conditions

In any general method for the design of 2-D IIR filters by the use of nonlinear programming techniques, we have to keep a few concerns in mind. For example, there is no formula or guideline for choosing the order of the filter, and the initial values for the coefficients of the transfer function. Hence we may have to try filters of different order and start the iterative procedure with different arbitrary values for the coefficients in order to improve the convergence rate. Besides, in each iterative step, we have to test if the interim transfer function is stable. Such problems in general take a lot of computational effort, with no guarantee that the effort will end with a stable filter function.

But notice that we have expressed (4.2) as the product of polynomials of the form denoted by D_1 and D_2 in (4.3) and (4.4). So the transfer function represents a cascade connection of filters of order no more than 2×2. There are many advantages in choosing such a structure for designing 2-D filters [34] as listed below:

1. The word length that guarantees the stability of the filter, when the coefficients of the filter are rounded is shorter than what is needed by a filter realized in the direct form given by (4.1).

2. The error in the frequency response due to quantization and finite word length of the coefficients is smaller.

3. The number of multiplications per output point is smaller than that required in the direct form filter of equivalent order. This advantage is particularly significant when symmetry conditions are imposed on the magnitude response of the filter.

4. It will be obvious in the following sections that assuring the stability of the filter is easier with the cascaded structure than with the direct form

4.2. DESIGN USING NONLINEAR PROGRAMMING

realization. It is perhaps the best reason why the cascade structure has been chosen by the three authors whose methods are discussed in the following sections.

In these methods, the vexing problem of assuring stability during the iterative process to minimize the error function is avoided by using an important result that is described below [30]. The strategy, based on this result, is to express the coefficients $d_{ij}^{(k)}$ of each component polynomial (i.e. for each value of k) in terms of another set of real and nonzero parameters $[q_l]$, according to the relationship between $d_{ij}^{(k)}$ and $[q_l]$. The 2-D polynomial containing any set of real values for $[q_l]$ is guaranteed to be a stable 2-D polynomial. When the optimum values for the parameters $[q_l]$ that minimize the error function (4.12) are found, the coefficients $d_{ij}^{(k)}$ are easily calculated. Now we elaborate on this result.

Let

$$D(s_1, s_2) = \sum_{j=0}^{I_2} \sum_{i=0}^{I_1} d_{ij} s_1^i s_2^j \qquad (4.15)$$

where s_1 and s_2 are two Laplace Transform variables and I_1 and I_2 are equal to either one or two. The authors [30] generate the polynomial (4.15) in the two variables from the system determinant of a lossless, frequency-independent, multiport network as shown in Fig. 4.1.

This network is assumed to be terminated by unit capacitors in the complex frequency variables s_1 and s_2 at n_1 and n_2 ports respectively. The input admittance is measured at another port, arbitrarily chosen without loss of generality as port 1 which is terminated by a unit resistor, so that the total number of ports is $1 + n_1 + n_2 = 1 + n$ as shown in the figure. Since the network is lossless and frequency independent, its admittance matrix $\mathbf{Y}$ is a real and skew-symmetric matrix given by

$$\mathbf{Y} = \left[\begin{array}{c|cccc} 0 & y_{12} & y_{13} & \cdots & y_{1n} \\ \hline -y_{12} & 0 & y_{23} & \cdots & y_{2n} \\ -y_{13} & -y_{23} & 0 & \cdots & y_{3n} \\ \vdots & \vdots & \vdots & & \vdots \\ -y_{1n} & -y_{2n} & -y_{3n} & \cdots & 0 \end{array}\right]$$

$$= \left[\begin{array}{c|c} \mathbf{Y}_{11} & \mathbf{Y}_{12} \\ \hline -\mathbf{Y}_{12}^T & \mathbf{Y}_{22} \end{array}\right] \qquad (4.16)$$

The admittance matrix of the network terminated by the resistor at port 1 and capacitors at the other $n_1 + n_2$ ports can be defined by

$$\widehat{\mathbf{Y}}(s_1, s_2, y_{kl}) = \mathbf{Y} + diag\left[1 \; \overbrace{s_1 \cdots s_1}^{n_1} \; \overbrace{s_2 \cdots s_2}^{n_2}\right] \qquad (4.17)$$

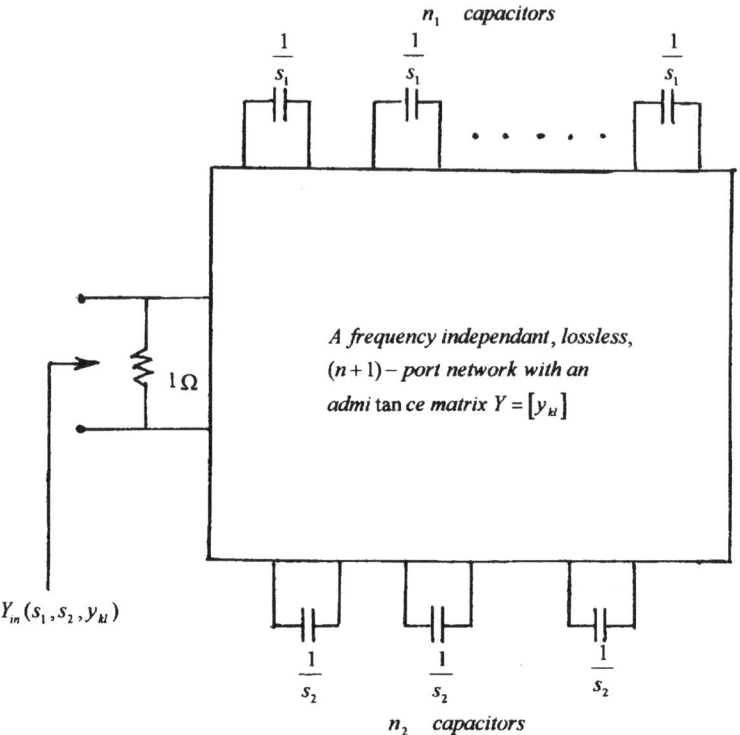

Fig. 4.1. A frequency-independent, lossless, multiport network, terminated by unit valued capacitors in two variables

where $diag \left[1 \overset{n_1}{\overleftrightarrow{s_1 \cdots s_1}} \overset{n_2}{\overleftrightarrow{s_2 \cdots s_2}} \right]$ represents a diagonal matrix in which the first element is a conductance of one mho, each of the next n_1 elements is an admittance s_1 and each of the last n_2 elements is an admittance s_2. Then, from network theory, we can derive the input admittance $\mathbf{Y}_{in}(s_1, s_2, y_{kl})$ at port 1 as

$$\begin{aligned}\mathbf{Y}_{in}(s_1, s_2, y_{kl}) &= 1 + \frac{\mathbf{Y}_{12} adj\left[\widehat{\mathbf{Y}}_{22}(s_1, s_2, y_{kl})\right] \mathbf{Y}_{12}^T}{\det\left[\widehat{\mathbf{Y}}_{22}(s_1, s_2, y_{kl})\right]} \\ &= 1 + \frac{M(s_1, s_2, y_{kl})}{N(s_1, s_2, y_{kl})}\end{aligned} \quad (4.18)$$

The determinant of the admittance matrix $\left[\widehat{\mathbf{Y}}(s_1, s_2, y_{kl})\right]$ in (4.17) is the polynomial $D(s_1, s_2, y_{kl})$. It is also given by $M(s_1, s_2, y_{kl}) + N(s_1, s_2, y_{kl})$. Since we have a lossless network terminated by a resistor, $D(s_1, s_2, y_{kl})$ is a strictly Hurwitz polynomial in two variables s_1 and s_2, for any set of real values of

4.2. DESIGN USING NONLINEAR PROGRAMMING

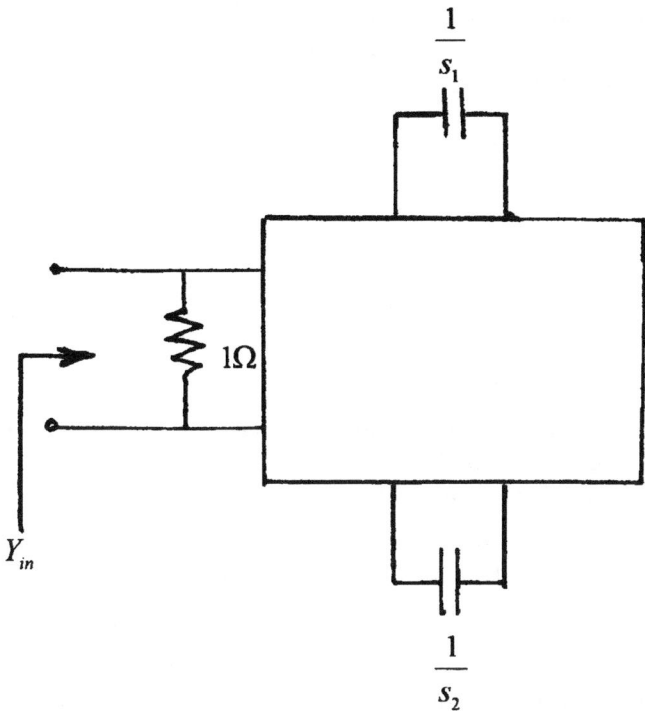

Fig. 4.2. Example of a simple two variable 3-port network

the parameters y_{kl}; $1 < k < l \leq n$. It means that if there exists a set of real parameters y_{kl}; $1 < k < l \leq n$, the polynomial $D(s_1, s_2, y_{kl})$ obtained as the determinant of the admittance matrix of the lossless, frequency - independent network terminated by unit capacitors s_1, s_2 and one input resistor, satisfies the condition

$$D(s_1, s_2, y_{kl}) \neq 0 \quad \text{for} \quad Re(s_1) \geq 0 \quad \text{and} \quad Re(s_2) \geq 0 \quad (4.19)$$

To illustrate this result, let us consider the case of a 3-port, lossless network as shown in Fig. 4.2. When it is terminated by one capacitor s_1, one capacitor s_2 and an input resistor of one ohm, ($n_1 = 1, n_2 = 1$), the admittance matrix of this network is

$$\widehat{\mathbf{Y}}(s_1, s_2, y_{kl}) = \begin{bmatrix} 1 & y_{12} & y_{13} \\ -y_{12} & s_1 & y_{23} \\ -y_{13} & -y_{23} & s_2 \end{bmatrix} \quad (4.20)$$

The determinant of this matrix is $y_{23}^2 + y_{13}^2 s_1 + y_{12}^2 s_2 + s_1 s_2$. If it is expressed in the form of (4.15), we have $D(s_1, s_2) = \sum_{j=0}^{I_1} \sum_{i=0}^{I_1} d_{ij} s_1^i s_2^j = d_{00} + d_{10} s_1 + d_{01} s_2 + d_{11} s_1 s_2$. Therefore we identify $d_{00} = y_{23}^2$, $d_{10} = y_{13}^2$, $d_{01} = y_{12}^2$ and

$d_{11} = 1$. If we use the notation $y_{23} = q_1$, $y_{13} = q_2$ and $y_{12} = q_3$, then for this case of $I_1 = 1$, $I_2 = 1$, the coefficients d_{ij} are expressed in terms of another set of parameters $[q_i]$ as given below:

Case I:

$$\begin{aligned} I_1 &= I_2 = 1 \\ d_{00} &= q_1^2 \\ d_{10} &= q_2^2 \\ d_{01} &= q_3^2 \\ d_{11} &= 1 \end{aligned}$$

Now consider a lossless network terminated by two capacitors s_1, one capacitor s_2, and one input resistor ($n_1 = 2, n_2 = 1$).

The admittance matrix of this network is

$$\widehat{\mathbf{Y}}(s_1, s_2, y_{kl}) = \begin{bmatrix} 1 & y_{12} & y_{13} & y_{14} \\ -y_{12} & s_1 & y_{23} & y_{24} \\ -y_{13} & y_{23} & s_1 & y_{34} \\ -y_{14} & -y_{24} & -y_{34} & s_2 \end{bmatrix}$$

By expanding its determinant, and denoting $y_{12} = q_1$, $y_{13} = q_2$, $y_{14} = q_3$, $y_{23} = q_4$, $y_{24} = q_5$ and $y_{34} = q_6$, we compare the coefficients of the terms. Then we get the expression for d_{ij} in terms of $[q_i]$ as follows:

Case II: $I_1 = 2$ and $I_2 = 1$

$$\begin{aligned} d_{00} &= (q_1 q_6 - q_2 q_5 + q_3 q_4)^2 \\ d_{10} &= q_5^2 + q_6^2 \\ d_{20} &= q_3^2 \\ d_{01} &= q_4^2 \\ d_{11} &= q_1^2 + q_2^2 \\ d_{21} &= 1 \end{aligned}$$

Similar results are obtained when $n_1 = I_1 = 1$, $n_2 = I_2 = 2$.

When $n_1 = n_2 = I_1 = I_2 = 2$, we get

Case III :

$$\begin{aligned} I_1 &= I_2 = 2 \\ d_{00} &= (q_5 q_{10} - q_6 q_9 + q_7 q_8)^2 \\ d_{10} &= (q_1 q_{10} - q_3 q_7 + q_4 q_6)^2 + (q_2 q_{10} - q_3 q_9 + q_4 q_8)^2 \\ d_{20} &= q_{10}^2 \\ d_{01} &= (q_1 q_8 - q_2 q_6 + q_3 q_5)^2 + (q_1 q_9 - q_2 q_7 + q_4 q_5)^2 \end{aligned}$$

4.2. DESIGN USING NONLINEAR PROGRAMMING

$$d_{11} = q_6^2 + q_7^2 + q_8^2 + q_9^2$$
$$d_{21} = q_3^2 + q_4^2$$
$$d_{02} = q_5^2$$
$$d_{12} = q_1^2 + q_2^2$$
$$d_{22} = 1$$

If the coefficients d_{ij}'s satisfy the above conditions, then

$$D(s_1, s_2)\Big|_{s_1 = \frac{(1-z_1^{-1})}{(1+z_1^{-1})},\ s_2 = \frac{(1-z_2^{-1})}{(1+z_2^{-1})}} = D(z_1^{-1}, z_2^{-1}) \neq 0$$
$$\text{for } |z_1| \geq 0 \text{ and } |z_2| \geq 0 \quad (4.21)$$

which means that the polynomial $D(z_1^{-1}, z_2^{-1})$ has no zeros outside the region:

$$\left\{ |z_1| \bigcup |z_2| \,;\, |z_1| \geq 1 \text{ and } |z_2| \geq 0 \right\}$$

Since $D(z_1^{-1}, z_2^{-1})$, generated from $D(s_1, s_2)$, by applying the above two bilinear transformations, takes the explicit form

$$\sum_{j=0}^{I_2} \sum_{i=0}^{I_1} d_{ij} \left[\frac{1-z_1^{-1}}{1+z_1^{-1}} \right]^i \left[\frac{1-z_2^{-1}}{1+z_2^{-1}} \right]^j \quad (4.22)$$

obviously it is not a polynomial. But we can multiply it by $(1+z_1^{-1})^{I_1}(1+z_2^{-1})^{I_2}$ to get it in the form of

$$\sum_{j=0}^{I_2} \sum_{i=0}^{I_1} d_{ij}(1 - z_1^{-1})(1 + z_1^{-1})^{I_1-i}(1 - z_2^{-1})^j(1 + z_2^{-1})^{I_2-j} \quad (4.23)$$

and the coefficients (4.23) can then be rearranged in the form of (4.3) or (4.4).

For designing a 2-D IIR digital filter with prescribed magnitude and constant group delay such that it is guaranteed to be stable, the authors of [1], assume the transfer function $H_d(z_1^{-1}, z_2^{-1})$ to be of the form (4.2) i.e. as a cascade of multiple lower order filters. Then $H_d(z_1^{-1}, z_2^{-1})$

$$= B \prod_{k=1}^{K} \frac{(1+z_1^{-1})^{J_1^{(k)}}(1+z_2^{-1})^{J_2^{(k)}} \sum_{j=0}^{J_2^{(k)}} \sum_{i=0}^{J_1^{(k)}} c_{ij}^{(k)} \left[\frac{1-z_1^{-1}}{1+z_1^{-1}}\right]^i \left[\frac{1-z_2^{-1}}{1+z_2^{-1}}\right]^j}{(1+z_1^{-1})^{I_1^{(k)}}(1+z_2^{-1})^{I_2^{(k)}} \sum_{j=0}^{I_2^{(k)}} \sum_{i=0}^{I_1^{(k)}} e_{ij}^{(k)} \left[\frac{1-z_1^{-1}}{1+z_1^{-1}}\right]^i \left[\frac{1-z_2^{-1}}{1+z_2^{-1}}\right]^j} =$$

$$B \prod_{k=1}^{K} \frac{\sum_{j=0}^{J_2(k)} \sum_{i=0}^{J_1^{(k)}} c_{ij}^{(k)}(1-z_1^{-1})^i(1+z_1)^{J_1^{(k)}-i}(1-z_2^{-1})^j(1+z_2^{-1})^{J_2^{(k)}-j}}{\sum_{j=0}^{I_2(k)} \sum_{i=0}^{I_1^{(k)}} e_{ij}^{(k)}(1-z_1^{-1})^i(1+z_1^{-1})^{I_1^{(k)}-i}(1-z_2^{-1})^j(1+z_2^{-1})^{I_2^{(k)}-j}} \quad (4.24)$$

The coefficients $e_{ij}^{(k)}$ are replaced by the parameters $q_l^{(k)}$ according to the stability conditions given above. Thus the elements of the parameter vector **b** are now modified to

$$\mathbf{b} = [((q_l^{(k)}, l = 1, 2, \ldots).k = 1, 2, \ldots, K), \\ (((c_{ij}^{(k)}, i = 0, 1, \ldots, J_1^{(k)}), j = 0, 1, \ldots, J_2^{(k)}), k = 1, 2, \ldots, K), B]^T \quad (4.25)$$

with $(i, j) \neq (0, 0)$. By finding the best values for these modified parameters in the denominator polynomials of (4.24), instead of $e_{ij}^{(k)}$, we avoid the need to check the stability of the transfer function after each iteration, since we have assured that it will be stable in all cases. Once we have found the values for the the parameters of $\mathbf{b}$, that minimize the error function (4.12)-(4.14), we calculate the original coefficients e_{ij} from $q_l^{(k)}$ and complete the approximation problem.

The authors [1] use the Fletcher and Powell optimization technique [10] to minimize the error function given by (4.12)-(4.14) in which $p = 2$. This least squares minimization method requires the calculation of the gradient vector of $E_m(\mathbf{b})$, E_{τ_1}, and E_{τ_2}. Derivation of the expressions for the gradient of the three component error functions $E_m(\mathbf{b})$, E_{τ_1} and E_{τ_2} as well as some details about choosing the corresponding relative weights R_m, R_{τ_1} and R_{τ_2} are given in [1].

4.2.3 Design Examples

We will consider a numerical example chosen by the authors of [1] for the design of a 2-D IIR filter with constant group delay. This example has been chosen by the authors of [19] and [21] also but with a slightly different passband region and group delays. We have chosen to include this example as a way for comparing the relative merits of the three methods proposed by these authors.

The filter is a lowpass circularly symmetric filter with the magnitude specification as given below:

$$H(\omega_1, \omega_2) = \begin{cases} 1.000 & \text{for} \quad 0 \leq r \leq 0.1\pi \\ 0.800 & \text{for} \quad 0.1\pi \leq r \leq 0.2\pi \\ 0.440 & \text{for} \quad 0.2\pi \leq r \leq 0.3\pi \\ 0.140 & \text{for} \quad 0.3\pi \leq r \leq 0.4\pi \\ 0.030 & \text{for} \quad 0.4\pi \leq r \leq 0.5\pi \\ 0.002 & \text{for} \quad 0.5\pi \leq r \leq 0.6\pi \\ 0.001 & \text{for} \quad 0.6\pi \leq r \leq \pi \end{cases} \quad (4.26)$$

where $r = \sqrt{\omega_1^2 + \omega_2^2}$ and the passband region is defined as the circular region: $r \leq 0.4\pi$. The group delays τ_1 and τ_2 are specified to be constant over the circular region: $r \leq 0.4\pi$.

Example 4.1

For the first example, the following parameters are chosen by the authors: $p = 2$, $\tau_1 = \tau_2 = 2$, $W(m, n) = 1$ and $w(m, n) = 1$ while the relative weights R_m and R_{τ_i} are chosen as $R_m = \alpha/E_m$ and $R_{\tau_i} = \frac{(1-\alpha)}{2E_{\tau_i}}$ where E_m and E_{τ_i} are defined as

$$E_m = \sum_{m=1}^{M}\sum_{n=1}^{N} W(m,n)(G(m,n) - H_d(m,n))^p \quad (4.27)$$

$$E_{\tau_i} = \sum_{m \in M_\tau}\sum_{n \in N_\tau} w(m,n)(\tau_i(m,n) - \tau_d(m,n))^p \quad (4.28)$$

and α is a factor that is used to vary the weighting factors R_m and R_{τ_i} as explained below. The discrete frequency samples are chosen by the authors in the entire upper half of the frequency plane i.e. at equally spaced points in the frequency region $\mathcal{R} = \{(\omega_1, \omega_2); -\pi \leq \omega_1 \leq \pi) \cup (0 \leq \omega_2 \leq \pi)\}$ i.e. at $|\omega_i/\pi| = 0.0, 0.1, 0.2, \ldots, 1.0$, $i = 1, 2$. It must be pointed out therefore that the magnitude function and group delays are not evaluated at all points exactly on the boundary of the circular region, during the optimization process. The authors also suggest choosing $\alpha = 1$ during the first five iterations and decreasing it by a step size of 0.1 every five successive iterations until it reduces to 0.5. The filter obtained by the authors is described by the following transfer function, $H_d(z_1^{-1}, z_2^{-1})$ (in which $I_1^{(1)} = I_2^{(2)} = J_1^{(1)} = J_2^{(2)} = 2$).

$$H_d(z_1^{-1}, z_2^{-1}) = C \frac{\begin{bmatrix} 1 & z_1^{-1} & z_1^{-2} \end{bmatrix} \mathbf{B}_0 \begin{bmatrix} 1 \\ z_2^{-1} \\ z_2^{-2} \end{bmatrix}}{\begin{bmatrix} 1 & z_1^{-1} & z_1^{-2} \end{bmatrix} \mathbf{A}_0 \begin{bmatrix} 1 \\ z_2^{-1} \\ z_2^{-2} \end{bmatrix}}$$

where

$$\mathbf{B}_0 = \begin{bmatrix} 1.0000 & 0.428998 & 0.470506 \\ 0.441454 & 0.316414 & 0.659983 \\ 0.372886 & 0.412246 & 0.168825 \end{bmatrix}$$

$$\mathbf{A}_0 = \begin{bmatrix} 1.0000 & -0.535537 & 0.008103 \\ -0.658556 & -0.119018 & 0.305373 \\ 0.081547 & 0.279834 & -0.202243 \end{bmatrix}$$

$$C = 0.035935 \quad (4.29)$$

Example 4.2

The second example, designed to realize the same magnitude function but with group delays $\tau_1 = \tau_2 = 4$ is described by the following transfer function in which $I_1^{(k)} = I_2^{(k)} = J_1^{(k)} = J_2^{(k)} = 2$ for $k = 1, 2$.

$$H_d(z_1^{-1}, z_2^{-1}) = 0.134044 \times 10^{-2} H_1(z_1^{-1}, z_2^{-1}) H_2(z_1^{-1}, z_2^{-1}) \quad (4.30)$$

where $H_1(z_1^{-1}, z_2^{-1})$ is given by

$$\frac{\begin{bmatrix} 1 & z_1^{-1} & z_1^{-2} \end{bmatrix} \mathbf{B}_1 \begin{bmatrix} 1 \\ z_2^{-1} \\ z_2^{-2} \end{bmatrix}}{\begin{bmatrix} 1 & z_1^{-1} & z_1^{-2} \end{bmatrix} \mathbf{A}_1 \begin{bmatrix} 1 \\ z_2^{-1} \\ z_2^{-2} \end{bmatrix}} \qquad (4.31)$$

where

$$\mathbf{B}_1 = \begin{bmatrix} 1.0 & 0.201054 & 0.600823 \\ 0.446691 & 1.162514 & 1.473812 \\ 0.570455 & 1.120600 & 0.463334 \end{bmatrix}$$

$$\mathbf{A}_1 = \begin{bmatrix} 1.0 & -0.490714 & 0.027371 \\ -0.536990 & -0.127213 & 0.236525 \\ 0.068811 & 0.211200 & -0.143242 \end{bmatrix}$$

and $H_2(z_1^{-1}, z_2^{-1})$ is given by

$$\frac{\begin{bmatrix} 1 & z_1^{-1} & z_1^{-2} \end{bmatrix} \mathbf{B}_2 \begin{bmatrix} 1 \\ z_2^{-1} \\ z_2^{-2} \end{bmatrix}}{\begin{bmatrix} 1 & z_1^{-1} & z_1^{-2} \end{bmatrix} \mathbf{A}_2 \begin{bmatrix} 1 \\ z_2^{-1} \\ z_2^{-2} \end{bmatrix}} \qquad (4.32)$$

where

$$\mathbf{B}_2 = \begin{bmatrix} 1.0 & 0.201054 & 0.600822 \\ 0.446690 & 1.162513 & 1.473811 \\ 0.570454 & 1.120600 & 0.463335 \end{bmatrix}$$

$$\mathbf{A}_2 = \begin{bmatrix} 1.0 & -0.491231 & 0.028179 \\ -0.538819 & -0.128680 & 0.236831 \\ 0.066576 & 0.209757 & -0.143378 \end{bmatrix}$$

Figure 4.3 and Fig. 4.4 show the magnitude and group delay $\tau_1(\omega_1, \omega_2)$ for the filter in Example 4.1, (the plot of $\tau_2(\omega_1, \omega_2)$ is similar) while Fig. 4.5 and Fig. 4.6 show a magnified view of the group delays $\tau_1(\omega_1, \omega_2)$ and $\tau_2(\omega_1, \omega_2)$ in only the circular passband region $r \leq 0.4\pi$.

It was mentioned that the authors of [1] used a single composite error function and using least squares minimization technique to minimize it, they were able to approximate both magnitude and group delays prescribed for an IIR filter. Hinamoto and Maekawa [19] approach the problem of designing a 2-D IIR filter in two stages. First they choose an all-pole transfer function $H_\phi(z_1^{-1}, z_2^{-1})$.

4.2. DESIGN USING NONLINEAR PROGRAMMING

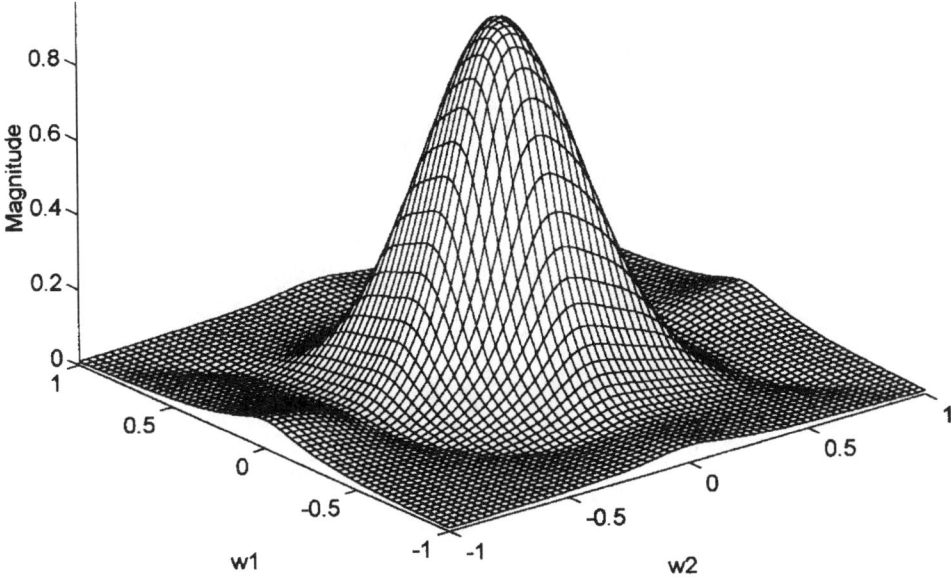

Fig. 4.3. Magnitude response of Aly-Fahmy filter

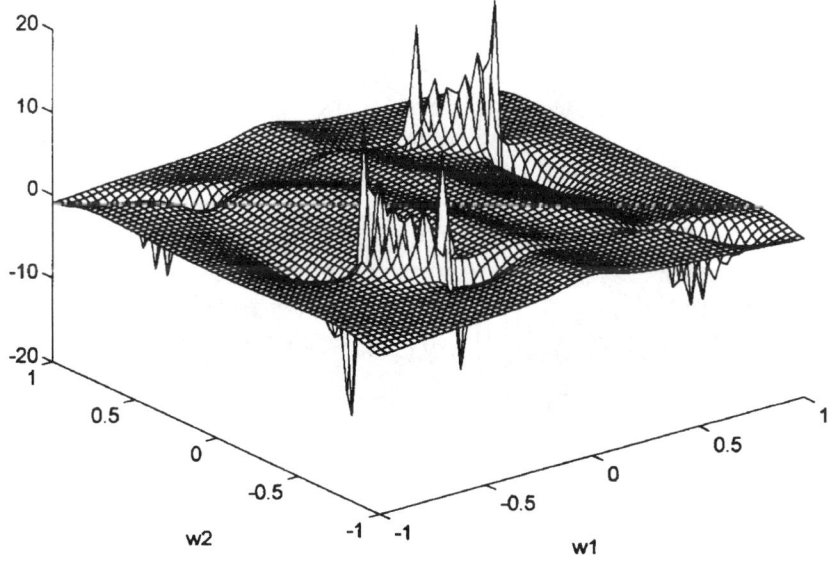

Fig. 4.4. Group delay response of Aly-Fahmy filter with respect to ω_1

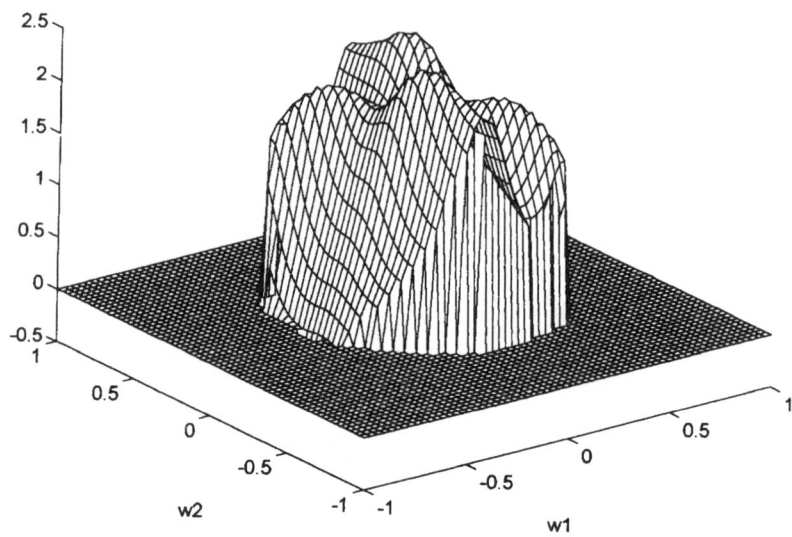

Fig. 4.5. Group delay response of Aly-Fahmy filter in the passband, with respect to ω_1

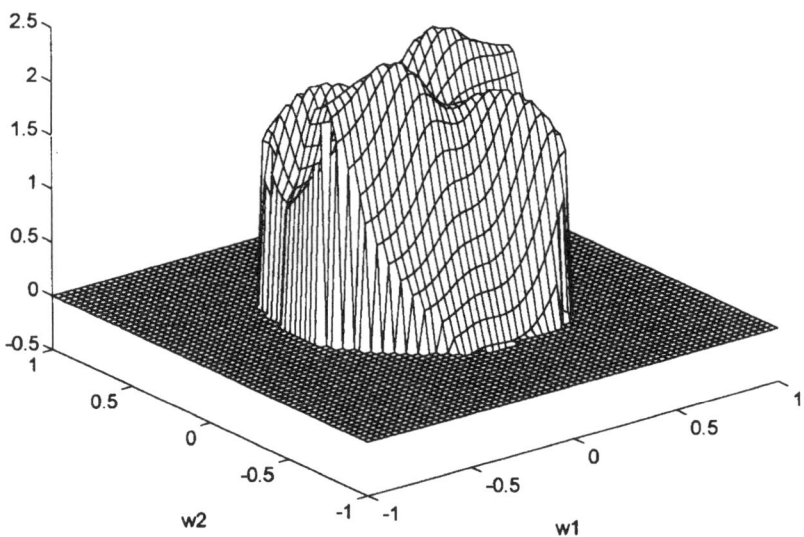

Fig. 4.6. Group delay response of Aly-Fahmy filter, in the passband, with respect to ω_2

4.2. DESIGN USING NONLINEAR PROGRAMMING

This function therefore has only a denominator of the same form as in (4.24) and the coefficients e_{ij} are replaced by the coefficients q_l as before, to assure stability. No constraint is imposed on this denominator that it be separable. Hence the parameter vector **b** is now defined by

$$\mathbf{b} = [q_l^{(k)}, \quad l=1,2,\ldots, k=1,2,\ldots,K]^T \tag{4.33}$$

Optimal values are found by using Davidon-Fletcher Powell method to approximate only the specified group delays τ_1 and τ_2, in the least squares sense ($p=2$). The performance index (error function) $J(\mathbf{b})$ is defined by them as

$$J(\mathbf{b}) = J_{\tau 1}(\mathbf{b}) + J_{\tau 2}(\mathbf{b}) \tag{4.34}$$

where

$$J_{\tau i}(\mathbf{b}) = \sum_{m \in M_\tau} \sum_{n \in N_\tau} w_i(m,n) [\tau_{di}(m,n) - \tau_i]^p \tag{4.35}$$

In the second stage, the above all-pole transfer function $H_\phi(z_1^{-1}, z_2^{-1})$ is multiplied by a mirror-image polynomial

$$N^\phi(z_1^{-1}, z_2^{-1}) = K_N \prod_{k=1}^{L} N_k(z_1^{-1}, z_2^{-1})$$

where L is the total number of sections cascaded. The coefficients of the mirror image polynomials are obtained in the second stage of optimization to approximate the prescribed magnitude response.

We have used 1-D mirror image polynomials in Chap. 2 in the design of IIR filters to approximate both magnitude and group delay. If a 2-D mirror image polynomial $N^\phi(z_1, z_2)$ is expressed as

$$N^\phi(z_1, z_2) = \sum_{j=-l}^{l} \sum_{i=-k}^{k} f_{ij} z_1^{-i} z_2^{-j}$$

then, it satisfies the conditions $f_{ij} = f_{-i,-j}$ and $f_{-i,j} = f_{i,-j}$ and the polynomial $N^\phi(e^{j\omega_1}, e^{j\omega_2})$ has zero phase. If this non-causal polynomial $N^\phi(z_1, z_2)$ is delayed by k and l units of time, it becomes a causal polynomial.

A typical (3×3) mirror image polynomial that is causal is given by

$$N_k(z_1^{-1}, z_2^{-1}) = \begin{bmatrix} 1 & z_1^{-1} & z_1^{-2} \end{bmatrix} \begin{bmatrix} b_{00}^{(k)} & b_{01}^{(k)} & b_{02}^{(k)} \\ b_{10}^{(k)} & b_{11}^{(k)} & b_{10}^{(k)} \\ b_{02}^{(k)} & b_{01}^{(k)} & b_{00}^{(k)} \end{bmatrix} \begin{bmatrix} 1 \\ z_2^{-1} \\ z_2^{-2} \end{bmatrix} \tag{4.36}$$

It can also be expressed in the following form

$$N_k(z_1^{-1}, z_2^{-1}) = z_1^{-1} z_2^{-1} \{ b_{11}^{(k)} + b_{01}^{(k)}(z_1 + z_1^{-1}) + b_{10}^{(k)}(z_2 + z_2^{-1}) + b_{02}^{(k)}(z_1^1 z_2^{-1} + z_1^{-1} z_2^1) + b_{00}^{(k)}(z_1^1 z_2^1 + z_1^{-1} z_2^{-1}) \} \tag{4.37}$$

It follows from (4.37) that

$$\begin{aligned}
N_k(e^{-j\omega_1}, e^{-j\omega_2}) &= e^{-j\omega_1}e^{-j\omega_2}\{b_{11}^{(k)} + 2b_{01}^{(k)}\cos(\omega_1) + 2b_{10}^{(k)}\cos(\omega_2) \\
&+ 2b_{02}^{(k)}\cos(\omega_1 - \omega_2) + 2b_{00}^{(k)}\cos(\omega_1 + \omega_2) \\
&= e^{-j\omega_1}e^{-j\omega_2}\{b_{11}^{(k)} + 2b_{01}^{(k)}\cos(\omega_1) + 2b_{10}^{(k)}\cos(\omega_2) \\
&+ 2(b_{02}^{(k)} + b_{00}^{(k)})\cos(\omega_1)\cos(\omega_2) + \\
&\quad 2(b_{02}^{(k)} - b_{00}^{(k)})\sin(\omega_1)\sin(\omega
\end{aligned} \quad (4.38)$$

If each of the polynomials $N_k(z_1^{-1}, z_2^{-1})$ is a mirror image polynomial, then the numerator

$$N^\phi(z_1^{-1}, z_2^{-1}) = K_N \prod_{k=1}^{L} N_k(z_1^{-1}, z_2^{-1})$$

is also a mirror image polynomial.

Hence the polynomial $N^\phi(e^{-j\omega_1}, e^{-j\omega_2})$ has a linear phase or a constant group delay added to that of $H_\phi(e^{-j\omega_1}, e^{-j\omega_2})$, when these two functions are multiplied. The result is $H_d(z_1^{-1}, z_2^{-1}) = H_\phi(z_1^{-1}, z_2^{-1}) N^\phi(z_1^{-1}, z_2^{-1}) =$

$$\frac{K_N \prod_{k=1}^{L} N_k(z_1^{-1}, z_2^{-1})}{\sum_{j=0}^{I_2(k)} \sum_{i=0}^{I_1^{(k)}} e_{ij}^{(k)}(1 - z_1^{-1})^i (1 + z_1^{-1})^{I_1^{(k)}-i}(1 - z_2^{-1})^j (1 + z_2^{-1})^{I_2^{(k)}-j}} \quad (4.39)$$

and the magnitude of the all-pole function is multiplied by the magnitude of the mirror image polynomial.

Only the coefficients of the numerator

$$N^\phi(z_1^{-1}, z_2^{-1}) = K_N \prod_{k}^{L} N_k(z_1^{-1}, z_2^{-1})$$

where N_k is shown in (4.37), are the unknown coefficients in the above equation. The number of independent coefficients is almost half the order of the polynomial because of their symmetry.

In the above equation, it can be assumed that $a_{00}^{(k)} = 1$ for all k, without loss of generality. A new parameter vector $\mathbf{b}$ is now defined as

$$\mathbf{b} = [(b_{00}^{(k)}, b_{01}^{(k)}, b_{10}^{(k)}, b_{02}^{(k)}); \quad k = 1, 2, \ldots, L), K_N]^T \quad (4.40)$$

Note the magnitude of the transfer function $H_d(e^{-j\omega_1}, e^{-j\omega_2})$ evaluated at the discrete frequencies has already been denoted by $H_d(m,n)$ and the magnitude specified is denoted by $G(m,n)$. Then an error function is defined as

$$J(\mathbf{b}) = \sum_{m=1}^{M} \sum_{n=1}^{N} W(m,n) E(m,n)^p \quad (4.41)$$

where

$$E(m,n) = H_d(m,n) - G(m,n) \quad (4.42)$$

and $W(m,n)$ is a weighting function chosen by the designer.

The authors [19] minimize the above error function using the Davidon-Fletcher-Powell method of optimization technique.

4.2. DESIGN USING NONLINEAR PROGRAMMING

Example 4.3

Here we consider one example of a 2-D lowpass filter designed by the authors [19]. The magnitude response specification is the same as given in Example 4.1 - except that the passband region chosen by the authors is a circle of radius $r = 0.3\pi$ (and not 0.4π). They have designed one filter with specified values $\tau_1 = \tau_2 = 1$ but we select the other filter with values for $\tau_1 = \tau_2 = 2$. We select this example so that the results can be compared with those for Example 4.1 and also Example 4.4 which will be discussed later. But the order of the filter designed by these authors is eight, whereas the order of the filter in Example 4.1 is four. So any comparison between the filters in Example 4.1 and Example 4.3 must consider these differences.

The transfer function of the lowpass circularly symmetric filter with group delays $\tau_1 = \tau_2 = 2$ and two 3×3 order sections is given by the following $H_d(z_1^{-1}, z_2^{-1})$.

$$\frac{10^{-2} \begin{bmatrix} 1 & z_1^{-1} & z_1^{-2} & z_1^{-3} & z_1^{-4} \end{bmatrix} \mathbf{B} \begin{bmatrix} 1 & z_2^{-1} & z_2^{-2} & z_2^{-3} & z_2^{-4} \end{bmatrix}^T}{D_1(z_1^{-1}, z_2^{-1}) D_2(z_1^{-1}, z_2^{-1})} \quad (4.43)$$

where

$$D_1(z_1^{-1}, z_2^{-1}) = \begin{bmatrix} 1 & z_1^{-1} & z_1^{-2} \end{bmatrix} \mathbf{D}_1 \begin{bmatrix} 1 \\ z_2^{-1} \\ z_2^{-2} \end{bmatrix}$$

$$\mathbf{D}_1 = \begin{bmatrix} 2.792030 & -2.512864 & 1.378853 \\ -2.512864 & 2.261611 & -1.240986 \\ 1.378853 & -1.240986 & 0.680952 \end{bmatrix}$$

and

$$D_2(z_1^{-1}, z_2^{-1}) = \begin{bmatrix} 1 & z_1^{-1} & z_1^{-2} \end{bmatrix} \mathbf{D}_2 \begin{bmatrix} 1 \\ z_2^{-1} \\ z_2^{-2} \end{bmatrix}$$

$$\mathbf{D}_2 = \begin{bmatrix} 2.489600 & -2.905738 & 0.916049 \\ -2.905738 & 3.391434 & -1.069167 \\ 0.916049 & -1.069167 & 0.337057 \end{bmatrix}$$

and the coefficient matrix $\mathbf{B}$ is given by

$$\begin{bmatrix} -1.517589 & 1.958025 & -0.783893 & 1.961028 & -1.517218 \\ 1.973989 & -0.481447 & 0.613957 & -0.481910 & 1.975659 \\ -0.264217 & 0.063123 & 3.658332 & 0.063123 & -0.264217 \\ 1.975659 & -0.481910 & 0.613957 & -0.481447 & 1.973989 \\ -1.517218 & 1.961028 & -0.783893 & 1.958025 & -1.517589 \end{bmatrix} \quad (4.44)$$

It is noticed that $b_{i,j} = b_{4-i, 4-j}$ $(i = 0, 1, 2, 3, 4; j = 0, 1, 2, 3, 4)$ in the above matrix $[\mathbf{B}]$ and therefore they are the coefficients of a mirror image polynomial.

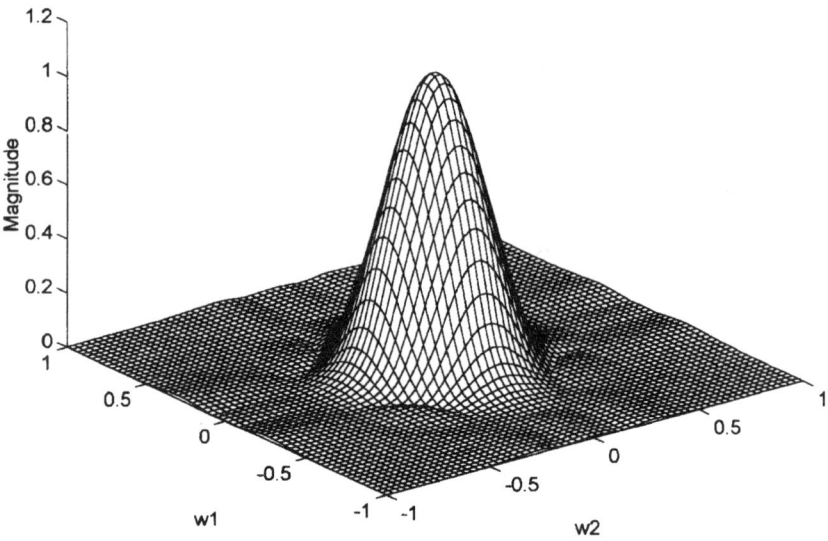

Fig. 4.7. Magnitude response of Hinamoto-Maekawa filter

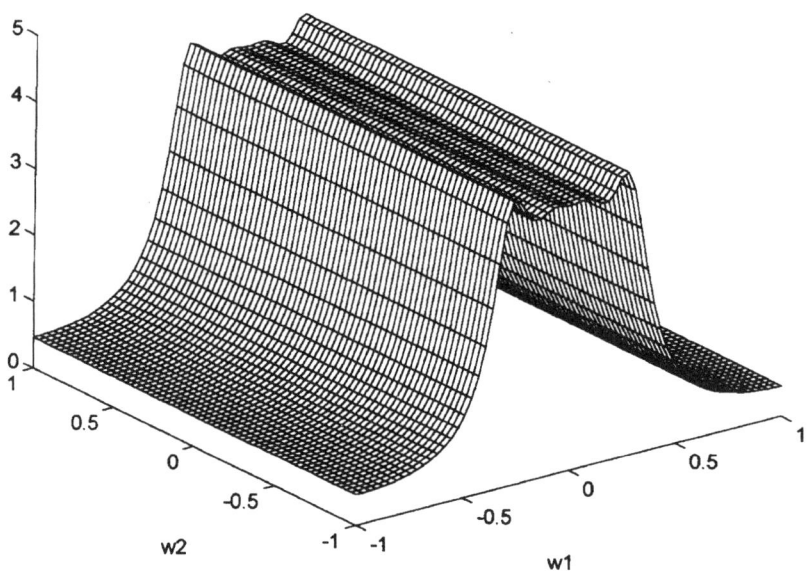

Fig. 4.8. Group delay response of Hinamoto-Maekawa filter, with respect to ω_1

4.2. DESIGN USING NONLINEAR PROGRAMMING

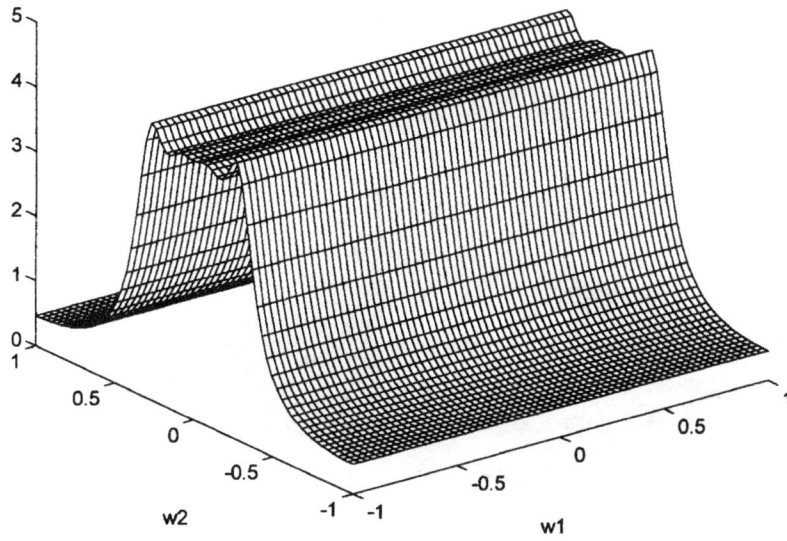

Fig. 4.9. Group delay response of Hinamoto-Maekawa filter, with respect to ω_2

The magnitude response of the above filter is shown in Fig. 4.7. The group delay of this filter with respect to ω_1 and ω_2 are plotted in Fig. 4.8 and Fig. 4.9 respectively.

The group delay responses depicted in Fig. 4.8 and Fig. 4.9 are similar to those obtained by using a separable denominator - as shown by the Example 4.4. It is not clear why the matrices in $\mathbf{D}_1$ and $\mathbf{D}_2$ shown in (4.43) happen to have a rank of one and hence the denominator polynomial is separable. Such a constraint was not imposed by the authors in finding the coefficients of the denominator. The group delays are magnified and plotted within the passband only in Fig. 4.10 and Fig. 4.11. It is pointed out that, the error function defined in (4.34) was used to minimize the group delays with respect to ω_1 and ω_2, and because the denominator is a separable polynomial, the plots in Fig. 4.8 and Fig. 4.9 show rectangular symmetry. Comparison of Fig. 4.10 and Fig. 4.11 with Fig. 4.5 and Fig. 4.6 show that the delay response for the filter in Example 4.3 has a better approximation to the constant value than the filter in Example 4.1.

The above two methods can be used to design a filter which has a prescribed magnitude response in its passband that is arbitrary in shape, since the optimization is carried over the entire upper half frequency plane: $-\pi \leq \omega_1 \leq \pi$, $0 \leq \omega_2 \leq \pi$. But, authors of both papers have chosen design examples of circularly symmetric filters only. Yet, they do not make use of the computational advantages that are available due to the symmetry. However, Hinamoto and Maekawa reduce the computational load by breaking up the design problem

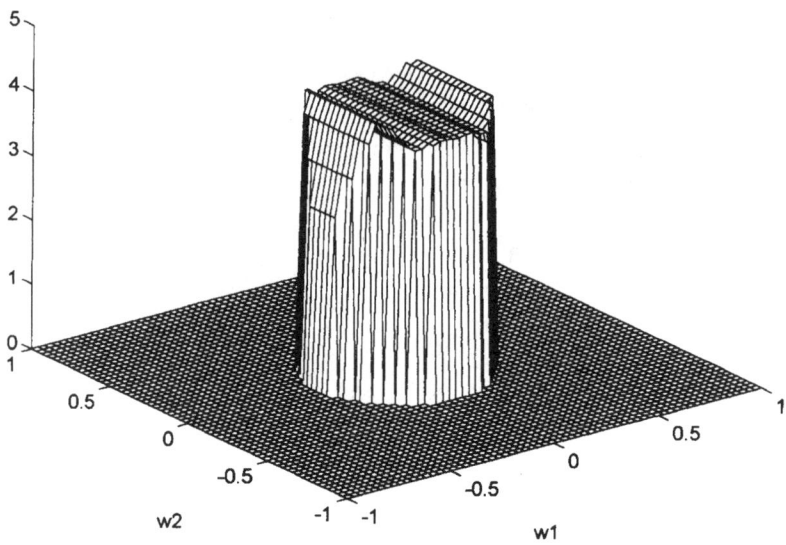

Fig. 4.10. Group delay response of Hinamoto-Maekawa, in the passband, with respect to ω_1

into two parts. They also choose a mirror image polynomial as the numerator polynomial, which has a reduced number of independent coefficients to be found by the optimization procedure.

A further modification to their design procedure was proposed by Kwan and Chan [21], in which properties of the circularly symmetric filters are fully utilized to reduce the computational load by a significant amount. Because the specification in this method is required to be a circularly symmetric filter response, it can not be applied for the design of filters with such passband or stopband regions like an elliptical region. However this method is applicable for the design of N-dimensional ($N \geq 2$) symmetric filters. We will describe this method for the design of only 2-D circularly symmetric filters in the following section.

If we denote $H(e^{-j\omega_1}, e^{-j\omega_2})$ by $H(\omega_1, \omega_2)$, then a circularly symmetric filter satisfies the following properties [29].

$|H(\omega_1, \omega_2)| = \left|H(\omega_1', \omega_2')\right|$ for all values of ω_1, ω_2 and ω_1', ω_2' such that $(\omega_1^2 + \omega_2^2)^{1/2} = (\omega_1'^2 + \omega_2'^2)^{1/2}$ i.e. the magnitude is the same for all points on any circle of radius $r < \pi$. This means that $|H(\omega_1, \omega_2)| = |H(\omega_1, -\omega_2)| = |H(-\omega_1, \omega_2)| = |H(-\omega_1, -\omega_2)|$ which represents a quadrantal symmetry. It also means that $|H(\omega_1, \omega_2| = |H(\omega_2, \omega_1)| = |H(-\omega_2 - \omega_1)| = |H(\omega_2, -\omega_1| = |H(-\omega_2, \omega_1)|$.

Hence one needs to choose discrete frequencies only in the region R defined by $(0 \leq \omega_1 \leq \pi)$ and $\omega_2 \leq \omega_1$ for optimizing the magnitude response; so the

4.2. DESIGN USING NONLINEAR PROGRAMMING

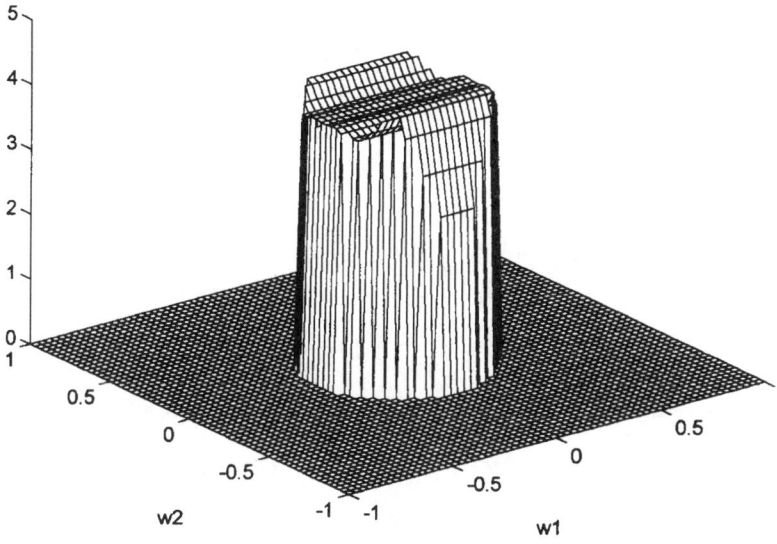

Fig. 4.11. Group delay response of Hinamoto-Maekawa filter, in the passband, with respect to ω_2

number of frequencies used for the evaluation in the approximation is reduced by a factor of four in comparison with the procedure of [19] - thereby saving a considerable amount of computation. In order to reduce the number of iterations to approximate the specified group delays of such a circularly symmetric filter, the authors choose the same two-step procedure as in [19] but the all-pole transfer function is assumed to be separable. That is, the all-pole transfer function is of the form

$$H_\phi(z_1^{-1}, z_2^{-1}) = \frac{1}{D(z_1^{-1}, z_2^{-1})} = \frac{1}{Q(z_1^{-1})Q(z_2^{-1})} \quad (4.45)$$

$Q(z_i^{-1})$ is either a polynomial such as

$$Q(z_i^{-1}) = \prod_{k=1}^{K}(d_0^{(k)} + d_1^{(k)}z_i^{-1} + d_2^{(k)}z_i^{-2}); \quad i = 1, 2 \quad (4.46)$$

or a polynomial such as

$$Q(z_i^{-1}) = (d_0^{(0)} + d_1^{(0)}z_i^{-1})\prod_{k=1}^{K}(d_0^{(k)} + d_1^{(k)}z_i^{-1} + d_2^{(k)}z_i^{-2}); \quad i = 1, 2 \quad (4.47)$$

In order to assure stability of (4.45), the coefficients in the above 1-D polynomials $Q(z_i^{-1})$ are expressed in terms of another set of real, nonzero parameters

q_l as given below

$$d_0^{(0)} = q_1^{(0)\,2} + 1$$
$$d_1^{(0)} = q_1^{(0)\,2} - 1$$

and

$$\begin{aligned}
d_0^{(k)} &= q_1^{(k)\,2} + q_2^{(k)\,2} + 1 \\
d_1^{(k)} &= 2(q_1^{(k)\,2} - 1) \\
d_2^{(k)} &= q_1^{(k)\,2} - q_2^{(k)\,2} + 1 \, for \, k = 1, 2, \ldots, K
\end{aligned} \qquad (4.48)$$

Therefore the parameter vector for approximating the given group delay is defined as

$$\mathbf{b} = [q_1^{(0)}, ((q_i^{(k)}, l = 1, 2), k = 1, 2, \ldots, K)]^T$$

The optimization is reduced to that of minimizing the error function in one frequency variable and hence the number of sample points to be chosen for minimizing the error function is considerably reduced. The error function chosen is defined as

$$J(\mathbf{b}) = \sum_{m=1}^{M_r} u_m [\tau_d(\omega_{1,m}) - \tau]^2 \qquad (4.49)$$

where $\tau = \tau_1 = \tau_2$ is the specified group delay which is a positive constant. After the error function is minimized by use of the Fletcher-Powell algorithm, the values for the coefficients $q_i^{(k)}$ are used to compute the coefficients of (4.46) and (4.47), using the above relationships.

The next step in the design procedure is to multiply the transfer function $H_\phi(z_1^{-1}, z_2^{-1})$ by a mirror image polynomial $N^\phi(z_1^{-1}, z_2^{-1})$ to get

$$H_d(z_1^{-1}, z_2^{-1}) = \frac{N^\phi(z_1^{-1}, z_2^{-1})}{Q(z_1^{-1})Q(z_2^{-1})} \qquad (4.50)$$

where $N^\phi(z_1^{-1}, z_2^{-1})$ is of the same form as defined earlier. The definition of the parameter vector, performance index and the method of minimizing the performance index are similar to those of [19] but with much smaller number of discrete frequency points chosen in the region R for evaluation of the approximation error, than the number of points chosen by the authors of [19].

Example 4.4

The same specifications as given in Example 4.3 are chosen for designing a lowpass filter following the procedure proposed by these authors [21]. However they use weighting functions $W(m, n)$ and $w(m, n)$ which are given explicitly by them and are different from those used in [19].

4.2. DESIGN USING NONLINEAR PROGRAMMING

The denominator polynomial for the design example is

$$D(z_1^{-1}, z_2^{-1}) = Q(z_1^{-1})Q(z_2^{-1})$$

where

$$\begin{aligned}Q(z_i^{-1}) &= (1.670937 - 1.503865 z_i^{-1} + 0.825198 z_i^{-2}) \\ &\quad \times (1.577846 - 1.841585 z_i^{-1} + 0.580569 z_i^{-2}); \ i = 1, 2\end{aligned} \quad (4.51)$$

The mirror image polynomial $N^\phi(z_1^{-1}, z_2^{-1})$ in the numerator is given by

$$10^{-2} \begin{bmatrix} 1 & z_1^{-1} & z_1^{-2} & z_1^{-3} & z_1^{-4} \end{bmatrix} \mathbf{B} \begin{bmatrix} 1 & z_2^{-1} & z_2^{-2} & z_2^{-3} & z_2^{-4} \end{bmatrix}^T$$

where the matrix $\mathbf{B}$ is given by (4.52). The coefficients have a symmetry which shows that the numerator is a mirror image polynomial.

$$\begin{bmatrix} 2.255632 & -2.910392 & 3.282184 & -2.910392 & 2.255632 \\ -2.910392 & 5.625357 & -4.034687 & 5.625357 & -2.910392 \\ 3.282184 & -4.034687 & 4.716775 & -4.034687 & 3.282184 \\ -2.910392 & 5.625357 & -4.034687 & 5.625357 & -2.910392 \\ 2.255632 & -2.910392 & 3.282184 & -2.910392 & 2.255632 \end{bmatrix} \quad (4.52)$$

Plots showing the magnitude and group delays of the above transfer function are found in Figs 4.12-4.14. The plots of the delays τ_1 and τ_2 for this filter are similar to those shown in Fig. 4.8 and Fig. 4.9. We notice that the plots of the group delays τ_1 and τ_2 show a rectangular symmetry because they are obtained by a separable denominator. Because the authors use Fletcher and Powell method to approximate the delay of the 1-D digital filter in the least squares sense, the delay exhibits some ripple in the passband and a peak at the edge of the passband and consequently this behavior is also seen in these figures.

Now, some general comments can be made in comparing the three methods described above.

Eq.(4.12) shows that the error function is a weighted sum of three error functions and depending on the relative weights R_m, R_{T_i}, $W(m,n)$ and $w(m,n)$, the rate of convergence of the error function may or may not be high. In contrast, the authors of [19] and [21], break up the problem into the minimization of two different error functions in sequence and hence improve the convergence rate. Figs. 4.12-4.14 look nearly the same as Figs. 4.7-4.11 respectively. But the authors Kwan and Chan [21] compare the number of iterations required for minimizing the error functions, using their method and the method of [19] and show that their method required much smaller number of iterations. They also compare the rms value of the errors as defined below for these design examples.

$$\epsilon_m = \frac{\left\{ \sum_{m=1}^{M''} u_m [|H_d(m)| - H(m)]^2 \right\}^{1/2}}{\left\{ \sum_{m=1}^{M''} H(m)^2 \right\}^{1/2}} \times 100 \quad (4.53)$$

210 CHAPTER 4. MAGNITUDE AND DELAY OF 2-D FILTERS

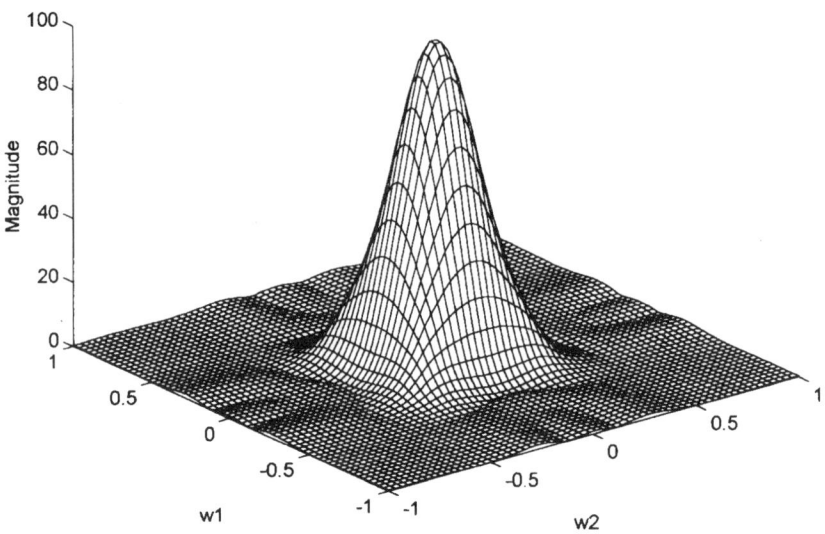

Fig. 4.12. Magnitude response of Kwan-Chan filter

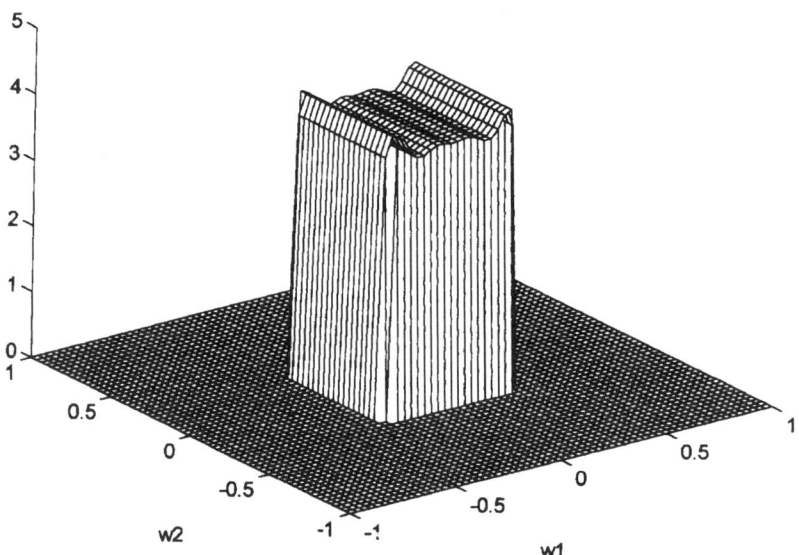

Fig. 4.13. Group delay response of Kwan-Chan filter, in the passband, with respect to ω_1

$$\epsilon_{\tau_i} = \frac{\left\{\sum_{m=1}^{M'} u_{\tau_i}[\tau_{d_i}(m) - \tau_i]^2\right\}^{1/2}}{\left\{\sum_{m=1}^{M'} \tau_i^2\right\}^{1/2}} \times 100 \qquad (4.54)$$

4.2. DESIGN USING NONLINEAR PROGRAMMING

where τ_i $(i = 1, 2)$ is the total group delay specified. The rms value of the corresponding errors used in [19] must consider the number of samples in the 2-D frequency plane. For the examples chosen, the above expressions can be used for comparison, because they all design a circularly symmetric lowpass filter. Hinamoto and Maekawa compute the rms values for their Example 4.3 and the Example 4.1 which show that their design procedure gives better results. The authors of [21] also make detailed comparison of the Examples 4.1, 4.3 and 4.4 using the order of the filter function given in the examples, values of the rms errors ϵ_m, ϵ_{τ_i}, the number of iterations required as the measures for comparison. Note that the radius of the passband in Example 4.1 is 0.4π, whereas the radius of the passband in Example 4.3 and 4.4 is 0.3π. Hence the magnitude response for the filters in Example 4.3 and 4.4 are the same but different from that for the filter in Example 4.1.

Further reduction in the computational complexity has been achieved by choosing the coefficients of $Q(z_i^{-1})$ without resorting to optimization [23]. There are two methods available for finding the coefficients of

$$H_\phi(z^{-1}) = \frac{1}{Q(z^{-1})} \qquad (4.55)$$

such that its group delay approximates a given constant either in the maximally flat sense or in the equiripple sense. In Chap. 2, we described Thiran's analytical formula for approximating a constant group delay in the maximally flat sense [33]. The formula was given by equation (2.35) and after normalizing its constant

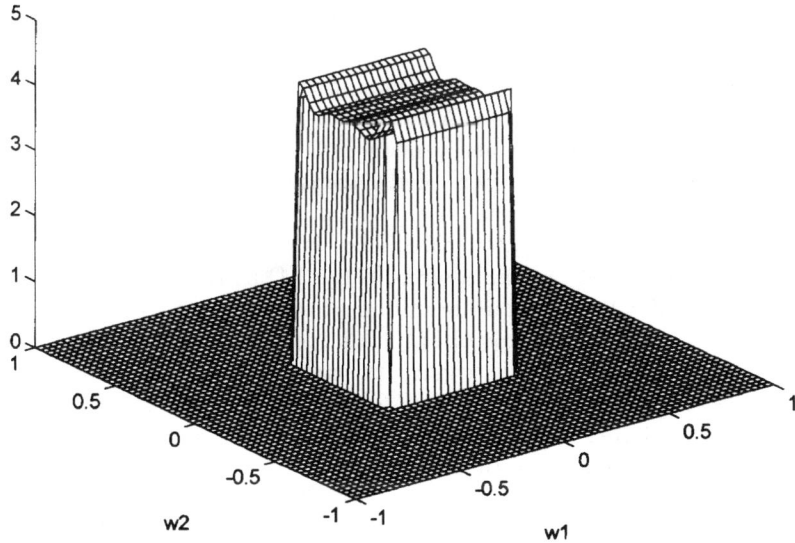

Fig. 4.14. Group delay response of Kwan-Chan filter, in the passband, with respect to ω_2

to unity, it is repeated below:

$$H_\phi(z^{-1}) = \frac{1}{Q(z^{-1})}$$
$$= \frac{1}{\sum_{k=0}^{N}\left[(-1)^k \cdot \binom{N}{k} \prod_{i=0}^{N} \frac{2\tau+i}{2\tau+k+i} z^{-k}\right]} \quad (4.56)$$

where N is the order of the filter, and τ is the desired group delay. After computing the coefficients of the denominator polynomial in (4.56), the authors choose a 1-D mirror image polynomial $N^\phi(z^{-1})$, to define

$$H(z^{-1}) = \frac{N^\phi(z^{-1})}{Q(z^{-1})} \quad (4.57)$$

Instead of approximating the magnitude of the above transfer function $H(z^{-1})$ to the specified magnitude, they set $N^\phi(z^{-1}) = H(z^{-1})Q(z^{-1})$ and approximate the magnitude of $H_d(z^{-1})Q(z^{-1})$ to the magnitude of $N^\phi(z^{-1})$ in the equiripple sense, by using Remez Exchange algorithm. (In the literature on system identification, or system modeling, both of these two approaches are used and the errors are known as modeling error and equation error respectively). Then they apply the modified McClellan's Transformation [24] *only* on the numerator polynomial $N^\phi(z^{-1})$ to generate the numerator polynomial $N^\phi(z_1^{-1}, z_2^{-1})$ of the 2-D IIR filter; while they generate the denominator polynomial as $Q(z_1^{-1}, z_2^{-1}) = Q(z_1^{-1})Q(z_2^{-1})$.

It should be pointed out that the 2-D transfer function

$$H_\phi(z_1^{-1}, z_2^{-1}) = \frac{1}{Q(z_1^{-1})Q(z_2^{-1})} \quad (4.58)$$

has a maximally flat group delay characteristic in the direction of both ω_1 and ω_2 axis; it is not zero in the stopband. Also it is noted that the modified McClellan Transformation maps any frequency in the 1-D ω-axis to a closed contour in the $\omega_1 - \omega_2$ plane, the parameters of the transformation being chosen to approximate these contours to circles or elliptic contours as closely as possible. The magnitude of the 1-D filter at any frequency is equal to the magnitude at all frequencies on the corresponding contour in the 2-D frequency plane. Hence when this modified McClellan's Transformation is applied only to the numerator polynomial $N(z^{-1})$, the magnitude of only this polynomial is mapped to the contours in the 2-D plane. So even though the parameters of the transformation are chosen to minimize the error in $N^\phi(z_1^{-1}, z_2^{-1})$ in the equiripple sense, the magnitude contributed by the function (4.58) would change the magnitude characteristics of the overall filter function

$$H_d(z_1^{-1}, z_2^{-1}) = \frac{N^\phi(z_1^{-1}, z_2^{-1})}{Q(z_1^{-1})Q(z_2^{-1})} \quad (4.59)$$

Since the above method [23] does not actually obtain an equiripple approximation to the magnitude response, it does not offer better results than those obtained from [21]. In [21], only one optimization using the Fletcher Powell method is needed to obtain the numerator, since the denominator is obtained by using Thiran's formula (4.56) i.e. without the need for optimization to approximate the prescribed constant group delays.

4.3 Design Using Linear Programming

Now we consider the use of Linear Programming for the approximation of both the magnitude and group delay of a 2-D IIR filter. The authors of [5] do so by extending their method for the design of 1-D IIR filters [6] which was described in section 2.5.5. In the following section, their method for designing 2-D IIR filters will be described.

Consider the transfer function

$$H_d(z_1^{-1}, z_2^{-1}) = \frac{\sum_i \sum_j b_{ij} z_1^{-i} z_2^{-j}}{\sum_i \sum_j d_{ij} z_1^{-i} z_2^{-j}} = \frac{N(z_1^{-1}, z_2^{-1})}{D(z_1^{-1}, z_2^{-1})}. \quad (4.60)$$

where we assume $d_{00} = 1$ without loss of generality. The coefficient arrays are chosen to represent either a nonsymmetric half plane [NSHP] filter or a quarter plane [QP] filter.

The index set or the region of support for the NSHP filter of order J is described by

$$\begin{array}{ll} 0 < j < J & ; i = 0 \\ -J < j < J ; 1 < i < J \end{array} \quad (4.61)$$

and the set for the QP filter is described by $0 \leq i < J$ and $0 \leq j < J$. A region of support for the NSHP filter is shown in Fig. 3.20.

The frequency response of this function is denoted in the form

$$H_d(e^{-j\omega_1}, e^{-j\omega_2}) = H_d(\omega_1, \omega_2) = \frac{\sum_i \sum_j b_{ij} e^{-j(\omega_1 i + \omega_2 j)}}{\sum_i \sum_j d_{ij} e^{-j(\omega_1 i + \omega_2 j)}} \quad (4.62)$$

As before, we use $G(e^{-j\omega_{1m}}, e^{-j\omega_{2n}}) = G(m, n)$ and $\tau_1(m, n)$, and $\tau_2(m, n)$ to represent the prescribed magnitude and group delay responses at the discrete points on the frequency plane $(\omega_{1m}, \omega_{2n})$, $m = 1, 2, \ldots, M$ and $n = 1, 2, \ldots, N$. By assuming $\tau_1 = \tau_2 = \tau$, to simplify the problem, we have

$$H(e^{-j\omega_{1m}}, e^{-j\omega_{2n}}) = G(m, n)e^{-j(\omega_{1m} + \omega_{2n})\tau} \quad (4.63)$$

The complex valued error between this and the function (4.62) evaluated at the same grid points on the frequency plane is defined as

$$\mathcal{E}(m, n) = G(m, n)e^{-j(\omega_{1m} + \omega_{2n})\tau} - \frac{\sum_i \sum_j b_{ij} e^{-j(\omega_{1m} i + \omega_{2n} j)}}{\sum_i \sum_j d_{ij} e^{-j(\omega_{1m} i + \omega_{2n} j)}} \quad (4.64)$$

In order to obtain linear equations in terms of the unknown coefficients b_{ij} and d_{ij}, the authors define a new equation error - which is complex-valued - by multiplying the above with $\sum_i \sum_j d_{ij} e^{-j(\omega_{1m} i + \omega_{2n} j)}$ and express the real and imaginary part of the numerator only, as $E_R(m,n)$ and $E_I(m,n)$ i.e.

$$\begin{aligned} E(m,n) &= \sum_i \sum_j d_{ij} e^{-j(\omega_{1m} i + \omega_{2n} j)} G(m,n) e^{-j(\omega_{1m} + \omega_{2n})\tau} \\ &\quad - \sum_i \sum_j b_{ij} e^{-j(\omega_{1m} i + \omega_{2n} j)} \\ &= E_R(m,n) + j E_I(m,n) \end{aligned} \qquad (4.65)$$

where

$$\begin{aligned} E_R(m,n) &= \sum_i \sum_j d_{ij} \{G(m,n) \cos[(i+\tau)\omega_{1m} + (j+\tau)\omega_{2n}]\} \\ &\quad + \sum_i \sum_j b_{ij} \{-\cos[\omega_{1m} i + \omega_{2n} j]\} \end{aligned} \qquad (4.66)$$

and

$$\begin{aligned} E_I(m,n) &= \sum_i \sum_j d_{ij} \{-G(m,n) \sin[(i+\tau)\omega_{1m} + (j+\tau)\omega_{2n}]\} \\ &\quad + \sum_i \sum_j b_{ij} \{\sin[\omega_{1m} i + \omega_{2n} j]\} \end{aligned} \qquad (4.67)$$

It is argued that the error (4.64) is minimized when the above two errors are minimized. Define a small positive variable ϵ, such that, for all m and n, the following inequalities are satisfied.

$$|E_R(m,n)| \leq \epsilon \qquad (4.68)$$
$$|E_I(m,n)| \leq \epsilon \qquad (4.69)$$

The above two inequalities are equivalent to the following

$$E_R(m,n) - \epsilon \leq 0 \qquad (4.70)$$
$$-E_R(m,n) - \epsilon \leq 0 \qquad (4.71)$$
$$E_I(m,n) - \epsilon \leq 0 \qquad (4.72)$$
$$-E_I(m,n) - \epsilon \leq 0 \qquad (4.73)$$

It can be seen that the above linear inequalities are in the form suitable for minimizing ϵ as much as possible. By letting $\zeta = -\epsilon$, the problem reduces to that of maximizing ζ as much as possible, subject to the constraint $\zeta \leq 0$. Indeed the Linear Programming problem is expressed as

$$maximize \quad g = \zeta \qquad (4.74)$$

subject to the inequalities

$$\sum_i \sum_j d_{ij} \{G(m,n) \cos[(i+\tau)\omega_{1m} + (j+\tau)\omega_{2n}]\}$$

4.3. DESIGN USING LINEAR PROGRAMMING

$$+ \sum_i \sum_j b_{ij} \{-\cos[\omega_{1m}i + \omega_{2n}j]\} + \varsigma \leq 0 \qquad (4.75)$$

$$- \sum_i \sum_j d_{ij} \{G(m,n)\cos[(i+\tau)\omega_{1m} + (j+\tau)\omega_{2n}]\}$$

$$- \sum_i \sum_j b_{ij} \{-\cos[\omega_{1m}i + \omega_{2n}j]\} + \varsigma \leq 0 \qquad (4.76)$$

$$- \sum_i \sum_j d_{ij} \{G(m,n)\sin[(i+\tau)\omega_{1m} + (j+\tau)\omega_{2n}]\}$$

$$+ \sum_i \sum_j b_{ij} \{\sin[\omega_{1m}i + \omega_{2n}j]\} + \varsigma \leq 0 \qquad (4.77)$$

$$\sum_i \sum_j d_{ij} \{G(m,n)\sin[(i+\tau)\omega_{1m} + (j+\tau)\omega_{2n}]\}$$

$$- \sum_i \sum_j b_{ij} \{\sin[\omega_{1m}i + \omega_{2n}j]\} + \varsigma \leq 0 \qquad (4.78)$$

for all $m = 1, 2, \ldots, M$ and $n = 1, 2, \ldots, N$

When the above Linear Programming problem is solved, it only means that the approximation of both the prescribed magnitude and group delays is achieved but there is no guarantee that the transfer function (4.60) is stable. The condition for stability used by the authors is given as

$$Re\left[D(z_1^{-1}, z_2^{-1})\right] > 0; \quad |z_1| = 1, |z_2| = 1 \qquad (4.79)$$

Note that the above condition is only a sufficient condition for stability of the transfer function, from which the authors arrive at the following linear inequality, with $b_{00} = 1$.

$$-\sum_i \sum_j d_{ij} \cos(\omega_1 i + \omega_2 j) \leq 1 - \delta \quad ; \qquad (4.80)$$

$$-\pi \leq \omega_1 \leq \pi, \quad 0 \leq \omega_2 \leq \pi \text{ and } (i + j \neq 0)$$

However, the above condition cannot be included as just one more inequality to be added for solving the Linear Programming problem, since ω_1 and ω_2 are continuous variables. So, we have to choose the same set of discrete values $(\omega_{1m}, \omega_{2n})$ for the two frequency variables and add inequalities for all $m = 1, 2, \ldots, M$ and $n = 1, 2, \ldots, N$. Hence the total number of inequalities to be considered becomes very large while we already have a large number of variables. So the computation time required by this method is significantly longer than that required for designing 1-D IIR filters, though the Linear Programming always finds the maximum value for ς if it exists. For these reasons this method may not be a preferred method for designing the 2-D IIR filters which approximate prescribed magnitude and group delay.

4.4 Design Using Singular Value Decomposition

In section 2.5.3, a method of designing a 1-D IIR filter to approximate the magnitude and group delay, using singular value decomposition(SVD) was discussed [12]. We will show that the same method can be extended for the design of 2-D IIR filters. For this purpose, let us assume that the given frequency response $G(e^{-j\omega_1}, e^{-j\omega_2}) = G(\omega_1, \omega_2)$ is such that it has a Fourier series representation of the form

$$\mathbf{T}_f(z_1^{-1}, z_2^{-1}) = \sum_{i=-\infty}^{\infty} \sum_{k=-\infty}^{\infty} T(i,k) z_1^{-i} z_2^{-k} \qquad (4.81)$$

where we have assumed that the coefficients are real and satisfy the condition for quadrantal symmetry i.e. $T(-i,-k) = T(i,k) = T(-i,k) = T(i,-k)$. Then we truncate the coefficient array by a rectangular window to get a 2-D polynomial

$$\mathbf{T}(z_1^{-1}, z_2^{-1}) = \sum_{i=-n_1}^{n_1} \sum_{k=-n_2}^{n_2} T(i,k) z_1^{-i} z_2^{-k} \qquad (4.82)$$

Hence it is a mirror image polynomial and non-causal but $T(e^{-j\omega_1}, e^{-j\omega_2})$ is a real function of ω_1 and ω_2. Equation (4.82) can be written in the form

$$\mathbf{T}(z_1^{-1}, z_2^{-1}) = Z_1^T \mathbf{R} Z_2 \qquad (4.83)$$

where $\mathbf{R}$ is given by

$$\begin{bmatrix} T(-n_1,-n_2) & T(-n_1,-n_2+1) & .. & T(-n_1,0) & .. & T(-n_1,n_2) \\ T(-n_1+1,-n_2) & ... & .. & T(-n_1+1,0) & .. & T(-n_1+1,n_2) \\ ... & ... & .. & ... & .. & ... \\ T(0,-n_2) & ... & .. & T(0,0) & .. & T(0,n_2) \\ ... & ... & .. & ... & .. & ... \\ T(n_1,-n_2) & ... & .. & T(n_1,0) & .. & T(n_1,n_2) \end{bmatrix}$$

$$(4.84)$$

and

$$Z_k^T = [\ z_k^{n_k}\ z_k^{n_k-1}\ \cdots\ z_k^1\ 1\ z_k^{-1}\ \cdots\ z_k^{-n_k}\ ]; \quad k = 1, 2 \qquad (4.85)$$

Let us assume that the matrix $\mathbf{R}$ has a rank $\rho > 0$. Then there exist two matrices $\mathbf{M}_1$ of size $\rho \times (2n_1+1)$ and $\mathbf{M}_2$ of size $\rho \times (2n_2+1)$ such that $\mathbf{R} = \mathbf{M}_1^T \mathbf{M}_2$. If $\mathbf{R} = USV^T$ is a singular value decomposition of $\mathbf{R}$, then we can identify

$$\mathbf{M}_1 = \sqrt{S} U^T \quad \text{and} \quad \mathbf{M}_2 = \sqrt{S} V^T \qquad (4.86)$$

or

$$\mathbf{F}_1(z_1^{-1}) = \mathbf{M}_1 Z_1 \quad \text{and} \quad \mathbf{F}_2(z_2^{-1}) = \mathbf{M}_2 Z_2 \qquad (4.87)$$

so that

$$\mathbf{T}(z_1^{-1}, z_2^{-1}) = \mathbf{F}_1^T(z_1^{-1}) \mathbf{F}_2(z_2^{-1})$$

The matrices $\mathbf{F}_1(z_1^{-1}) = \mathbf{M}_1 Z_1$ and $\mathbf{F}_2(z_2^{-1}) = \mathbf{M}_2 Z_2$ can be put in the form

$$\mathbf{F}_1(z_1^{-1}) = \sum_{i=-n_1}^{n_1} M_{1\rho}(i) z_1^{-i} \text{ and } \mathbf{F}_2(z_2^{-1}) = \sum_{i=-n_2}^{n_2} M_{2\rho}(i) z_2^{-i} \qquad (4.88)$$

where each of $M_{1\rho}(i)$ and $M_{2\rho}(i)$ is a $\rho \times 1$ matrix or a column vector for each i.

The matrices $\mathbf{M}_1$ and $\mathbf{M}_2$ are of size $\rho \times (2n_1+1)$ and $\rho \times (2n_2+1)$ respectively and can be represented by the following two column matrices.

$$[M_{1\rho}(-n_1) \quad M_{1\rho}(-n_1+1) \cdots M_{1\rho}(-1) \quad M_{1\rho}(0) \cdots M_{1\rho}(n_1)]$$

and

$$[M_{2\rho}(-n_2) \quad M_{2\rho}(-n_2+1) \cdots M_{2\rho}(-1) \quad M_{2\rho}(0) \cdots M_{2\rho}(n_2)] \qquad (4.89)$$

The importance of the above result is that $\mathbf{F}_1(z_1^{-1})$ and $\mathbf{F}_2(z_2^{-1})$ are 1-D FIR digital filters with one input and ρ outputs.

Next we shift the non-causal 2-D FIR filter to the form

$$\mathbf{T}_g(z_1^{-1}, z_2^{-1}) = z_1^{-m_1} z_2^{-m_2} \mathbf{T}(z_1^{-1}, z_2^{-1}) = \mathbf{G}_1^T(z_1^{-1}) \mathbf{G}_2(z_2^{-1}) \qquad (4.90)$$

where

$$\mathbf{G}_1(z_1^{-1}) = z_1^{-m_1} \mathbf{F}_1(z_1^{-1}) \quad \text{and} \quad \mathbf{G}_2(z_2^{-1}) = z_2^{-m_2} \mathbf{F}_2(z_2^{-1}) \qquad (4.91)$$

and m_1 and m_2 are chosen to be some arbitrary positive integers such that $0 < m_1 < n_1$ and $0 < m_2 < n_2$. It is seen that the frequency response of $\mathbf{G}_1(e^{-j\omega_1})$ and $\mathbf{G}_2(e^{-j\omega_2})$ have the same magnitudes as $\mathbf{F}_1(e^{-j\omega_1})$ and $\mathbf{F}_2(e^{-j\omega_2})$ respectively and they are all real functions; but $\mathbf{G}_1(e^{-j\omega_1})$ and $\mathbf{G}_2(e^{-j\omega_2})$ contain additional group delays of m_1 and m_2 in the ω_1 and ω_2 directions respectively.

It is also pointed out that $M_j(-i) = M_j(i)$, since we have assumed $T(i,k)$ has quadrantal symmetry. Now we are ready to make the significant observation that the problem of approximating the magnitude of a 2-D IIR filter is reduced to that of finding the approximation for 1-D FIR filters: $\mathbf{G}_1(z_1^{-1})$ and $\mathbf{G}_2(z_2^{-1})$. The method for approximating the magnitude response of a 1-D IIR filter by a 1-D FIR filter has been briefly mentioned when we described Method 2.5 in section 2.5.3. The method for designing the 1-D filters will be described in greater detail in the following section. Because of this important result, the need for assuring stability of a 2-D IIR filter is avoided and the total computational time is slightly more than that required for the design of a 1-D IIR filter.

4.5 Chebyshev Approximation Theory

In (4.81), it was implicitly assumed that $\mathbf{T}(z_1^{-1}, z_2^{-1})$ is square-integrable on the unit bi-disc $T^2 = \{(z_1, z_2) : |z_1| = 1, |z_2| = 1\}$. Therefore, from the Fourier series theory and Parseval's Theorem, we can make the following two statements:

Statement 1.

$$\frac{1}{(2\pi)^2} \int_{-\pi}^{\pi} \int_{-\pi}^{\pi} \left| \mathbf{T}(e^{-j\omega_1}, e^{-j\omega_2}) \right|^2 d\omega_1 d\omega_2 \quad (4.92)$$

$$= \sum_{i=-\infty}^{\infty} \sum_{k=-\infty}^{\infty} |T(i,k)|^2 < \infty$$

The Fourier series coefficients $T(i,k)$ can be computed from

$$T(i,k) = \frac{1}{(2\pi)^2} \int_{-\pi}^{\pi} \int_{-\pi}^{\pi} \mathbf{T}(e^{-j\omega_1}, e^{-j\omega_2}) e^{j(i\omega_1 + k\omega_2)} d\omega_1 d\omega_2 \quad (4.93)$$

Statement 2.
If we define the finite summation of the series by

$$\mathbf{T}_{mn}(z_1^{-1}, z_2^{-1}) = \sum_{i=-m}^{m} \sum_{k=-n}^{n} T(i,k) z_1^{-1} z_2^{-1} \quad (4.94)$$

then $\mathbf{T}_{mn}(z_1^{-1}, z_2^{-1})$ converges to $\mathbf{T}(z_1^{-1}, z_2^{-1})$ in the sense

$$\lim_{(m,n) \to (\infty, \infty)} \left\{ \int_{-\pi}^{\pi} \int_{-\pi}^{\pi} \left| \mathbf{T}(e^{-j\omega_1}, e^{-j\omega_2}) - \mathbf{T}_{mn}(e^{-j\omega_1}, e^{-j\omega_2}) \right|^2 d\omega_1 d\omega_2 \right\}$$

$$= 0$$

except at points of discontinuity, where Gibbs oscillations would occur. Hence Fourier series expansion approximates the given function in the least squares sense i.e. the least squared error approaches zero as $(m,n) \to (\infty, \infty)$.

In [13], the authors raise the following question for which no answer has been provided before. What are the necessary and sufficient conditions under which the Chebyshev norm of the error function converges to zero as $(m,n) \to (\infty, \infty)$, i.e. what are necessary and sufficient conditions to be satisfied by $\mathbf{T}(z_1^{-1}, z_2^{-1})$ such that

$$\lim_{(m,n) \to (\infty, \infty)} \{ \|\mathbf{T} - \mathbf{T}_{mn}\|_c \} = 0 \quad (4.95)$$

where the Chebyshev norm $\|E\|_c$ of an error function E_c is defined by

$$\|E\|_c = \sup \left\{ \left| \mathbf{T}(z_1^{-1}, z_2^{-1}) \right| : (z_1, z_2) \in T^2 \right\} \quad (4.96)$$

They arrive at a partial answer to this question. They derive only sufficient conditions to answer the above question and these conditions are stated below as a theorem.

Theorem 3 *If the following three conditions are satisfied by* $\mathbf{T}(e^{-j\omega_1}, e^{-j\omega_2})$, *then (4.95) is true.*

4.5. CHEBYSHEV APPROXIMATION THEORY

$$\int_{-\pi}^{\pi}\int_{-\pi}^{\pi}\left|\frac{\partial^2 \mathbf{T}(e^{-j\omega_1}, e^{-j\omega_2})}{\partial\omega_1\partial\omega_2}\right|^2 \partial\omega_1\partial\omega_2 < \infty \quad (4.97)$$

$$\int_{-\pi}^{\pi}\left|\frac{\partial \mathbf{T}(e^{-j\omega_1}, e^{-j\omega_2})}{\partial\omega_1}\right|^2 \partial\omega_1 < \infty \quad (4.98)$$

$$\int_{-\pi}^{\pi}\left|\frac{\partial \mathbf{T}(e^{-j\omega_1}, e^{-j\omega_2})}{\partial\omega_2}\right|^2 \partial\omega_2 < \infty \quad (4.99)$$

The proof consists of two parts. First, it is shown that if the above three conditions are satisfied by $\mathbf{T}(z_1^{-1}, z_2^{-1})$, then $T(i,k)$ is absolutely summable i.e. $\sum_{i=-\infty}^{\infty}\sum_{k=-\infty}^{\infty} |T(i,k)| < \infty$. Then it is shown that if $T(i,k)$ is absolutely summable, (4.95) is satisfied. Before the proof is provided, however, it is necessary to prove the following lemma:

Lemma 4 *If the Fourier series coefficients $T(i,k)$ used in (4.81) can be written as $T(i,k) = g_{ik}h_{ik}$, with g_{ik} and h_{ik} satisfying*

$$\sum_{i=-\infty}^{\infty}\sum_{k=-\infty}^{\infty} |g_{ik}|^2 < \infty \quad \text{and} \quad \sum_{i=-\infty}^{\infty}\sum_{k=-\infty}^{\infty} |h_{ik}|^2 < \infty \quad (4.100)$$

then

$$\sum_{i=-\infty}^{\infty}\sum_{k=-\infty}^{\infty} |T(i,k)| \leq \sqrt{\sum_{i=-\infty}^{\infty}\sum_{k=-\infty}^{\infty} |g_{ik}|^2}\sqrt{\sum_{i=-\infty}^{\infty}\sum_{k=-\infty}^{\infty} |h_{ik}|^2} < \infty \quad (4.101)$$

Proof. Let $Q_{in+k} = T(i,k)$ for $|i| \leq m$ and $|k| \leq n$ where m, n are some positive integers. Then $\{Q_{in+k}\}$ is a 2-sided 1-D sequence, constructed from the 2-D sequence $\{T(i,k)\}$. Applying the well-known Schwartz inequality to this $\{Q_{in+k}\}$, we get

$$\sum_{i=-m}^{m}\sum_{k=-n}^{n} |T(i,k)| \leq \sqrt{\sum_{i=-m}^{m}\sum_{k=-n}^{n} |g_{ik}|^2}\sqrt{\sum_{i=-m}^{m}\sum_{k=-n}^{n} |h_{ik}|^2} \quad (4.102)$$

Since the two terms on the right side converge as $(m,n) \to (\infty, \infty)$, the term on the left side also converge as $(m,n) \to (\infty, \infty)$ which completes the proof for the above Lemma.

Now we prove that if (4.97)-(4.99) are satisfied, then $T(i,k)$ is absolutely summable.

Proof. We derive

$$\frac{\partial^2 \mathbf{T}(e^{-j\omega_1}, e^{-j\omega_2})}{\partial\omega_1\partial\omega_2} = \sum_{i=-\infty}^{\infty}\sum_{k=-\infty}^{\infty} ikT(i,k)e^{-j(i\omega_1+k\omega_2)}$$

$$= \sum_{i=-\infty}^{\infty}\sum_{k=-\infty}^{\infty} w_{ik}e^{-j(i\omega_1+k\omega_2)}$$

where
$$\sum_{i=-\infty}^{\infty}\sum_{k=-\infty}^{\infty}|w_{ik}|^2 < \infty \qquad (4.103)$$

which follows from (4.97) and the Parseval's Theorem. Similarly by using the other two conditions, we have

$$\frac{\partial \mathbf{T}(e^{-j\omega_1},1)}{\partial(j\omega_1)} = -\sum_{i=-\infty}^{\infty} iT(i,0)e^{-j(i\omega_1)} = \sum_{i=-\infty}^{\infty} g_i e^{-j(i\omega_1)} \qquad (4.104)$$

$$\frac{\partial \mathbf{T}(1,e^{-j\omega_2})}{\partial(j\omega_2)} = -\sum_{i=-\infty}^{\infty} kT(0,k)e^{-j(i\omega_2)} = \sum_{i=-\infty}^{\infty} h_k e^{-j(i\omega_2)} \qquad (4.105)$$

Using (4.98) and (4.99), we obtain

$$\sum_{i=-\infty}^{\infty} |g_i|^2 < \infty \quad \text{and} \quad \sum_{k=-\infty}^{\infty} |h_k|^2 < \infty \qquad (4.106)$$

From the above expressions, we can see that

$$T(i,0) = -\frac{g_i}{i} \quad for \quad i \neq 0$$

$$T(0,k) = -\frac{h_k}{k} \quad for \quad k \neq 0 \qquad (4.107)$$

$$T(i,k) = \frac{w_{ik}}{ik} \quad for \quad i \neq 0, k \neq 0 \qquad (4.108)$$

and we substitute the above relations in (4.109)

$$\sum_{i=-\infty}^{\infty}\sum_{k=-\infty}^{\infty} |T(i,k)| = |T(0,0)| + \sum_{i \neq 0} |T(i,0)|$$
$$+ \sum_{k \neq 0} |T(0,k)| + \sum_{i \neq 0}\sum_{k \neq 0} |T(i,k)|$$

From Lemma 2, we have

$$\sum_{i \neq 0} |T(i,0)| = \sum_{i \neq 0} \left|\frac{g_i}{i}\right| \leq \sqrt{\sum_{i \neq 0} i^{-2}} \sqrt{\sum_{i \neq 0} |g_i|^2} < \infty$$

$$\sum_{k \neq 0} |T(0,k)| = \sum_{k \neq 0} \left|\frac{h_k}{k}\right| \leq \sqrt{\sum_{k \neq 0} k^{-2}} \sqrt{\sum_{k \neq 0} |h_k|^2} < \infty$$

$$\sum_{i \neq 0}\sum_{k \neq 0} |T(i,k)| = \sum_{i \neq 0}\sum_{k \neq 0} \left|\frac{w_{ik}}{ik}\right|$$
$$\leq \sqrt{\sum_{i \neq 0}\sum_{k \neq 0}(ik)^{-2}} \sqrt{\sum_{i \neq 0}\sum_{k \neq 0}|w_{ik}|^2} < \infty \qquad (4.109)$$

4.5. CHEBYSHEV APPROXIMATION THEORY

Thus we complete the proof that

$$\sum_{i=-\infty}^{\infty}\sum_{k=-\infty}^{\infty}|T(i,k)| < \infty \qquad (4.110)$$

Next we consider the second part of the proof. Assume that (4.110) is satisfied. Then

$$\begin{aligned}\|\mathbf{T}-\mathbf{T}_{mn}\|_c &= \left\|\sum_R\sum T(i,k)e^{-j(i\omega_1+k\omega_2)}\right\|_c \qquad (4.111)\\ &\leq \sum_R\sum |T(i,k)|\end{aligned}$$

where the index set $R = \{(i,k) : i > m \text{ or } k > n\}$. When (4.110) is satisfied, it is clear that as $(m,n) \to (\infty,\infty)$, we have shown $\|\mathbf{T}-\mathbf{T}_{mn}\|_c \to 0$. So we have proved that (4.97)-(4.99) are sufficient conditions for the convergence of $T(z_1^{-1}, z_2^{-1})$ in the Chebyshev sense. They are conditions in the frequency domain and hence very useful, because they lead to methods of designing filters in terms of specifications in the frequency domain and not in terms of the desired spatial sequences $h(m,n)$. Since the conditions are only sufficient, there may be many different classes of functions that would satisfy these conditions. The authors in [13] have proposed one such class of functions as described below, under the following assumptions: (1) the Fourier series coefficients obtained from the given frequency domain specifications satisfy the conditions $T(-i,-k) = T(i,k)$ (2) the boundaries of the passband and stopband regions in the $\omega_1 - \omega_2$ plane are two closed contours defined by $\Phi_p(\rho_p, \psi_p)$ and $\Phi_s(\rho_s, \psi_s)$ respectively. Any point in the $\omega_1 - \omega_2$ plane is related to its equivalent polar coordinates ρ and ψ by $\omega_1 = \rho\cos\psi$ and $\omega_2 = \rho\sin\psi$ or $\rho = \sqrt{\omega_1^2 + \omega_2^2}$ and $\psi = \tan^{-1}\left(\frac{\omega_1}{\omega_2}\right)$ (3) there is only one passband region and one stopband region in the region $R_2 = \{(\omega_1,\omega_2); |\omega_1| \leq \pi, |\omega_2| \leq \pi\}$ and that Φ_p and Φ_s are continuous functions of ρ and ψ (4) $|\Phi_p - \Phi_s| > 0$ for all points in R_2 and (5) $|T(e^{-j\omega_1}, e^{-j\omega_2})| = 1$ in the passband region and is zero in the stopband region.

Since the magnitude in the transition band is not specified in the above conditions, this freedom is exploited by the authors to choose $T(e^{-j\omega_1}, e^{-j\omega_2})$ in the transition band region such that with the above assumptions, it satisfies the conditions (4.97)-(4.99). The function chosen is

$$T(e^{-j\omega_1}, e^{-j\omega_2}) = \frac{1}{2}\left\{1 + \cos\left(\frac{\rho - \Phi_p(\rho_p,\psi_p)}{\Phi_p(\rho_p,\psi_p) - \Phi_s(\rho_s,\psi_s)}\pi\right)\right\} \qquad (4.112)$$

where ρ and ψ are in the transition band region. For a fixed value of ψ, e.g. when $\psi = 0$, $|\mathbf{T}(e^{-j\omega_1}, e^{-j\omega_2})|$ is a continuous function of ρ as shown in Fig. 4.15. Its second derivative is shown in Fig. 4.16.

As we traverse along any closed contour in the region R_2, the corresponding polar coordinate ψ is found to be a continuous variable but the other coordinate ρ may or may not be continuous. For example, the boundary of a rectangular

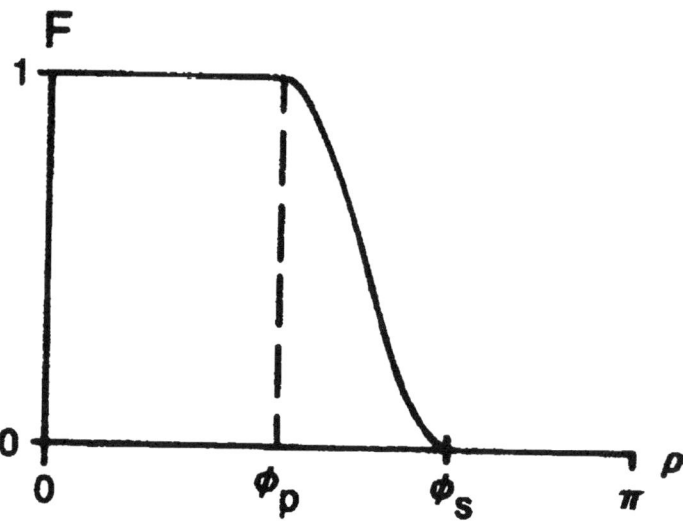

Fig. 4.15. Plot of the function in Equation (4.116) (By permission from IEEE)

region has a discontinuity in ρ at its corners, whereas that of a circular or elliptic region is continuous. For practical design purposes, if the transition

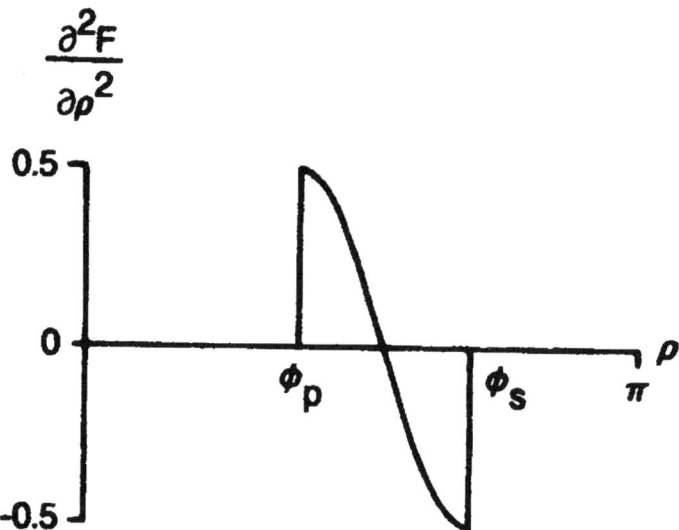

Fig. 4.16. Second derivative of the function in equation (4.116)(By permission from IEEE)

4.5. CHEBYSHEV APPROXIMATION THEORY

band is specified to be bounded by rectangular region, the corners of the region can be rounded by small circles as suggested in [13].

It can be easily proved that the conditions (4.97)-(4.99) are equivalent to the following three quantities being square integrable.

$$\frac{\partial^2 T(e^{-j\omega_1}, e^{-j\omega_2})}{\partial \rho^2}, \frac{\partial^2 T(e^{-j\omega_1}, e^{-j\omega_2})}{\partial \psi^2} \text{ and } \frac{\partial^2 T(e^{-j\omega_1}, e^{-j\omega_2})}{\partial \rho \partial \psi} \quad (4.113)$$

When both polar coordinates defining the boundaries of the transition region are continuous, and the magnitude in that region is given by (4.112), the authors have shown that (4.97)-(4.99) are satisfied by the function $\mathbf{T}(z_1^{-1}, z_2^{-1})$, subject to the assumptions made in the previous paragraph.

With the above theory assuring that the Fourier series expansion of $\mathbf{T}(z_1^{-1}, z_2^{-1})$ converges in the Chebyshev sense, the design procedure is straightforward. First it is shown [13] that the theory of 2-D Discrete Fourier Transform can be applied to compute the Fourier series coefficients of $\mathbf{T}_{mn}(z_1^{-1}, z_2^{-1})$ by using the FFT algorithm. Then the design of 2-D IIR filters that approximate the specified magnitude and constant group delay is carried out by application of singular value decomposition and model reduction techniques. Though these techniques have been mentioned in Chap. 2, a more detailed description is given below:

After we have found the Fourier series coefficients $T(i,k)$ of $\mathbf{T}_{mn}(z_1^{-1}, z_2^{-1})$, we shift them by m and n to get a strictly causal FIR filter function $H(z_1^{-1}, z_2^{-1}) = z_1^{-m} z_2^{-n} T_{mn}(z_1^{-1}, z_2^{-1})$. Let us consider

$$H_c = \begin{bmatrix} h(0,0) & h(0,1) & \ldots & h(0,2n) \\ h(1,0) & h(1,1) & \ldots & h(1,2n) \\ \ldots & \ldots & \ldots & \ldots \\ h(2m,0) & h(2m,1) & \ldots & h(2m,2n) \end{bmatrix} \quad (4.114)$$

so that the transfer function of the causal 2-D FIR filter is represented by $H(z_1^{-1}, z_2^{-1}) = Z_1^T H_c Z_2$ where

$$\begin{aligned} Z_1^T &= \begin{bmatrix} 1 & z_1^{-1} & z_1^{-2} & \ldots & z_1^{-2m} \end{bmatrix} \\ Z_2 &= \begin{bmatrix} 1 & z_2^{-1} & z_2^{-2} & \ldots & z_2^{-2n} \end{bmatrix}^T \end{aligned} \quad (4.115)$$

If we assume that the matrix H_c has a rank $r_h > 0$, then as in the previous section, we find a singular value decomposition $H_c = U^T S V$. Therefore we identify two matrices: a matrix $\mathbf{M}_1$ of size $r_h \times (2m+1)$ and a matrix $\mathbf{M}_2$ of size $r_h \times (2n+1)$ such that $\mathbf{M}_1^T = U^T \sqrt{S}$ and $M_2 = \sqrt{S} V$. Therefore,

$$H(z_1^{-1}, z_2^{-1}) = Z_1^T \mathbf{M}_1^T \mathbf{M}_2 Z_2 = \mathbf{H}_1(z_1)^T \mathbf{H}_2(z_2) \quad (4.116)$$

where

$$\mathbf{H}_1(z_1) = \sum_{k=0}^{2m} M_1(k) z_1^{-k} \quad (4.117)$$

$$\mathbf{H}_2(z_2) = \sum_{k=0}^{2n} M_2(k) z_2^{-k} \quad (4.118)$$

where $M_1(k)$ and $M_2(k)$ are of the same form as (4.89). So we have a separable transfer function. Since each of the transfer functions is a polynomial in one variable and not a ratio of polynomials, the balanced model reduction procedure is considerably simplified. For the same reason, a state space realization $\{A_i, B_i, C_i, D_i\}$ in a special form as shown below for each $H_i(z_i)$, $(i = 1, 2)$ can be defined.

$$A_i = \begin{bmatrix} 0 & 0 & \cdots & \cdots & 0 \\ 1 & 0 & 0 & \cdots & \cdots \\ 0 & 1 & 0 & \cdots & \cdots \\ \cdots & \cdots & \cdots & \cdots & \cdots \\ 0 & \cdots & \cdots & 1 & 0 \end{bmatrix} \qquad (4.119)$$

$$B_i = \begin{bmatrix} 1 \\ 0 \\ \cdots \\ \cdots \\ 0 \end{bmatrix} \quad \text{and} \quad D_i = M_i(0) \qquad (4.120)$$

$$C_1 = [M_1(1) \quad \cdots \quad \cdots \quad \cdots \quad M_1(2m)] \qquad (4.121)$$
$$C_2 = [M_2(1) \quad \cdots \quad \cdots \quad \cdots \quad M_2(2m)] \qquad (4.122)$$

Then, $H_i(z_i) = D_i + C_i(z_i I - A_i)^{-1} B_i$ for $i = 1, 2$ are each in the form of a polynomial, with the above choice for the state space realization. Now we show how a simple model reduction method can be developed to get a 2-D IIR filter transfer function, for the above special form of the state space realization. Let us define the observability Grammian P_1 and the controllability Grammian Q_i for $H_i(z_i)$ as follows:

$$P_i = \sum_{k=0}^{\infty} (A_i^T)^k C_i^T C_i A_i^k \qquad (4.123)$$

$$Q_i = \sum_{k=0}^{\infty} A_i^k B_i B_i^T (A_i^T)^k; \quad i = 1, 2 \qquad (4.124)$$

By substituting the state space realization into the above expressions, it can be shown that $Q_1 = I_{2m}$ and $Q_2 = I_{2n}$. It is also known that P_i admits a singular value decomposition $P_i = T_i^T \Sigma_i^2 T_i$ where Σ_i is a diagonal matrix with Hankel singular values of $H_i(z_i)$ in descending order. The matrix T_i is a real orthogonal matrix because P_i is known to be symmetric and positive definite, though in general T_i is not unique. Then the balanced realization is computed from

$$A_{ib} = T_i A_i T_i^T; \quad B_i = T_i B_i; \quad C_{ib} = C_i T_i^T$$

$$D_{ib} = D_i; \qquad i = 1, 2$$

The reduced order models can be now derived by direct truncation

$$A_{ir} = [\, I_{ir} \ 0 \,] A_{ib} \begin{bmatrix} I_{ir} \\ 0 \end{bmatrix}, \quad B_{ir} = [I_{ir} \ 0 \ B_{ib}] \qquad (4.125)$$

4.6. DESIGN USING DIGITAL SPECTRAL TRANSFORMATION

$$C_{ir} = C_{ib} \begin{bmatrix} I_{ir} \\ 0 \end{bmatrix}, \text{ and } D_{ir} = D_i \qquad (4.126)$$

From the above matrices, we construct $\hat{H}_i(z_i) = D_{ir} + C_{ir}(z_i I - A_{ir})^{-1} B_{ir}$; $i = 1, 2$. Then the 2-D IIR filter transfer function of reduced order is given in the separable form by

$$H_r(z_1, z_2) = \hat{H}_1^T(z_1) \hat{H}_2(z_2) \qquad (4.127)$$

This is a brief outline of the procedure for obtaining the balanced model reduction of a 2-D FIR filter given in the form (4.117), which is obtained from (4.114). This is the same method that should be used for obtaining a reduced order 2-D separable IIR filter from (4.91). More details of the model reduction techniques for the general case are found in text books on modern control theory.

4.6 Design Using Digital Spectral Transformation

Several types of frequency transformations and digital spectral transformations have been used in this book, to reduce the approximation problem in one frequency domain to that of another frequency domain where the solution is relatively easy or where the solution already exists. When this solution is more analytical than numerical, it is always preferred over strictly a numerical approach that provides the solution. Well-known examples are the frequency transformations that transform a lowpass analog prototype filter to an analog, highpass, bandpass and bandstop filter. These are transformations from one Laplace transform frequency variable to another Laplace transform frequency variable $[p = f(s)]$ for the approximation of continuous time (analog) filter functions. There is a similar class of frequency transformations in the z-transform for one dimensional digital filters, which were proposed by Constantinides [7]. These transformations were discussed in Chap. 1. In the same chapter, we introduced the bilinear transform which transforms the frequency response of a one-dimensional analog filter to that of a one-dimensional digital filter. In Chap. 3, we introduced a transformation to get the transfer function $H(s_1, s_2)$ of a 2-D analog filter from the transfer function $H(s)$ of a 1-D analog filter and also the use of two bilinear transforms $s_1 = \frac{z_1-1}{z_1+1}$ and $s_2 = \frac{z_2-1}{z_2+1}$ to transform the function $H(s_1, s_2)$ to a 2-D digital filter function $H(z_1, z_2)$. Then there is a set of digital spectral transformations which transforms a 2-D digital filter lowpass function to another set of 2-D digital filters [28]. This type of transformation was also discussed at the end of Chap. 3. A good survey of some properties of all these transformations is found in [3].

Now, we consider the application of a digital spectral transform (DST) proposed by [14, 15] for designing two-dimensional IIR digital filters from one-dimensional IIR digital filters. This is the Method 4.3 that was discussed in

Chap. 3. The design procedure based on these transformations has many advantages compared to the digital spectral transformations proposed by [4, 11] and a few other methods which were described in the previous chapter. The advantages are listed below:

1. The 2-D, IIR filters obtained by employing the DSTs proposed in [14, 15] are always stable, provided that the 1-D IIR filter chosen for the transformation is stable.

2. The method can be used for designing lowpass filters with a circular region as well as an elliptic region for the passband (or the stopband).

3. By use of the spectral transformations listed in [28], many other classes of filter responses can be realized.

4. The method does not involve any interpolation for computing the filter output, as required in [4].

5. The transform contains only a few unknown parameters that determine the passband region and the search for their best values does not require any linear programming or complicated nonlinear programming.

6. The 2-D filter realizes the same type of magnitude response characteristic in the passband region as the 1-D digital filter. Though the DST is a separable transformation, the 2-D filter function generated from it is not separable. In contrast, the transfer functions obtained in [21] and [23] have separable denominators. Such filters with separable denominator polynomials can only realize a subclass of specifications.

7. Most significantly it is the only digital spectral transformation (DST) that has been shown to exhibit almost the same group delay response characteristic in the passband region of the 2-D digital filter as the 1-D digital filter-besides preserving the magnitude response. If the 1-D digital filter has a maximally flat magnitude response as well as a constant group delay in the passband, it is shown that the DST proposed in [14]-[15] yields a 2-D IIR filter that has a maximally flat magnitude response in the passband region that is a very close approximation to the specified circular or elliptic region in the $\omega_1 - \omega_2$ plane and also has nearly the same constant group delay along the ω_1 and ω_2 axis.

None of the other methods described in this chapter and in the previous chapter have all of these advantages as the method based on the two DST's proposed here. This method will be described below in detail.

Let the transfer function of a 1-D stable IIR filter be denoted $H(z)$. The digital spectral transformation proposed by [14] is

$$z = z_1' z_2' = \left(\frac{z_1 + a}{az_1 + 1}\right)\left(\frac{z_2 + b}{bz_2 + 1}\right); \quad |a| < 1; \quad |b| < 1 \quad (4.128)$$

4.6. DESIGN USING DIGITAL SPECTRAL TRANSFORMATION

It is important to note that in this method z, z_1 and z_2 are assumed to be *delay operators* so the conditions $|a| < 1$ and $|b| < 1$ assure that the transformations $G_1(z_1) = \left(\frac{z_1 + a}{az_1 + 1}\right)$ and $G_2(z_2) = \left(\frac{z_2 + b}{bz_2 + 1}\right)$ are stable functions. Then it can be easily proved [14] that

$$H(z_1, z_2) = H(G_1(z_1)G_2(z_2)) \qquad (4.129)$$

is also stable in the $z_1 - z_2$ plane. It may also be noted that $G_1(z_1)$ and $G_2(z_2)$ are allpass functions and points on the unit bi-disc : $\{(z_1, z_2); |z_1| = 1 \text{ and } |z_2| = 1|\}$ map to points on the unit circle $|z| = 1$. In particular, the origin in the $\omega_1 - \omega_2$ plane (i.e. $z_1 = 1$ and $z_2 = 1$) maps to $\omega = 0$ for all values of a and b. The general mapping relationship between points on the ω-axis and those in the $\omega_1 - \omega_2$ plane is derived from (4.128) as

$$\omega = \omega_1 + \omega_2 - 2\left(\tan^{-1}\frac{a\sin\omega_1}{a\cos\omega_1 + 1} + \tan^{-1}\frac{b\sin\omega_2}{b\cos\omega_2 + 1}\right) \qquad (4.130)$$

For fixed values of the parameters a and b, we see that the above maps a frequency on the ω-axis to a closed contour in the $\omega_1 - \omega_2$ plane. As illustration, we fixed the values of $a = 0.61$, $b = 0.61$ and choosing some sample values for $\omega = 0.1\pi, 0.2\pi, 0.3\pi, \ldots, 0.9\pi$, we plotted the resulting contours in Fig. 4.17; the same plot in the first quadrant only is shown in Fig. 4.18.

There is a striking resemblance of these plots to those obtained from the McClellan's transformation [24] used in designing circularly symmetric 2-D FIR filters. But the McClellan's transformation chosen for designing circular filters has four-quadrant symmetry, whereas the transformation (4.128) has a symmetry only along the diagonal line defined by $\omega_1 + \omega_2 = 0$. As the values of a and b are changed, the contour corresponding to any given value of ω changes and this degree of freedom is used to find the optimal values for these two parameters such that the transformation gives a contour in the $\omega_1 - \omega_2$ plane of the 2-D filter which approximates the boundary of the given circular (or elliptical) passband region (in the first and third quadrant) with the least amount of deviation.

When we substitute $\omega_2 = 0$, $\omega_1 = W^{(2)}$ and $\omega_2 = W^{(2)}$, $\omega_1 = 0$ in (4.130), we get

$$W^{(1)} = W^{(2)} - 2\tan^{-1}\left(\frac{a\sin W^{(2)}}{a\cos W^{(2)} + 1}\right)$$

$$W^{(1)} = W^{(2)} - 2\tan^{-1}\left(\frac{b\sin W^{(2)}}{b\cos W^{(2)} + 1}\right) \qquad (4.131)$$

where $W^{(1)}$ is the frequency on the ω-axis which maps into these two points $(W^{(2)}, 0)$ on the ω_1 axis and $(0, W^{(2)})$ on the ω_2 axis. So in the case of a circular passband, it is obvious that $a = b$. Therefore (4.130) reduces to

$$\omega = \omega_1 + \omega_2 - 2\left(\tan^{-1}\frac{a\sin\omega_1}{a\cos\omega_1 + 1} + \tan^{-1}\frac{a\sin\omega_2}{a\cos\omega_2 + 1}\right) \qquad (4.132)$$

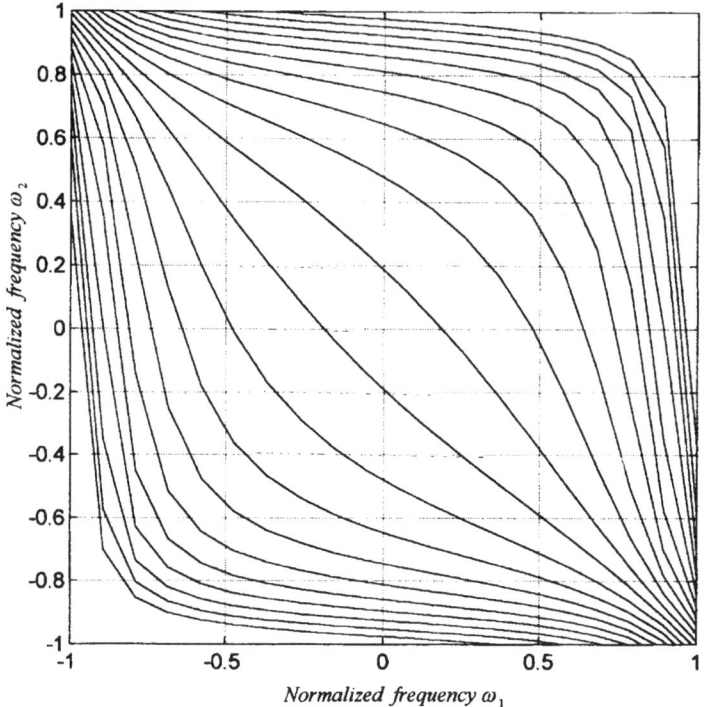

Fig. 4.17. Mapping of the constant magnitude contours, under the Digital Spectral Transformation

In the case of an elliptic passband region the semi-minor axis and semi-major axis $W_1^{(2)}$ and $W_2^{(2)}$ are different but they are related by the following equation.

$$\left(\frac{\omega_1}{W_1^{(2)}}\right)^2 + \left(\frac{\omega_2}{W_2^{(2)}}\right)^2 = 1 \tag{4.133}$$

When the mapping relationship for the intersection of the elliptic contour with the ω_1 and ω_2 axes are used, we get

$$W^{(1)} = W_1^{(2)} - 2\tan^{-1}\frac{a \sin W_1^{(2)}}{1 + a \cos W_1^{(2)}} \tag{4.134}$$

$$W^{(1)} = W_2^{(2)} - 2\tan^{-1}\frac{b \sin W_2^{(2)}}{1 + b \cos W_2^{(2)}} \tag{4.135}$$

Hence we conclude that in this case $a \neq b$. But once a value for a is chosen, b is constrained by the above two relationships and therefore, we need to search for the optimum value of only the parameter a in the interval $|a| < 1$. With

4.6. DESIGN USING DIGITAL SPECTRAL TRANSFORMATION

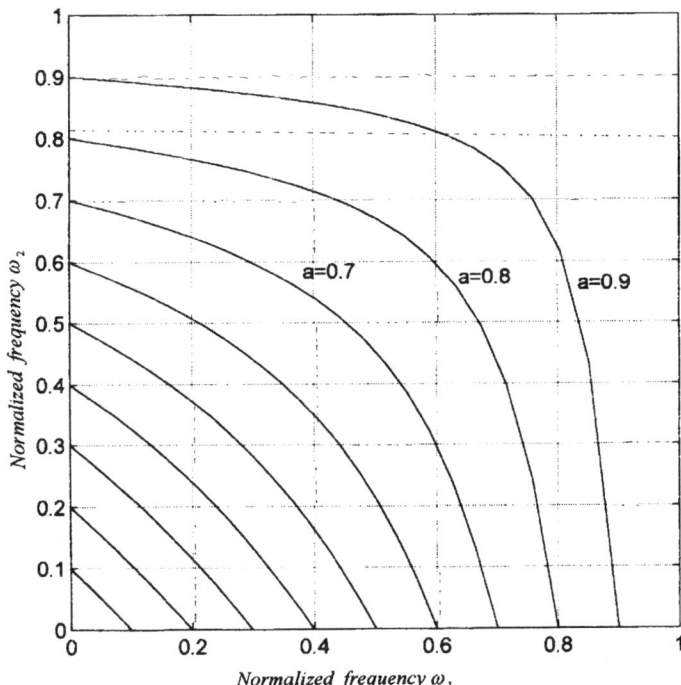

Fig. 4.18. Mapping of the constant magnitude contours in the first quadrant

the optimum value so chosen, we get a contour that approximates a circle or ellipse in the first and third quadrant of the $\omega_1 - \omega_2$ plane. By cascading the resulting transfer function $H(G_1(z_1)G_2(z_2))$ with $H(G_1(z_1)G_2(z_2^{-1}))$, which has a frequency response that approximates a circle or an ellipse in the second and fourth quadrant so that the product of the transfer functions $H(G_1(z_1)G_2(z_2))$ and $H(G_1(z_1)G_2(z_2^{-1}))$ approximates the full circle or ellipse that defines the boundary of the passband. It is pointed out that we do not find a need for cascading more than these two functions, whereas the mapping of the rotated filter used by [8] is such (Compare Fig. 4.17 and Fig. 3.22) that several of these filters rotated by different angles β_k are found necessary to get a fairly good approximation to a circularly symmetric passband. Hence the order of the final filter and consequently the number of multiplications per output sample is much higher in [8] as well as in the method of [25]. These methods were discussed in Chap. 3.

There is one modification of the above product that is necessary. When two complex numbers $z_1' = e^{j\omega_1'}$ and $z_2' = e^{j\omega_2'}$ are multiplied, $z = z_1'z_2'$ represents a mapping $\omega = \omega_1' + \omega_2'$. This means that every point on the ω axis maps to a straight line $\omega_2' = -\omega_1' + \omega$ in the $\omega_1 - \omega_2$ plane which has the effect of rotating the piece-wise constant frequency response of a 1-D digital filter by 135°. (See Fig. 3.14b in Chap. 3). Because the frequency response of a digital filter is periodic, such a rotation causes the appearance of undesirable portions of the adjacent frequency responses at the four corners of the $2\pi \times 2\pi$ region $R_4 : \{(\omega_1, \omega_2); |\omega_1| \leq \pi, |\omega_2| \leq \pi\}$, when $H(G_1(z_1)G_2(z_2))$ and $H(G_1(z_1)G_2(z_2^{-1}))$ are multiplied.(This spill over effect was pointed out - in Chap. 3 - in the design of fan filters also.) Therefore we cascade their product with a rectangular 2-D lowpass IIR filter $R(z_1, z_2) = \overline{H}(z_1)(\overline{H}(z_2)$, where $\overline{H}(z)$ is a 1-D lowpass IIR filter with a cutoff frequency $= \max (W_1^{(2)}, W_2^{(2)})$; it is designed to have a constant group delay also in the passband chosen. The 2-D lowpass filter $R(z_1, z_2)$ has a rectangular passband which should considerably attenuate the undesirable portions of the frequency response at the four corners. Yet, it must be pointed out that when the passband radius of the circularly symmetric lowpass filter or the semi-major axis of the elliptically symmetric filter is more than 0.9π, the attenuation provided by the rectangular filter $R(z_1, z_2)$ is not sufficient enough to remove the undesirable frequency response around the corners. Hence this design is limited to filters with a frequency response that lies within a circle of radius less than 0.9π. It is also found that when the radius of the circular passband region is less than about 0.55π, the approximation of the actual contour to a circle does not give good results. In that case, we design a filter $T(z_1, z_2)$ with a radius twice the specified radius. Then $H(z_1, z_2) = T(z_1^2, z_2^2)$ realizes a circular filter with half the radius of $T(z_1, z_2)$, which is therefore equal to the specified radius for the circular passband region.

It should now become clear that the design procedure based on the above method is relatively simple and computationally easier, compared to the methods described in the preceding sections and in Chap. 3. Let us first consider the design of a circularly symmetric lowpass filter with a passband region of radius $W^{(2)}$. We choose an initial value for a ($|a| = |b| < 1$) and from the constraints (4.131), we obtain the corresponding bandwidth $W^{(1)}$ of the 1-D lowpass filter. With these values, the closed contour in the $\omega_1 - \omega_2$ plane no doubt passes through the intersection points $(W^{(2)}, 0)$ and $(0, W^{(2)})$, but the rest of the contour may not match the points exactly on the circle with the radius $W^{(2)}$. So we compute N discrete points $(\omega_{1i}, \omega_{2i})$, $i = 1, 2, \ldots, N$ distributed at equal angular intervals on the actual contour, using the mapping function (4.130), and calculate the radial distance $\sqrt{\omega_{1i}^2 + \omega_{2i}^2}$ from the origin. The difference between this radius and the ideal radius $W^{(2)}$ gives the amount of deviation at each point. Because of symmetry, it is sufficient to choose the N discrete points on the contour in the first quadrant only and summing the errors at these N points, we define an average normalized error E' in percentage as

$$E' = \frac{\sum_{i=0}^{N} \left| W^{(2)} - \sqrt{\omega_{1i}^2 + \omega_{2i}^2} \right|}{N.W^{(2)}} \times 100 \qquad (4.136)$$

Next we change the value of a by a small step size and repeat the iteration until a minimum value for the error E' is obtained. The value for $a(= b)$ that corresponds to this minimum error is used to define the digital spectral transformation (4.128). The frequency response of the 1-D filter is the same as the

4.6. DESIGN USING DIGITAL SPECTRAL TRANSFORMATION

intersection of the frequency response of the 2-D filter with the ω_1 or ω_2 axes. Hence we design a lowpass 1-D IIR filter that simultaneously satisfies the magnitude and the group delay specifications which correspond to the specifications in the corresponding region in the $\omega_1 - \omega_2$ plane. Several methods were described in Chap. 2 for designing such 1-D IIR filters. Then we apply the digital spectral transformation with the optimal value for $a(=b)$, to this 1-D IIR filter and finish the design of the 2-D IIR filter as explained above. Indeed one can generate a look-up table for the optimum values of a that map circles of different radius in the $\omega_1 - \omega_2$ plane to points on the frequency axis ω of the 1-D filter and the corresponding minimum value of E' that is achieved. From this table, one can easily choose the optimum value for a, when a radius for the circular lowpass filter is given. This data can also be used to design 2-D highpass filters with the circle of given radius as the edge of the passband. In Table 4.1, some typical values for the radius of the 2-D circular filter, the optimal values of a that are recommended to minimize the error and the minimum value of the error E' are listed.

Table 4.1. Optimal values of a for different specified values of radius

Radius	Optimum value of a	Error E'
0.50π	0.90	12.00 %
0.55π	0.90	9.9 %
0.60π	0.90	7.6 %
0.65π	0.90	4.9%
0.70π	0.90	1.7 %
0.75π	0.61	1.3 %
0.80π	0.49	1.3 %
0.85π	0.42	1.4 %
0.90π	0.39	1.4 %
0.95π	0.36	1.4 %

Of course one could also choose an rms value for the error in the form (4.137).

$$E = \frac{\sqrt{\sum_{i=0}^{N} \left| W^{(2)} - \sqrt{\omega_{1i}^2 + \omega_{2i}^2} \right|^2}}{N.W^{(2)}} \qquad (4.137)$$

Let us now consider the case of a lowpass filter which has an elliptic region for the passband, the ellipse being defined by (4.133). In this case $a \neq b$. But after choosing an initial value for a, the constraint equations (4.135) are used to determine the value of $W^{(1)}$ and b. After finding these values, we compute N discrete points distributed on the contour which is described by (4.130) and which approximates the ideal ellipse. But the error in this case is defined as

$$E' = \frac{\sum_{i=1}^{N} \left| 1 - \left\{ \left(\frac{\omega_{1i}}{W_1^{(2)}}\right)^2 + \left(\frac{\omega_{2i}}{W_2^{(2)}}\right)^2 \right\} \right|}{N} \times 100 \qquad (4.138)$$

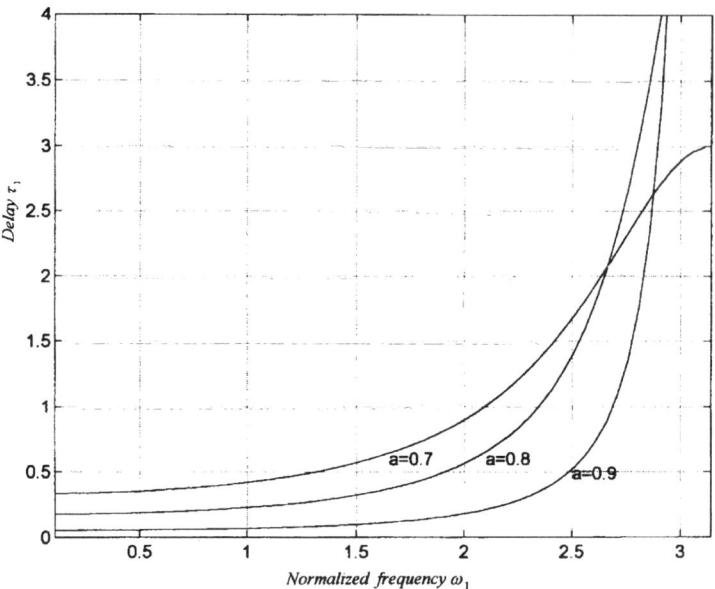

Fig. 4.19. Plot of the group delay τ_1 as a function of ω_1

Next we change the value of a by a small step size and repeat the procedure until the error E' reaches its minimum value. The rest of the design procedure is the same as for the circular filter. (The authors tried to apply the above method to design a fan filter but the results are not very good.) The authors in [32] have shown that using the DST proposed in (4.128), a 1-D filter having a constant group delay response is mapped to a 2-D filter with nearly a constant group delay. To show this result, assume that the 1-D filter has a constant group delay of τ. Hence the corresponding phase response is given by $\omega\tau$. When the mapping function (4.130) is substituted for ω and the result is differentiated with respect to ω_1 and ω_2 we get the following expressions for the two group delays.

$$\tau_1 = -\frac{\partial \theta(\omega_1,\omega_2)}{\partial \omega_1} = -\tau\left[1 - 2\frac{a^2 + a\cos\omega_1}{a^2 + 2a\cos\omega_1 + 1}\right]$$

$$\tau_2 = -\frac{\partial \theta(\omega_1,\omega_2)}{\partial \omega_2} = -\tau\left[1 - 2\frac{b^2 + b\cos\omega_2}{b^2 + 2b\cos\omega_2 + 1}\right]$$

Plots of the group delay τ_1 for some typical values of a are shown in Fig. 4.19. They show that over a passband region of the 2-D lowpass filter with a radius $r \leq 0.55\pi$, the group delay is nearly constant. A similar result holds good for the group delay τ_2 as well.

4.6. DESIGN USING DIGITAL SPECTRAL TRANSFORMATION

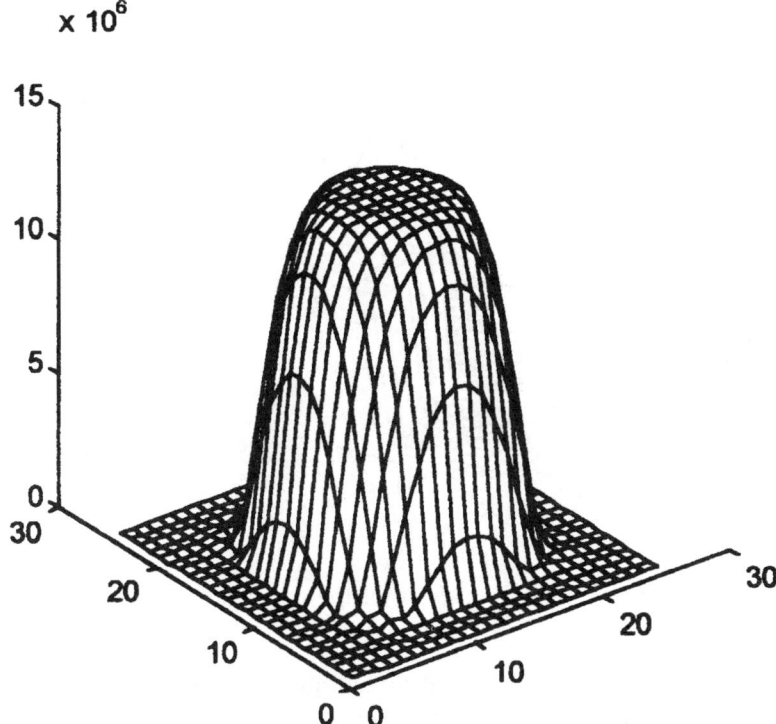

Fig. 4.20. Magnitude response of the circularly symmetric lowpass filter

4.6.1 Example 4.5

We choose an example for the design of a lowpass IIR filter having a circular passband region of radius of 0.825π radians. The value chosen for a and b is 0.46. The corresponding 1-D lowpass filter is required to have a bandwidth $W^{(1)} = 0.2\pi$ radians. A 12th order, 1-D maximally flat IIR filter with a bandwidth of 0.2π and a constant group delay of 9 sample periods is designed using the method described in sections 2.4.3 and 2.4.4. Fig. 4.20 shows the magnitude response of the 2-D digital filter obtained from this choice. Of special interest are the Figs. 4.21-4.22, which show the group delay τ_1 and τ_2- in only one quarter of the $\omega_1 - \omega_2$ plane. When they are compared with the Fig. 4.5-4.6, and Figs. 4.10-4.11, it would seem that the method of [14] is very attractive for designing 2-D lowpass filters, producing a very good magnitude response and a constant group delay behavior.

As another example, a 2-D IIR filter with an elliptical passband region is considered-for which the semi-major axis is 0.925π and the semi-minor is 0.7π radians. The plots of magnitude, and the group delays τ_1 and τ_2 are shown in Figs. 4.23-4.25.

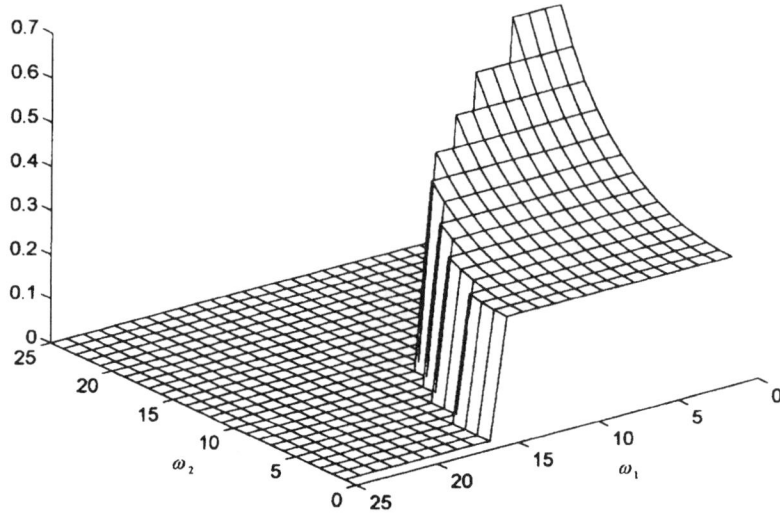

Fig. 4.21. Group delay τ_1 in one quadrant of the passband

A few years after the digital spectral transformation (4.128) was proposed in [14], two more transformations have been reported. The transformation sug-

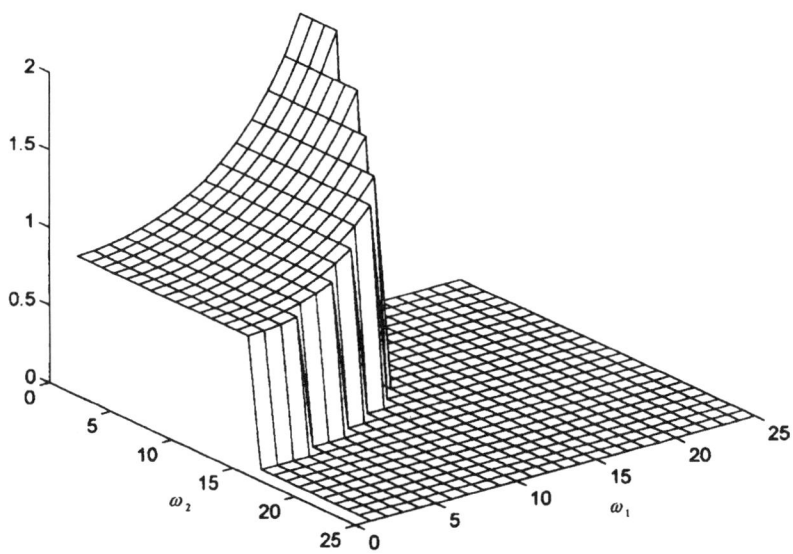

Fig. 4.22. Group delay τ_2 in one quadrant of the passband

4.6. DESIGN USING DIGITAL SPECTRAL TRANSFORMATION

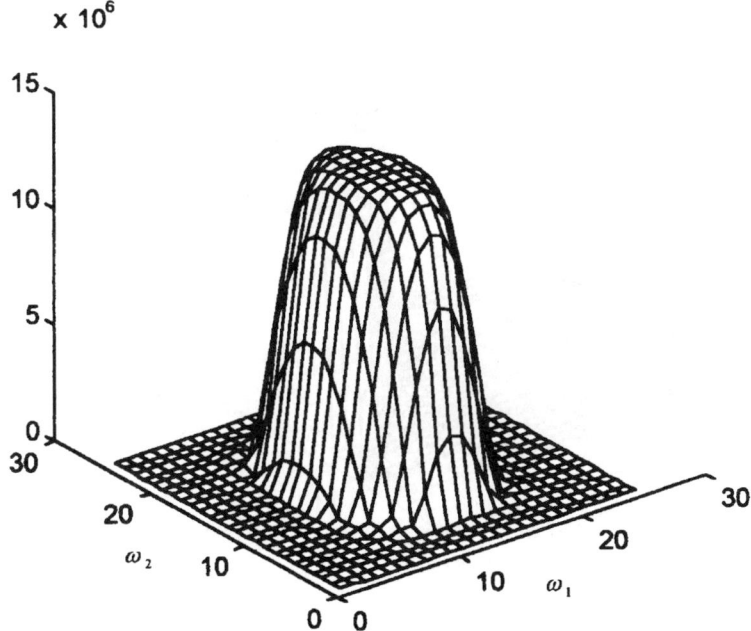

Fig. 4.23. Magnitude response of an elliptic passband filter

gested in [22] is given as

$$z = \frac{1 + a(z_1 + z_2 - z_1 z_2)}{-a(1 - z_1 - z_2) + z_1 z_2} \tag{4.139}$$

which is stable when $0 < a < 1/3$. (Note that the variables z, z_1 and z_2 in (4.139) and (4.140) denote delays.) The design procedure and the results given in [22] are almost the same as in [14]. The other transformation proposed more recently in [15] is

$$z = z_1 z_2 \frac{(1 + a_1 z_1^{-1} + a_2 z_1^{-2})(1 + b_1 z_2^{-1} + b_2 z_2^{-2})}{(1 + a_1 z_1 + a_2 z_1^2)(1 + b_1 z_2 + b_2 z_2^2)} \tag{4.140}$$

The above transformation can also be written in the form

$$z = \frac{(z_1 + a_1 + a_2 z_1^{-1})(z_2 + b_1 + b_2 z_2^{-1})}{(1 + a_1 z_1 + a_2 z_1^2)(1 + b_1 z_2 + b_2 z_2^2)} \tag{4.141}$$

which shows that it is not obtained simply by extending the first order allpass functions of (4.128) to second order allpass functions. This transformation is much different from Goodman's DST given by equations (3.142-3.143).

236   CHAPTER 4. MAGNITUDE AND DELAY OF 2-D FILTERS

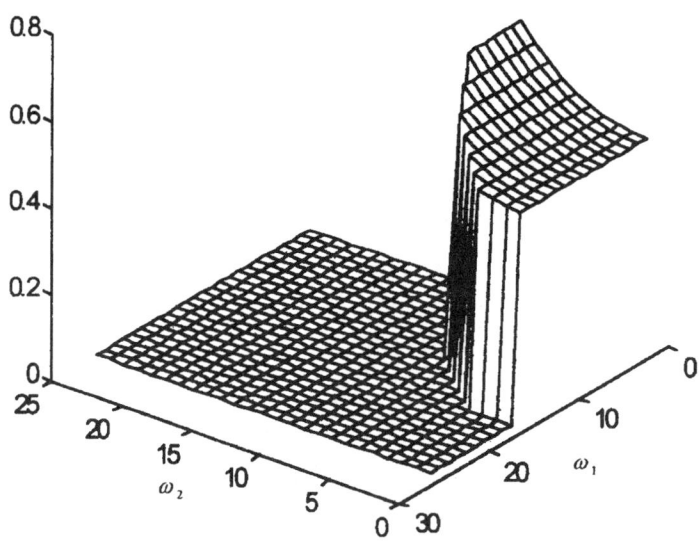

Fig. 4.24. Group delay τ_1 in one quadrant of the elliptic passband

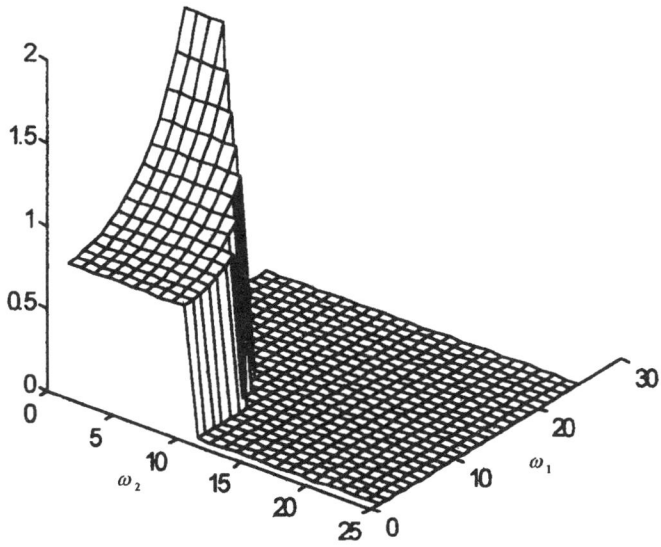

Fig. 4.25. Group delay τ_2 in one quadrant of the elliptic passband

4.6. DESIGN USING DIGITAL SPECTRAL TRANSFORMATION

The mapping function for this Digital Spectral Transformation (4.140) is given by

$$\omega = \omega_1 + \omega_2 - 2\tan^{-1}\left[\frac{a_1 \sin \omega_1 + a_2 \sin 2\omega_1}{1 + a_1 \cos \omega_1 + a_2 \cos 2\omega_1}\right]$$

$$-2\tan^{-1}\left[\frac{b_1 \sin \omega_2 + b_2 \sin 2\omega_2}{1 + b_1 \cos \omega_2 + b_2 \cos 2\omega_2}\right] \quad (4.142)$$

The design procedure is again the same as used in [14]. But in the case of a circular lowpass filter, we get $a_1 = b_1$ and $a_2 = b_2$. In the case of an elliptically symmetric filter, $a_1 \neq b_1$, $a_2 \neq b_2$, but there is one constraint between the four unknown parameters similar to (4.133). In addition, they are required to satisfy the two pairs of constraints: $|a_2| > 1$, $|a_1| < 1 + a_2$ and $|b_2| > 1$, $|b_1| < 1 + b_2$ in order to assure that the 2-D IIR filter is stable. Hence in using this digital spectral transformation for designing the 2-D IIR filter, a simple iterative procedure is required to find the best values for the unknown parameters that minimizes the error function E' subject to these constraints.

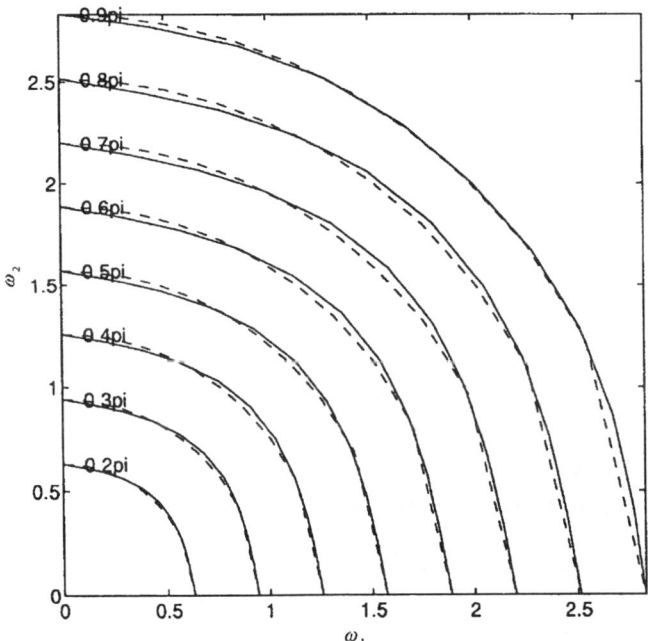

Fig. 4.26. Contours of circles with different radius and best approximations of actual contours

In Fig. 4.26, we have plotted the circular contours with different radius, (in dotted lines) and the actual contours obtained with the optimal values for

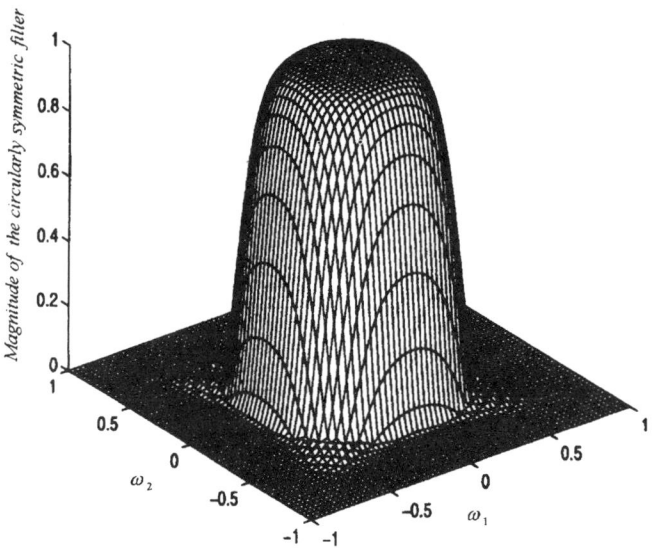

Fig. 4.27. Magnitude response of the circularly symmetric lowpass filter in Example 4.6

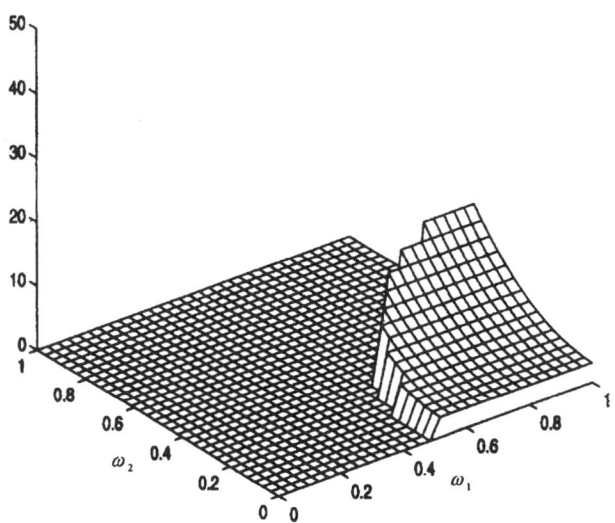

Fig. 4.28. Group delay τ_1 in one quadrant of the passband of the filter in Example 4.6

4.6. DESIGN USING DIGITAL SPECTRAL TRANSFORMATION

the parameters $a_1(=b_1)$ and $a_2(=b_2)$ which approximate them with the least

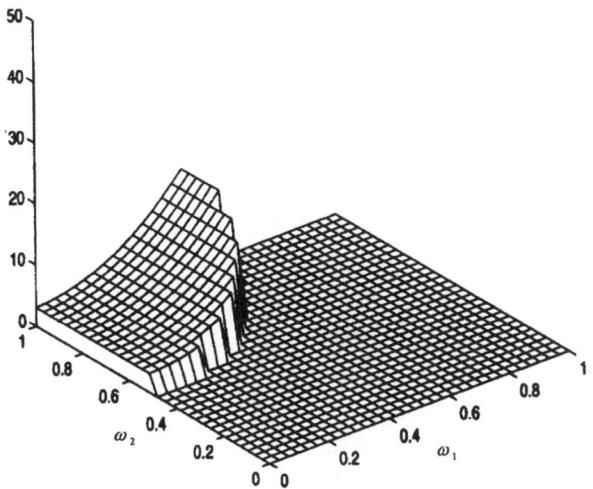

Fig. 4.29. Group delay τ_2 in one quadrant of the passband of the filter in Example 4.6

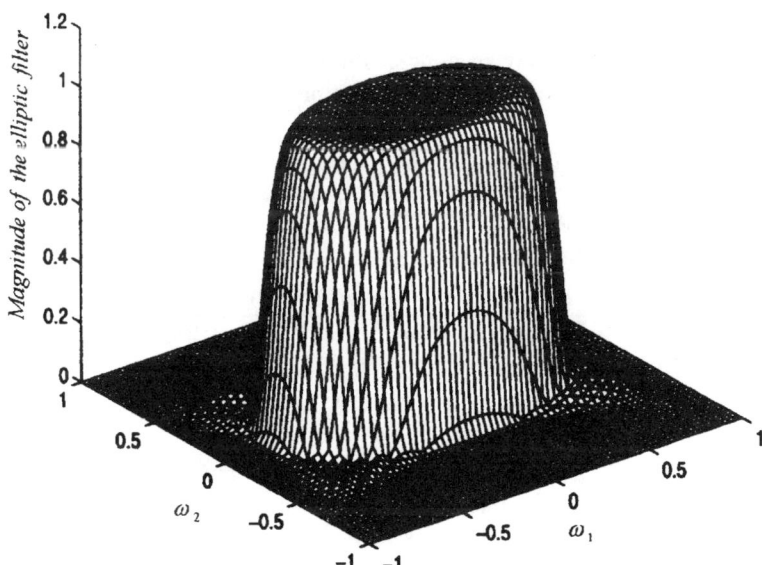

Fig. 4.30. Magnitude response of the elliptic passband filter

amount of error E'. The optimum values a_1 and a_2 and the corresponding values for the minimum error E' are listed in Table 4.2. In this Table, the last column shows the minimum error obtained by [14] for the filters with the same radius as listed in the first column. The optimum values of the parameter $a = b$ of that DST are given in Table 4.1. It is evident that the transformation given in [15] yields considerably lower values for the error E' than the Digital Spectral Transformation given in [14].

Table 4.2. Optimal values of a_1 and a_2, and the two error values compared

Radius	Optimum value of a_1	Optimum value of a_2	Error E'	Error E' from [14]
0.2π	0.5035	0.1525	0.9026%	NA
0.3π	0.2025	0.2115	0.9713%	NA
0.4π	0.6023	0.0926	0.9472%	NA
0.5π	0.4402	0.0986	0.9801%	12.%
0.6π	0.4297	0.0824	0.9805%	7.6%
0.7π	0.4121	0.0699	0.9597%	1.7%
0.8π	0.3768	0.0795	0.8065%	1.3%
0.9π	0.3836	0.1102	0.8861%	1.4%

4.6.2 Example 4.6

The magnitude and group delay responses obtained by the application of the DST [15] to meet the same specifications as Example 4.5, are shown in Figs. 4.27-4.29.

The mapping relationship between the group delay τ of the 1-D filter and $\tau_1 = -\frac{\partial(\tau\phi(\omega_1,\omega_2))}{\partial \omega_1}$ and $\tau_2 = -\frac{\partial(\tau\phi(\omega_1,\omega_2))}{\partial \omega_2}$ of the 2-D filter are derived from (4.142) and are given below and soon after Fig. 4.28.

$$\tau_1 = -\tau\left[1 - 2\left(\frac{a_1^2 + 2a_2^2 + a_1(1+3a_2)\cos\omega_1 + 2a_2\cos 2\omega_1}{1 + a_1^2 + a_2^2 + 2a_1(1+a_2)\cos\omega_1 + 2a_2\cos 2\omega_1}\right)\right]$$

$$\tau_2 = -\tau\left[1 - 2\left(\frac{b_1^2 + 2b_2^2 + b_1(1+3b_2)\cos\omega_2 + 2b_2\cos 2\omega_2}{1 + b_1^2 + b_2^2 + 2b_1(1+b_2)\cos\omega_2 + 2b_2\cos 2\omega_2}\right)\right] \quad (4.143)$$

When they are plotted, they are found to be nearly constant for the values of ω_1 and ω_2 up to 0.55π - just like the plots in Fig. 4.19. Similar results are obtained in the case of elliptic filters as shown in Figs. 4.30-4.32. The 1-D lowpass filter with its magnitude response shown in Fig. 2.9 was used to design a 2-D highpass filter, by choosing the best DST that maps the 1-D cutoff frequency to a circular contour in the $\omega_1 - \omega_2$ plane. We obtain a 2-D highpass filter which has zero phase and its magnitude is plotted in Fig. 4.33.

4.7. CONCLUSION

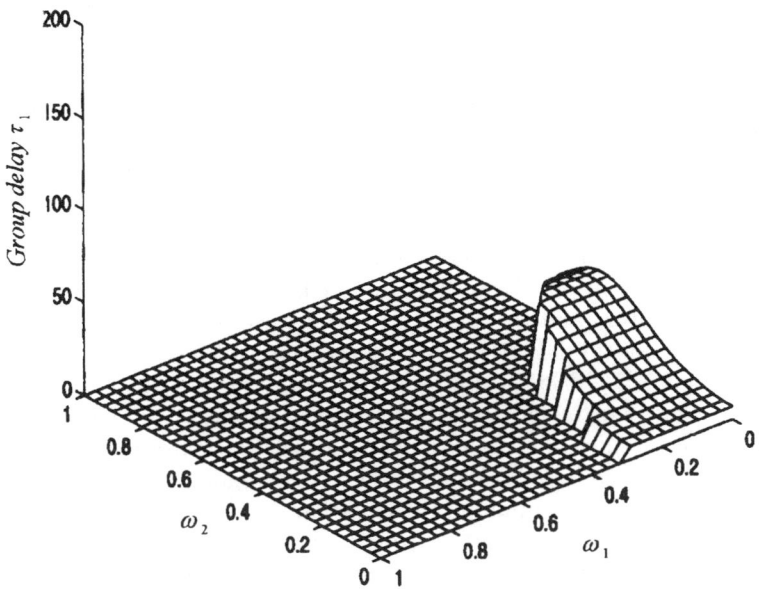

Fig. 4.31. Group delay τ_1 in one quadrant of the elliptic passband filter

4.7 Conclusion

Several important methods reported recently, for the design of two-dimensional IIR filters that approximate both the prescribed magnitude and the group delay response have been described in this chapter. If any comparison of these methods is to be made, the following criteria should be considered. Analytical methods are always preferable over methods that are heavily dependent on computer-aided programming methods. If the analytical methods reduce the problem to that of approximating a one-dimensional digital or analog filter, it is an advantage to make use of the rich body of literature on the design of such (analog and digital) filters. Hence the use of digital spectral transformation is an attractive approach compared to the computer-aided approximation methods. If programming is required, it should be easy to estimate the optimal or sub-optimal order of the 2-D IIR filter, the algorithm should converge as fast as possible and should guarantee that the 2-D filter is stable. The filter should require a minimum number of multiplications and its realization should be very easy. The method should be applicable for the design of filters with different kinds of boundaries defining the passband and stopband regions like the circular, elliptical regions, as well as fan filters and so on. The design method should be able to meet the given attenuation in the specified passband region as well as the stopband region. The design should be suitable for approximating both magnitude and group delay response. In terms of these criteria, the best methods appear to be those proposed by [19, 14],and [15, 16].

More detailed comparison of the three methods described in the above sections needs to be made before any definitive conclusions can be deduced, regarding their relative merits in terms of all the criteria. Further research needs to be directed towards theoretically more elegant and computationally more efficient methods. It has been pointed out that 2-D IIR digital filter functions with a separable denominator polynomial can not be used to design filters which have steep rolloff characteristics in the transition band [29]. The transfer functions obtained by the application of the DST proposed in [14] and [15], contain non-separable denominators and the rolloff characteristics in their transition band is the same as that of the 1-D IIR lowpass filter from which they are generated. Hence designing a 1-D IIR lowpass filter with the specified rolloff characteristics will meet the same characteristics in the transition band of the resulting 2-D IIR filter. In this method, we can also meet the specification of a radius for the stopband region and a minimum attenuation in it, by finding the corresponding stopband frequency and the minimum attenuation A_s of the 1-D, IIR filter. All filters - except filters designed by linear programming - discussed in this chapter are quarter plane filters or combination of such filters with regions of support in all the four quadrants. It would seem reasonable to conjecture that IIR filters with a non-symmetrical half-plane support (NSHP) should be able to realize a

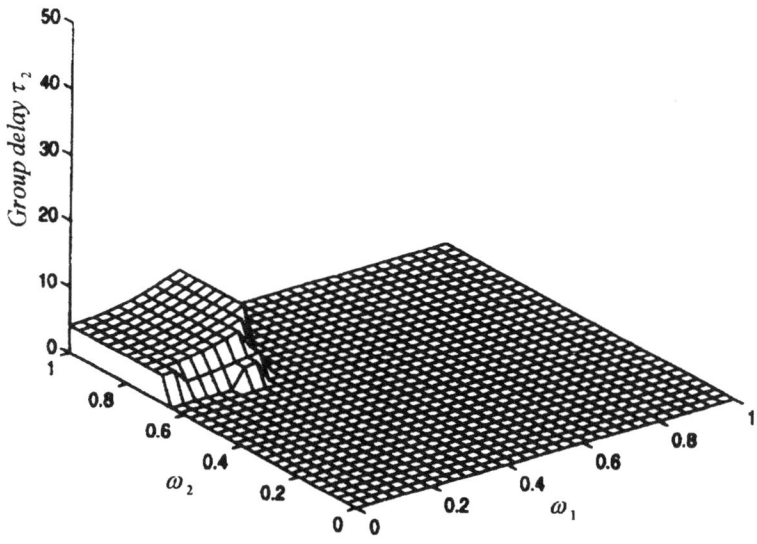

Fig. 4.32. Group delay τ_2 in one quadrant of the elliptic passband filter

4.7. CONCLUSION

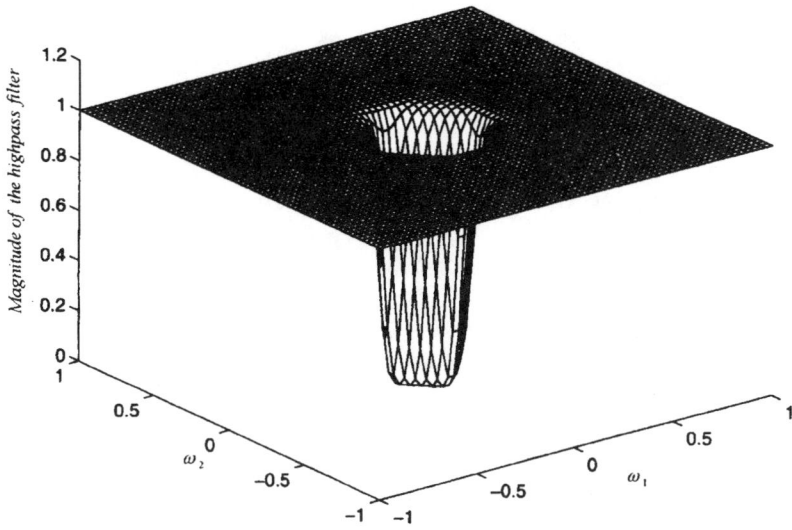

Fig. 4.33. Magnitude response of a 2-D highpass filter

wider class of frequency domain specifications than the filters discussed in this chapter and in the previous chapter. In Chap. 3, the fan filter discussed in section 3.3.6 has a nonsymmetrical half plane support (NSHP) but the magnitude response shown in Fig. 3.21 is not better than the response shown in Fig. 3.19. Therefore, new techniques for the design of 2-D NSHP filters [31] need to be investigated to extend the design methodology of stable 2-D IIR filters.

Bibliography

[1] S.A.H. Aly and M.M. Fahmy, "Design of two-dimensional recursive digital filters with specified magnitude and group delay characteristics," IEEE Trans. on Circuits and Systems, CAS-25, pp. 908-915, November 1978.

[2] J.A. Cadzow, "Recursive digital filter synthesis via gradient based algorithms," IEEE Trans. on Acoustics, Speech and Signal Processing, ASSP-24, pp. 349-355, October 1976.

[3] S. Chakrabarti and S.K. Mitra, "Design of two-dimensional digital filters via spectral transformations," Proc. IEEE, vol. 65, no. 6, pp. 905-914, June 1977.

[4] H. Chang and J.K. Aggarwal, " Design of 2-dimensional recursive filters by interpolation," Proc. IEEE Int'l Symposium on Circuits and Systems, pp. 369-372, April 1976.

[5] A.T. Chottera and G.A. Jullien, "Design of two-dimensional recursive digital filters using Linear Programming," IEEE Trans. on Circuits and Systems CAS-29, pp. 817-826, December 1982.

[6] A.T. Chottera and G.A. Jullien, "A Linear Programming approach to recursive digital filter design with linear phase," IEEE Trans. on Circuits and Systems, CAS-29, pp. 139-149, December 1982.

[7] A.G. Constantinides, "Spectral transformations for digital filters," Proc. IEE, vol. 117, pp. 1585-1590, 1970.

[8] J.M. Costa and A.N. Venetsanopolous, "Design of circularly symmetric two-dimensional recursive filters," IEEE Trans. on Acoustics, Speech and Signal Processing, ASSP-22, pp. 432-443, December 1974.

[9] R. Fletcher, *Practical Methods of Optimization*, (2nd edition) Chichester, John Wiley, 1987

[10] R. Fletcher and M.J.D. Powell, "A rapidly convergent descent method for minimization," Comput. Journal, vol. 6, no. 2, pp. 163-168, 1963.

[11] D.M. Goodman, "A design technique for circularly symmetric lowpass filters," IEEE Trans. on Acoustics, Speech and Signal Processing, ASSP-26, pp. 290-303, August 1978.

[12] G. Gu and B.A. Shenoi, "A novel approach to the synthesis of recursive digital filters with linear phase," IEEE Trans. on Circuits and Systems, vol. 38, pp. 602-612, June 1991.

[13] G. Gu, B.A. Shenoi and C. Zhang, "Synthesis of 2-D linear phase digital filters," IEEE Trans. on Circuits and Systems, vol. 37, pp. 1499-1508, December 1990.

[14] L. Harn and B.A. Shenoi, "Design of stable two-dimensional IIR filters using digital spectral transformations," IEEE Trans. on Circuits and Systems, CAS-33, pp. 483-490, May 1986.

[15] Rajamohana Hegde and B.A. Shenoi, "Design of 2-D IIR filters using a new digital spectral transformation," Proc. IEEE Int'l Symposium on Circuits and Systems, vol II, pp. 344-347, May 1995.

[16] -ibid-, "A unified approach to the design of 2-D IIR digital filters with flat passband magnitude and delay characteristics," Proc. IEEE Int'l Sympo. on Circuits and Systems, vol. I, pp. 765-768, 1997.

[17] Rajamohana Hegde and B.A. Shenoi, "Magnitude approximation of IIR digital filters with constant group delay response," Proc. IEEE Int'l Sympo. on Circuits and Systems, vol. IV, pp. 2200-2203, 1997.

[18] Rajamohana Hegde and B.A. Shenoi, "Magnitude approximation of digital filters with specified degrees of flatness and constant group delay characteristics," IEEE Trans. on Circuits and Systems: Part II. (To be published)

[19] T. Hinamoto and S. Maekawa, "Design of two-dimensional recursive digital filters using mirror image polynomials," IEEE Trans. on Circuits and Systems, CAS-33, pp. 750-758, August 1986.

[20] T.S. Huang, J.W. Burnett, and A.G. Deczky, "The importance of phase in image processing filters," IEEE Trans. on Acoustics, Speech and Signal Processing, ASSP-23, pp. 529-542, December 1975.

[21] H.K. Kwan and C.L. Chan, "Multidimensional spherically symmetric recursive digital filter design satisfying prescribed magnitude and constant group delay response," Proc.IEE, Pt.G: Electronic Circuits and Systems, pp. 187-193, August 1987.

[22] H.K. Kwan and C.L. Chan, "Design of two-dimensional circularly symmetric IIR digital low-pass filters by spectral transformation," Proc.IEEE Asian Electronics Conference, Hong Kong, pp. 29-33, 1987.

[23] H.K. Kwan and C.L. Chan, "Design of linear phase circularly symmetric two-dimensional recursive digital filters," IEEE Trans. on Circuits and Systems, CAS-36, pp. 1023-1029, July 1989.

[24] J.S. Lim, *Two-Dimensional Signal and Image Processing*, Prentice- Hall, 1990.

[25] G.V. Mendonca, A. Antoniou and A.N. Venetsanoplous, "Design of two-dimensional pseudorotated digital filters satisfying prescribed specifications," IEEE Trans. on Circuits and Systems, CAS-34, pp. 1-10, January 1987.

[26] K. Nishikawa and R.M. Mersereau, "Design of circularly symmetric two-dimensional IIR lowpass digital filters with constant group delay using McClellan Transformation," Trans. IEICE, vol. E75-A, pp. 830-836, July 1992.

[27] T.Q. Nguyen, T.I. Laakso and R.D. Koilpillai, "Eigenfilter approach for the design of allpass filters approximating a given phase response," IEEE Trans. on Signal Processing, vol. 42, pp. 2257-2263, September 1994.

[28] N.A. Pendergrass, S.K. Mitra and E.I. Jury, "Spectral transformations for two-dimensional digital filters," IEEE Trans. on Circuits and Systems, vol. CAS-23, pp. 26-35, January 1976.

[29] P.K. Rajan and M.N.S. Swamy, "Quadrantal symmetry associated with two-dimensional digital transfer functions," IEEE Trans. on Circuits and Systems, CAS-25, pp. 340-343, June 1978.

[30] P.A. Ramamoorthy and L.T. Bruton, "Frequency domain approximation of stable multi-dimensional discrete recursive filters," Proc.of Int'l Symp. on Circuits and Systems, pp. 654-657, April 1977.

[31] P.A. Ramamoorthy and L.T. Bruton, Chapter 3: Design of Two-Dimensional Recursive Filters, *Digital Signal Processing*. (T.S.Huang, Editor) Springer-Verlag, 1981

[32] B.A. Shenoi and P. Misra, "Design of two-dimensional IIR digital filters with linear phase," IEEE Trans. on Circuits and Systems-II, vol. 42, pp. 124-129, February 1995.

[33] J.P. Thiran, "Recursive digital filters with maximally flat group delay," IEEE Trans. on Circuit Theory, vol. CT-18, pp. 659-663, November 1971.

[34] A.N. Venetsanopolous, Chapter 12: Computer-Aided Design of Two-Dimensional Digital Filters, *Multidimensional Systems-Techniques and Applications* (S.G.Tzafestas, Editor), Marcel Dekker Inc, 1986.

Index

Aliasing, 44
Approximation, 1
    Butterworth approximation, 9
    Chebyshev approximation, 18, 217
    Computer-aided approximation, 64
    Elliptic approximation, 27
    Equiripple approximation, 4, 7, 8, 9
    Inverse Chebyshev approximation, 23
    Linear phase approximation, 4,8
    Least squares approximation, 4, 8, 31
    Maximally flat approximation, 9
    Min-max approximation, 4, 7
Attenuation, 9

Bilinear Transformation, 54, 71, 147, 160

Cauer filters, 28
Characteristic function, 9
Chebyshev norm, 111, 218
Chebyshev Rational Function, 27, 28, 29
Circularly symmetric filters, 173
Computer-aided approximation, 64
Cutoff frequency, 11, 14
Commensurate, distributed networks, 104, 117, 119
Complementary Operation, 145
Complete elliptic integral, 28

Delay
    Approximation, 71, 187
    Equalizers, 4
    Group delay, 73, 83
    Maximally flat group delay, 74, 87, 98
Digital Filters, 2
    1-D digital filters, 2, 4, 41
        IIR filters, 41, 71, 72, 187
        Recursive filters, 110
        FIR filters, 41, 72
    2-D digital filters, 2
        FIR filters, 227
        IIR filters, 137, 173, 175, 187, 196
        Recursive filters, 110
Digital Spectral Transformation, 60, 62, 153, 182, 225, 226

Eigen Filters, 151, 165

Filters, 2
    allpass filters, 71, 74, 84
    analog and digital filters, 2
    anti-aliasing filters, 45
    bandpass filter, 2, 3, 34, 36, 108
    bandstop filter, 2, 3, 34, 38, 108
    Butterworth filter, 6, 14, 31, 133, 179
        prototype, lowpass filter, 11, 34
    Cauer filters, 28
    Chebyshev filter, 20, 21, 22, 31, 33, 161, 179
    Chebyshev II filter, 23
    Eigen filters, 104, 122, 129
    Elliptic filters, 27, 31, 33, 161, 179
    Fan filters, 151, 165
    highpass filters, 2, 3, 34, 108
    Inverse Chebyshev filter, 23, 26, 31, 161
    lowpass filter, 2, 3, 34, 108, 161
    Pulse shaping filters, 99

Fletcher-Powell Optimization, 84, 196
Fourier Transform, 1, 41
    Discrete Fourier Transform, 223
    Discrete-Time, Fourier Transform, (DTFT), 42

Half plane symmetry, 141, 148, 149, 179
Hankel matrix, 112

Impulse Invariant Transformation,
Intersymbol interference, 3, 86, 100

Jacobian elliptic sine function, 28, 31

Laurent series, 87
Linear phase approximation, 121
Linear Programming, 104, 113, 117, 173, 213, 214, 215
Loss function, 9

Maclaurin series, 6
Magnitude, 1
    Approximation, 1, 71, 187
    Compensators, 3
    Maximally flat magnitude, 87, 98
McClellan's Transformation, 212, 227
Mirror Image Polynomial, 83, 85

Nonessential singularities of the second kind, 177
Nonlinear Optimization, 83
Nonlinaer Programming, 173, 188, 190
Non-Symmetric Half Plane (NSHP) filters, 169, 171, 213
Nyquist frequency, 46

Passband, 2, 3, 27, 137, 138
    Rectangular passband, 138
Peak-constrained least squares (PCLS) Optimization, 85
Polynomials
    Butterworth polynomials, 12
    Chebyshev polynomials, 18
    Mirror Image Polynomials 83, 85
Pulse shaping filters, 99

Quadrantal symmetry, 141, 149

Rayleigh's Principle, 122, 123, 126
Reconstruction Formula, 44
Recursive Quadratic Programming, 85
Remez Exchange Algorithm, 81, 88, 97, 109, 127, 212

Sampling Theorem, 45
Signals
    one-dimensional, discrete-time, 1
    two-dimensional, discrete-time, 1
Singular Value Decomposition, 104, 110, 216
Stabilty conditions, 190
Stopband, 2, 3, 27, 137, 138
    stopband frequency, 11
    equiripple stopband, 95

Taylor series, 6, 75
Transition band, 3
Transmission zeros, 29